Chambers Nuclear Energy and Radiation Dictionary

EDITOR

Professor P. M. B. Walker, CBE, FRSE

Chambers

EDINBURGH NEW YORK TORONTO

Published 1992 by W & R Chambers Ltd,
43–45 Annandale Street, Edinburgh EH7 4AZ

British Library Cataloguing in Publication Data
Chambers nuclear energy and radiation dictionary.
1. Earth Sciences
I. Walker, Peter M. B.
550

ISBN 0 550 13246–5 Hbk
ISBN 0 550 13247–3 Pbk

Cover design by John Marshall

Printed in England by Clays Ltd, St Ives plc

Contents

How the dictionary was made

Chambers Nuclear Energy and Radiation Dictionary was compiled and designed on a COMPAQ 386 personal computer. The original database was made with the INMAGIC library retrieval software from Head Computers Ltd. The text was set using the Xerox VENTURA desktop publishing system and the drawings made with the Micrografx DESIGNER graphics program.

Preface

Chambers Nuclear Energy and Radiation Dictionary is designed to make it easier for those who are concerned about nuclear power and radiation to learn more about nuclear energy and to come to an informed opinion about a subject which is of great concern to a very wide public. It has a number of unique features and was developed from the database of the acclaimed *Chambers Science and Technology Dictionary* with many new and revised definitions in this important subject.

A unique feature of this book is the inclusion of 11 chapters preceding the dictionary proper. These chapters describe in simple language and with copious illustrations those special features which make nuclear energy such an important and controversial subject.

The first two describe the properties of the atomic nucleus which make nuclear energy possible and then the problems which have to be overcome in harnessing this energy. The next two chapters discuss the many different kinds of power stations which rely on the splitting of the atomic nucleus for their energy and then the methods with which it is hoped that the energy from the fusion of atomic nuclei may produce power in the next century. There are then two chapters on nuclear safety and on the production and enrichment of uranium fuel, together with methods for its eventual disposal. These are followed by a chapter on nuclear bombs of various kinds and one on how nuclear and other forms of radiation can be detected. There is then a chapter which relates the radiation resulting from nuclear fission to other kinds of radiation with which we are more familiar. Because nuclear radiation can be damaging to living organisms, the next chapter discusses some basic biology and in particular the subject of cancer. Finally, the biological effects of radiation are described before comparing the amounts of man-made radiation to that which comes naturally from outer space and from the rocks beneath us. This then leads to the radiation limits which are determined by the various regulatory authorities and the kinds of evidence upon which their decisions are based.

Within the dictionary part of this book there are also illustrated panels on special subjects, such as *food irradiation, non-ionizing radiation, radioactive dating* and *risk assessment.* There are also panels on other related subjects like *atomic structure, particle accelerators* and *lasers.*

Arrangement

This book is fully cross-referenced so that many terms in the dictionary can be used as an index to sections within the first 11 chapters. The entries in the dictionary are strictly alphabetical with single letter entries occurring at the beginning of each letter. Panels within the dictionary occur either on the same page as the parent entry or on the pages immediately following with the alphabetical entry stating 'See panel on p. 0'. Cross-references to headwords which themselves refer to panels use the style 'See **trace elements analysis** p. 205' and cross-references from the dictionary to the earlier chapters are shown as 'p. 26' the end of an entry.

Italic and Bold

Italic is used for:

(1) alternative forms of, or alternative names for, the headword usually after 'also' at the end of an entry;
(2) the expanded form of an abbreviated headword, provided that the expanded form is not found as a headword elsewhere;
(3) terms derived from the headword, often after 'adj.' or 'pl.';
(4) variables in mathematical formulae;
(5) for emphasis.

Bold is used for:

> cross-references, either after 'see', 'cf.' etc. or in the **body of the entry**. It is also used after 'Abbrev. for' when the expanded form can be found as a headword elsewhere.

Bold italic is used in the panels to highlight a term explained within that article.

Tables

Much information about elements important in the nuclear industry is presented as tables within the text of the dictionary part of the book. There are also tables for 'SI units' and 'Fundamental Dynamical Units'. The latter gives comparisons with the older scientific units found in some texts.

Acknowledgements

I would like to express my grateful thanks to Dr David E. Watt of the Environmental, Health and Safety Services of St Andrews University for reading much of the material and checking, in particular, Chapters 8 to 11. Professor John Evans, Director of the Medical Research Council's Human Genetics Unit in Edinburgh and Dr Jed Adams, Director of the Medical Research Council's Radiobiology Unit at Harwell also provided me with useful papers and references.

I am also very grateful to Dr J. R. Stanbridge and his colleagues at British Nuclear Fuels plc for their comments on the technical matters in Chapter 3. Dr D. C. W. Sanderson of the Scottish Universities Research and Reactor Centre kindly sent me his reports on airborne radiometric surveying and food irradiation. Philip Jones, Librarian at the Harwell Laboratories of A.E.A. Technology allowed me to consult material in the library and suggested relevant books, and J. H. C. Maple, Head of Press and Public Relations at the JET fusion laboratories, provided further useful information.

I would also like to acknowledge information gained from many books and articles with special thanks to the following:

D. W. Devins, *Energy; Its Physical Impact on the Environment,* Wiley, New York, is a source book for the technology of all forms of energy production, not just nuclear energy but also conventional and renewable sources. It is written at university level and is invaluable for comparing different energy strategies in a quantitative scientific way.

W. Marshall has edited a compendious three volume work, *Nuclear Power Technology,* Clarendon Press, Oxford, 1983, containing chapters written at a highly technical level on a wide range of relevant subjects, although some of them are not now of primary importance.

A more accessible one volume book is H. A. Cole's *Understanding Nuclear Power,* Gower, Brookfield, Vermont, 1988, which covers much of the ground and is written from the informed point of view of a participant in much of the relevant work.

K. S. Krane's *Nuclear Physics,* Wiley, New York, 1988, is an excellent university text book covering much of the field.

Finally, H. Morland's *The Secret that Exploded*, Random House, New York, 1981, is an account of one man's pilgrimage round the unclassified parts of the US nuclear arsenals in search of information on the construction of the hydrogen bomb. Fig. 7.4 is taken from that book.

As a biologist, I freely acknowledge my debt to all these individuals but I must emphasise that any opinions and comments are mine and mine alone. I have tried to provide the basic factual information which readers may interpret in their own image.

Peter M. B. Walker
1992

Contributors

Extensive use has been made of the entries in *Chambers Science and Technology Dictionary* and I would like to thank all those who contributed to individual sections of that dictionary. In particular my thanks are due to Professor H. W. Wilson, the former Director of the Scottish Universities Research and Reactor Centre, who edited and contributed the Nuclear Engineering entries in the dictionary. All the figures in this book have been drawn in electronic format by myself and are based on the best information which I have been able to find.

P.M.B.W.

A CHRONOLOGY OF NUCLEAR POWER

Nuclear Energy And Radiation

1671 Newton studies the colours refracted by a prism and decides that each colour has an inherent difference.

1798 Uranium discovered by Klaproth.
1800 Infrared radiation discovered by its heating effect on a thermometer by William Herschel, who considers it to be similar to visible light.

1860 Gustav Kirchoff states that the ratio of absorption to emission for energy at any wavelength is the same for all bodies at the same temperature. Later he develops the concept of the black body, leading to Planck's discovery of quantization of radiation.

1864 Clerk Maxwell discovers that electromagnetic waves are propagated at the same velocity as light.

1868 Helium discovered in the spectrum of the Sun by Jules Janssen and Norman Lockyer.

Biology And Technology

1796 Vaccination for smallpox discovered by Edward Jenner, a country doctor.

1859 Charles Darwin's '*The Origin of Species by Means of Natural Selection*' published.

First industrial exploitation of oil at Titusville, Pennsylvania, USA.

1862 Louis Pasteur disproves the idea of the spontaneous generation of bacteria from inorganic matter.

1865 Publication by Gregor Mendel of the laws of heredity.

1873 First electric power distribution system installed in Vienna by Hippolyte Fontaine.
1876 The telephone invented by Alexander Graham Bell.

The idea of chemical potential discovered by Josiah Willard Gibbs and with it the extension of thermodynamics to chemistry.

The four-cycle internal combustion engine built by Nikolaus Otto, Gottlieb Daimler and Wilhelm Maybach.
1878 Louis Pasteur publishes his great

A CHRONOLOGY OF NUCLEAR POWER

Nuclear Energy And Radiation

1887 Heinrich Hertz discovers the photoelectric effect and confirms the existence of electromagnetic waves, possessing the same properties as light.

1895 X-rays discovered by Wilhelm Röntgen.

1896 Natural radioactivity of uranium discovered by H. Becquerel.

1897 The electron discovered and characterized by J.J. Thompson.

1898 Radium and polonium discovered by Marie and Pierre Curie.

1899 The term 'electron' is coined by Hendrik Lorentz, following his studies of electromagnetism; he also shows the identity between electrons and cathode rays.

1900 Gamma rays discovered by Paul Villard.

First indications given by Max Planck that radiation may be quantized.

1905 The quantum nature of the photon

Biology And Technology

work on microbiology and the term 'microbe' is coined by Charles Sédillot.

1880 The invention of the punched-card machine for storing and analysing data by Charles Hollerith.

1882 Construction of the first industrial alternator by Sebastian de Ferrranti.

1884 Invention of the electrical transformer by Lucien Gaulard.

Invention of the steam turbine by Charles Parsons.

1885 August Weissman in studying heredity affirms that acquired characters cannot be inherited.

1896 Guglielmo Marconi patents his system of radiotelegraphy.

1900 The rediscovery of Mendel's Laws by Hugo de Vries.

1902 The application of Mendel's Laws to animals by William Bateson and Lucien Cuénot, and the demonstration that the chromosomes behaved in the same way as Mendel's genes by Theodor Boveri and W.S. Sutton.

1903 Konstantine Tsiolkovski publishes his 'Exploration of Outer Space by Reaction Machines', the first description of how rockets work.

A CHRONOLOGY OF NUCLEAR POWER

Nuclear Energy And Radiation

demonstrated by Albert Einstein.

1911 The deduction by Ernest Rutherford from alpha-particle recoil that an atom has a small dense nucleus.
1913 Model of the radioactive atom described by Niels Bohr.

1919 The first atom split in the transmutation of nitrogen into oxygen with alpha particles by Ernest Rutherford.
1920 Neutron postulated by Rutherford.

1928 Positron postulated by Paul Dirac.
1931 Cyclotron built by Ernest Lawrence and Stanley Livingstone.
1932 Electrostatic particle accelerator built by J.D. Cockroft and E.T.S. Walton and used to bombard lithium with protons causing first artificial nuclear disintegration.
Neutron discovered by James Chadwick.
Deuterium (heavy hydrogen) discovered by Harold Urey.
Positron detected by Carl Anderson.

1934 Irène and Frédéric Joliot produce radioisotopes by alpha-particle bombardment.

Biology And Technology

1906 The discovery of chromatography by Mikhail Tsvet, a method of separating a mixture of substances.
1910 The work of Thomas Morgan on the fly *Drosophila melanogaster* finally demonstrates that the genes lie linearly along the chromosomes.

1913 The invention of the hot cathode X-ray tube by William Coolidge.
The invention of the nuclear particle counter by Hans Geiger.
1915 The theory of continental movement proposed by Alfred Wegener.

1920 The continuity between ultraviolet and X-rays in the electromagnetic spectrum established independently by Fernand Holweck and Jean Thibaud.
1926 The discovery that X-rays produce mutations in Drosophila by Hermann Muller.

1933 The discovery of the electron microscope by Ernst Brüche, M. Knoll and Ernst Ruska.

A CHRONOLOGY OF NUCLEAR POWER

Nuclear Energy And Radiation

Enrico Fermi produces radioisotopes by neutron bombardment.

Fermi produces neptunium by neutron bombardment of uranium, thus causing fission but without recognizing it.

1938 Fermi's experiments repeated by Otto Hahn and Fritz Strassman, also without recognizing fission.

Lise Meitner and Otto Frisch explain Hahn and Strassman's results as being nuclear fission and postulate that mass has been converted into energy.

1939 Independent confirmation of nuclear fission by several laboratories and the yield of neutrons from uranium fission quantified.

1940 Plutonium discovered in neutron-bombarded uranium by Glenn Seaborg.

Uranium-235 separated from uranium-238 by mass spectrometry.

1942 The Chicago Pile (CP-1) built by the Metallurgical Laboratory of the University of Chicago at Stagg Field, Illinois, USA first goes critical at 3:25pm on 2 December. It is dismantled in March 1943 to make way for CP-2, an improved design. Both used carbon as moderator and natural uranium metal as fuel.

1944 The first heavy-water moderated reactor (CP-3) goes critical at Chicago, Illinois, USA.

1945 Detonation of the first atomic bomb, code named Trinity, on 16 July.

Nuclear weapons first used on 6 and 10 August against Japan.

First nuclear reactor built outside the US goes critical at Chalk River, Canada.

1947 First nuclear reactor in Western Europe built at Harwell, Oxfordshire, UK.

Uranium fuel first manufactured in UK.

1953 Pressurized-water reactor with highly enriched uranium as fuel for use as a submarine propulsion unit, but built on

Biology And Technology

1942 As the result of irradiating the fungus *Neurospora* with X-rays and studying the mutations produced, George Beadle and E.L. Tatum show that one enzyme was made by one gene.

1948 The discovery of the transistor by John Bardeen, Walter Brattain and William Shockley.

1953 The discovery of the structure of the genetic material, DNA, by Francis Crick and James Watson, based on the

A CHRONOLOGY OF NUCLEAR POWER

Nuclear Energy And Radiation

land, goes critical.

1954 USS *Nautilus* using pressurized-water reactor launched.

1956 UK graphite-moderated, water-cooled nuclear power station commissioned at Calder Hall, Cumbria, UK.

1957 Shippingport demonstration pressurized-water reactor goes critical.

Windscale reactor fire.

1960-3 Full-scale commercial power stations commissioned in USA, all water-cooled.

1962 First full-scale commercial nuclear power station commissioned in the UK at Berkeley, Gloucestershire, using Magnox fuel arrangement.

1975 Prototype fast reactor generates electricity at Dounreay, Caithness, UK.

Browns Ferry reactor fire, Alabama, USA.

1976 Commercial Advanced Gas-cooled Reactor commissioned at Hunterston, Ayrshire, UK.

1978 The fusion project, Joint European Torus, started at Culham, Oxford, UK.

1979 Three Mile Island accident, Pennsylvania, USA.

1986 Chernobyl reactor disaster, Ukraine, USSR.

1991 The Joint European Torus fusion reactor, achieves a peak output of 14.3 Mw for two seconds in its first experiments using tritium.

Biology And Technology

structural studies of Rosalind Franklin and Maurice Wilkins.

1954 The invention of the maser by Charles Townes, which then suggested the laser.

1957 SPUTNIK, the first artificial satellite, placed in orbit by USSR, 4 October.

1960 The first laser constructed by Theodore Maiman.

1966 The genetic code elucidated by Marshall Nirenberg and others.

1969 Neil Armstrong and Edwin Aldrin land on the Moon on 21 July.

1971 The first microprocessor of 2300 transistors built.

1978 The first medical images demonstrated from nuclear magnetic resonance.

A SHORT GLOSSARY OF NUCLEAR TERMS

Specialist groups develop their own jargon and this was very obvious in the early development of nuclear bombs and nuclear power when most of the work was done in conditions of secrecy. A few of these terms have come into general science, others are highly specialized and almost forgotten. This short glossary defines some of these terms as an introduction to the following chapters, although most of them will also be found in the dictionary part of the book.

Two terms are very widely used in nuclear physics although they are not SI units or their derivatives:

barn symbol σ, measures the likelihood of one kind of nuclear particle interacting with another, especially the chance of a neutron being captured by or causing fission in a nucleus. It can be thought of as an area centred on a nucleus; the bigger the area the greater the likelihood of the neutron being captured. One barn equals 10^{-24} cm^2 and the number of barns determines what is called the *fission* or *capture cross-section*. Neutron capture cross-sections can range from 0.01 to millions of barns. It is said that the term originated because, although it is a very small area, to a neutron nearby 'it is as big as a barn door'. A view supported by the existence of a little-used but much smaller relative called a ***shed*** and equal to 10^{-24} barn.

fermi symbol *fm*, is easier to understand; it is exactly the same as the SI unit *femtometre* (10^{-15} metre) with the same symbol. Fermi is a far more elegant name, commemorating one of the great pioneers of nuclear physics. The radius of the proton is about 1.2 fermi.

The next group of terms concern nuclear energy, primarily nuclear reactors:

chain reaction the cycle of events in which more than one neutron is produced by the fission of a nucleus and these neutrons go on to cause fission in further nuclei. If the neutrons from a first fission can cause on average more than one subsequent fission, the chain reaction is *divergent* and enormous energy will be produced very rapidly. If less than one subsequent fission occurs the chain is *convergent* and the amount of power produced will quickly fall.

critical the condition in which the number of neutrons produced is just sufficient, after accounting for all losses, to cause exactly the same number of neutrons to be produced at the next generation of fissions, i.e. the reactor is in a steady state. If more are produced the assembly is *supercritical*; if less *subcritical*.

reactivity, symbol *k*, measures the overall state of a reactor and is equal to one when the system is just critical. Smaller changes can be expressed in ***niles*** (0.01 *k*) or even ***milliniles*** (0.00001 *k*).

Neutrons can be delineated by two pairs of terms, one describing their origin and the other their velocity.

prompt neutrons are produced directly from fission in the fuel (often uranium-235).

delayed neutrons come from the fission of the unstable elements made when a uranium nucleus breaks up. If criticality is maintained by prompt neutrons alone, such a reactor would be ***prompt critical*** and would have an ***excess reactivity*** equal to the contribution of the delayed neutrons, about 0.0065 *k*. The latter defines another unit of reactivity called a ***dollar*** ($1.00), equal here to 0.0065 but depending on the fissile isotope present. There are also ***cents***.

fast neutrons move at around the velocity at which they are emitted during fission, i.e. about 20000 km per sec with an energy of about one million electron-volts.

thermal neutrons have been slowed down to velocities which match those encountered in ordinary atoms at the working temperature of a reactor.

Reactors change their reactivity with time in response to alterations in control rod position, coolant flow, temperature and so forth. Thefollowing are relevant:

inhour is a curious unit which is derived from inverse hour (h^{-1}). If the neutron flux has a period of half an hour it will have a rate of change of reactivity of two inhours. There is an important *inhour equation*.

worth or *reactivity worth*, is a measure of the effect of altering the conditions within a reactor, thus the complete withdrawal of a control rod could be said to be worth a dollar or even a few cents. In the UK it would be more usual to express it in reactivity (*k*).

Two important terms define the kind of isotopes which can provide nuclear energy:

fertile describes isotopes only able to produce one neutron on fission but in doing so are *converted* to a fissile isotope. They have an even number atomic weight. Uranium-238 and thorium-232 are fertile isotopes.

fissile describes isotopes able to produce more than one neutron during fission. They can sustain a chain reaction and always have an odd number atomic weight. Uranium-235, plutonium-239, uranium-233 and plutonium-241 are fissile isotopes.

There are also terms confined apparently just to bomb-making, such as:

shake which is the lifetime of a fast neutron before it collides with another fissionable nucleus in an uncontrolled chain reaction. It is taken as 10^{-12} seconds. Used as in 'it takes seven shakes for a 100 kiloton bomb to release 99% of its energy'.

urchin is the neutron source needed to prime early fission bombs and is situated at their centre.

Early reactors were described with several special terms:

pile was the early name for the whole ensemble, as in a pile of graphite bricks.

slug was the name for the separate pieces of uranium fuel.

coupon described the arrangement of slug, moderator and structural components which could be tried out and changed in the early piles.

lattice is the more modern term for the geometric arrangement of fuel, moderator and coolant and is still in use.

There are many other more specialist terms particularly for fuel handling and storage, e.g. canal, coffin, pond. These are defined in the dictionary.

Chapter 1

Atoms and Neutrons

This chapter provides some of the information needed to understand how various nuclear devices work. It is therefore a preliminary to a discussion of nuclear reactors, bombs and nuclear safety. It is concerned with the structure of the atom and the way that the elements can be given a shorthand name which describes their physical properties. It is also concerned with the behaviour and properties of the key atomic particle of the nuclear age. This is the *neutron*, a common product of the 'splitting' of the atom and the intermediary in releasing its energy.

The structure of the atom

There are commonly believed to be 92 naturally occurring elements in the universe, although about 10 grams of a 93rd, plutonium-244, may be present on Earth because of the presence of its disintegration products and tracks in meteorites. Further, plutonium-244 is the only isotope heavier than uranium with a half-life comparable to the age of the Earth at about 100 million years. Each element can be distinguished from the others on the basis of its chemical properties. Iron is different from the gas, chlorine, which are both different from the carbon found in diamond or, in another commoner form, graphite. They differ because each has a different number of electrons in orbit round a central nucleus. These electrons can be shared with electrons from another element to form molecules of a chemical compound with properties very different from its parents. A good example is table salt, sodium chloride, which is made from one atom of sodium, a highly reactive metal which will explode if put in water and one atom of chlorine, a greenish-yellow poisonous gas.

Each electron has a small mass and a negative electrical charge. The negative electrons are in orbit round a nucleus with an equal number of positive protons to maintain an electrically neutral atom. The reason why attraction between the positive and negative charges does not cause the atom to collapse is because the orbit of each electron is defined by special quantum-mechanical considerations. The inner electrons must maintain their least energetic position, which is distant from the nucleus (see Fig. 1.1a).

There are therefore up to 92 (perhaps 94) protons in the nuclei of the natural elements and the same number of electrons revolving round them. This was believed to be the structure of the atom in the 1920s. A third component, the *neutron* was discovered by James Chadwick in 1932. The neutron has almost the same mass as a proton but carries no charge and one or more are present in all nuclei except the lightest, hydrogen. The identification of an atom therefore requires not only the *atomic number*, i.e. the number of protons (and electrons), but also the number of neutrons. The convention is to add the number of protons and neutrons to give the *mass number* and both together are called *nucleons*. One way of writing this is $^{3}_{2}$He, where He is the shorthand name for

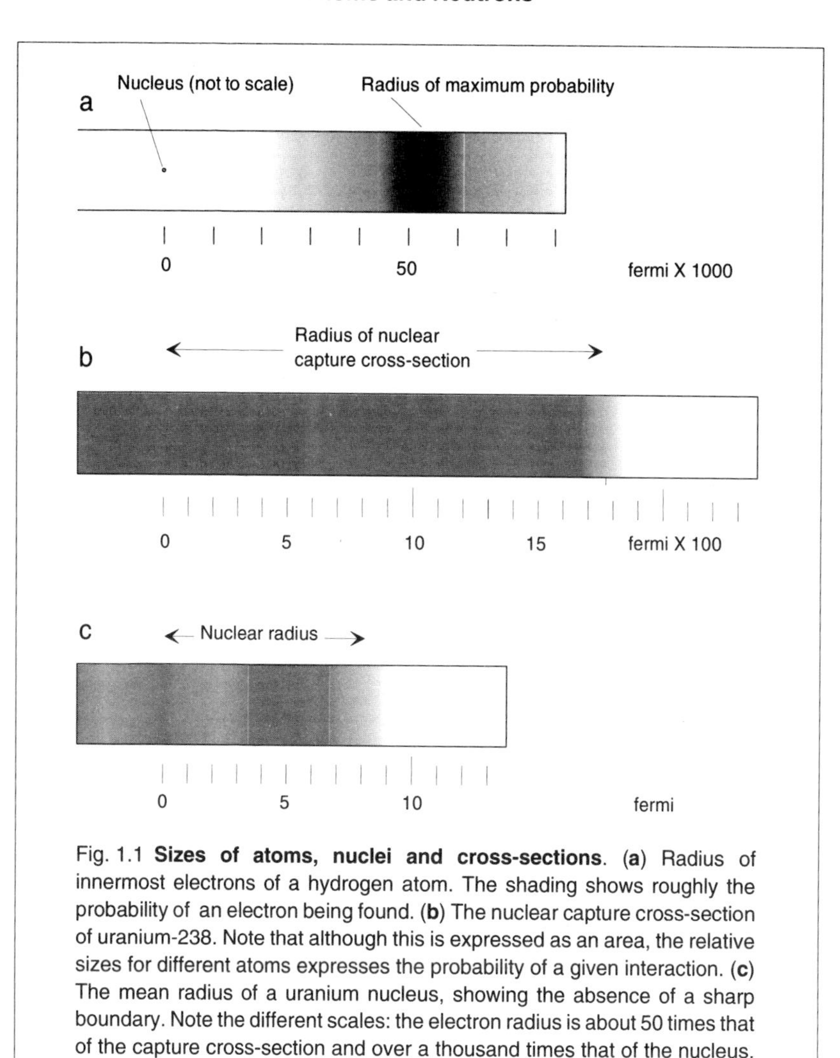

Fig. 1.1 **Sizes of atoms, nuclei and cross-sections**. (**a**) Radius of innermost electrons of a hydrogen atom. The shading shows roughly the probability of an electron being found. (**b**) The nuclear capture cross-section of uranium-238. Note that although this is expressed as an area, the relative sizes for different atoms expresses the probability of a given interaction. (**c**) The mean radius of a uranium nucleus, showing the absence of a sharp boundary. Note the different scales: the electron radius is about 50 times that of the capture cross-section and over a thousand times that of the nucleus. Dimensions are in fermi, 10^{-15} metre.

the element, Helium, the upper number 3 is the number of nucleons, and the lower number, 2, is the number of protons, although the latter is usually left out. Other examples are $^{238}_{92}U$, for one kind of uranium and $^{12}_{6}C$ for carbon. The alternative names, e.g. uranium-238 and carbon-12 are used in most of this book.

If the atomic number is preserved, but the number of neutrons in an element's nucleus changes so that the element has a different mass number, then the element will be a different *isotope*. They are often called *radioisotopes* if they disintegrate spontaneously (see panel, p. 10). These spontaneous disintegrations cause the element to lose mass, which is converted into energy and detected as radioactivity. A good example is hydrogen, the lightest element and a gas at normal temperatures.

atoms and isotopes

The ***atomic mass unit*** (symbol u) which is defined as one twelfth of the mass of the commonest isotope of carbon:

$$1 \text{ u} = 1.6598 \times 10^{-27} \text{ kg} .$$

In this system the mass of the hydrogen atom, one proton and one electron is very nearly but not quite 1. Similarly, a proton alone has a mass of 1.007 82 u and a neutron 1.008 67 u. The mass of an electron is about one two thousandth of that of a proton.

The ***mass number*** (abbrev. *A*) of an atom is the total number of protons and neutrons in its nucleus. Electrons have a mass number of 0.

The ***atomic number*** (abbrev. *Z*) of an atom is the same as the number of protons in its nucleus. These carry a positive electrical charge which is exactly balanced by the number of electrons surrounding the nucleus. An atom of an element, *X*, can therefore be defined as follows,

$${}^{A}_{Z}X_{N},$$

where *N* is the number of neutrons and is equal to *A* – *Z*. Because every element, *X* has a unique atomic number, both *Z* and *N* are redundant and usually omitted. For clarity, an isotope of an element can be written as *Name* -*A*, e.g. Uranium-235. This method is used in this book.

Isotopes of an element (*A* varying but not *Z*) can be of two kinds, *stable* and *unstable*; stable isotopes do not spontaneously disintegrate and an element can have one stable isotope or several (tin has 10). Unstable isotopes (also called *radioisotopes*) will disintegrate and are themselves the result of the radioactive decay of elements with higher mass numbers. Two other less common terms are sometimes used: ***isobar*** refers to elements sharing the same mass number (having the same *A*) and ***isotone*** refers to elements with the same number of neutrons (the same *N*).

The stability of an atom depends on the number of protons and neutrons in its nucleus. For mass numbers (A) below about 40, stable isotopes must have about the same number of protons as neutrons, a ratio near one; for mass numbers between 40 and 200 this ratio declines to about 0.6 and above 200 all elements are radioactive.

Unstable isotopes can change to other forms in several ways, depending on whether their proton to neutron ratio is high or lower than the stable figure. If it is higher, the isotope will lose a proton:

$$\text{proton} \Rightarrow \text{neutron} + \beta^{+} + \nu,$$

where β^{+} stands for a positive beta particle, i.e. an electron with a positive charge (called a *positron*), and ν is a **neutrino**, a massless uncharged particle which is lost; the element will change to the next lower element in the periodic table. If the ratio is lower, the isotope will lose a neutron:

$$\text{neutron} \Rightarrow \text{proton} + \beta^{-} + \bar{\nu},$$

where β^{-} is an ordinary beta particle, i.e. an electron, and $\bar{\nu}$ an antineutrino; the element changes to another of its isotopes. Heavy elements can also lose excess positive charge by losing two protons and two neutrons:

$${}^{A}_{Z}X_{N} \Rightarrow {}^{A-4}_{Z-2}X_{N-2} + {}^{4}_{2}\text{He}_{2} ,$$

where the He nucleus is the alpha particle. Conversions from one isotope to another or from one element to another must obey the following rule: the mass number, neutron number and atomic number (therefore the electric charge) must be the same on each side of the equation before and after conversion. The atomic number of the electron is taken as –1 (positron +1).

Its commonest isotope is simply ${}^{1}_{1}H$, but there are two others ${}^{2}_{1}H$ or deuterium which does not spontaneously disintegrate and ${}^{3}_{1}H$ or tritium, which is unstable and therefore radioactive.

Half-life

Half-life ($\tau_{1/2}$) defines the rate at which an unstable isotope will decay and is the time required for half the amount of isotope to change into its daughter product, which may itself be unstable. This time can vary from microseconds or less to thousands of years. Decay is a random process and it is impossible to predict the exact time at which any single nucleus will disintegrate.

The size of the atomic nucleus

The atomic nucleus is not like a marble with a hard surface which allows its size to be measured accurately. Rather it appears to have fuzzy edges with its mean density relatively constant up to a certain radius and then dropping off. Nuclear diameters can be measured in several ways including determining how high-energy electrons are scattered by different sized nuclei. The results are rather consistent: nucleons (protons + neutrons) do not cluster near the centre but are spread fairly uniformly out to a radius depending on the value of *A*, the mass number. The relation is as follows:

$$R = R_0 A^{1/3},$$

where R is the mean radius in fermis (fm) and the constant $R_0 \approx 1.2$ fm. The fermi is a non-SI unit, very convenient for nuclear dimensions and equal to one femtometre (10^{-15} m). Fig. 1.1c illustrates the size of one nucleus.

The Coulomb force and nuclear binding

As we have seen, electrons and protons attract each other by virtue of their opposite electrical charges and the protons in a nucleus repel those in other nuclei because they all carry positive charges. The electrical force of attraction and repulsion is called the *Coulomb force* and well describes the properties of atoms. What keeps the protons and neutrons together in the nucleus? This is the *strong nuclear force* carried by **gluons**, which is now understood by theoretical physicists but is outside the scope of this book. It does have, however, one profoundly important property: it is only effective over very small distances. If two nuclei are to have any chance of fusing together they must come almost as close as protons and neutrons are already in the nucleus.

Nuclear fission

Binding energy

The binding energy is the key factor which allows energy to be extracted from atomic nuclei when they either fuse or fission and is the *difference* between the mass energy of a nucleus and the mass energy of the protons, neutrons and electrons of which it is made. The panel on p. 13 defines mass energy in the appropriate units and shows how the binding energy can be calculated but the essential result is given in Fig. 1.2. This is a plot of the binding energy per nucleon (*B/A*) against the mass number (*A*) and shows a number of important features. Although *B/A* stays fairly constant around a value of eight for most elements, for those with mass numbers below about 30, *B/A* increases with higher values of A, but for elements with *A* above 100, the reverse occurs. This means that for the light elements fusing two nuclei will produce energy, while for heavy elements fission to give lighter elements gives energy. Iron-56, the most stable element, separates the two regions of the binding energy curve. It also shows why fusion reactions release more energy for a given mass of fuel than fission reactions.

Fission

Natural uranium contains 0.72 per cent of an isotope, uranium-235 which has a unique property among natural isotopes. Bombarding its nuclei with neutrons causes them to form a compound nucleus, uranium-236 which then immediately disintegrates or *fissions* into two other nuclei, emitting mostly neutrons in the process. Because the mass of all these fission products is less than the parent nucleus and its bombarding neutron, energy must be released and most of this appears as the kinetic energy of the two fission nuclei as they fly apart. Then as they collide with other nuclei they will slow down and convert their kinetic energy into heat which can eventually provide steam in a nuclear power plant. Some additional energy is released in the radioactive decay of the fission product nuclei (see p. 18).

A small part of this excess energy is in the form of fresh neutrons expelled during the process. Because there is on average, two and a half neutrons produced for the one needed to initiate the conversion, a *chain reaction* can occur, in which the fresh neutrons collide with more nuclei and initiate a further round. If that was all that there was to it, highly enriched uranium-235 would have exploded long ago. These neutrons, however, have too high an energy for many of them to cause another nucleus of uranium-235 to fission, i.e. their *fission cross-sections* (see below) are too small for high-energy neutrons. Many will therefore easily escape from a small mass of fissionable isotope. To produce useful power these neutrons must lose energy by colliding with the nuclei of substances, which do not themselves absorb many neutrons, until their energies match the nuclear energy levels of the uranium-235 fuel. Such substances have a low molecular weight and are called *moderators,* and include graphite and water. What happens when neutrons of different energies collide with a nucleus is the key to understanding how nuclear devices work and depends on the *cross-section* of the nucleus and involves *resonance.*

Fission and capture cross-sections

The *nuclear cross-section* is a measure of how close a nuclear particle must come to a nucleus before it can interact with it. It measures the *probability* of such an interaction. The cross-section (abbrev. σ) has the dimensions of an area and is measured in a unit, peculiar to nuclear physicists, called a *barn,* which has the value of 10^{-24} cm^2 (see Fig. 1.1b). For neutrons, σ_f refers to the fission, σ_a to the capture and σ_{tot} to the combined cross-section.

Neutrons can have a wide range of different velocities depending on their origin and history. It follows that their kinetic energy, half the *mass* times *velocity* squared and measured in electron-volts (eV), can also vary widely. In nuclear reactions we are concerned with a range from a few million electron-volts, the

binding energy

The *mass energy* of a nucleus or any other particle is the product of its mass and energy, and is defined by the mass-energy equation which Einstein deduced from the Special Theory of Relativity:

$$E = mc^2,$$

where E is the energy, m the mass and c the speed of light. In nuclear physics it is convenient to define m in terms of u, the atomic mass unit, when $u.c^2$ becomes 931.50 MeV (million electron-volts).

To a high degree of accuracy the mass energy of a particular nucleus is equal to the mass energy of the atom less that of its electrons which can be simply expressed as the mass energy of hydrogen-1 multiplied by the number of protons plus the mass energy of all the neutrons. A nucleus X will have Z protons and N neutrons and so its binding energy will be Z times the mass of a hydrogen-1 atom plus N times the mass of a neutron less the mass of the combined nucleus X which will have $Z + N$ (= A) nucleons. All of this being multiplied by c^2 as shown:

$$B = \left[Zm(^1\text{H}) + Nm(neutron) - m(^A\text{X})\right] c^2 ,$$

where m is the mass of the particle within the following brackets. The figure below shows the relation between binding energy and mass number.

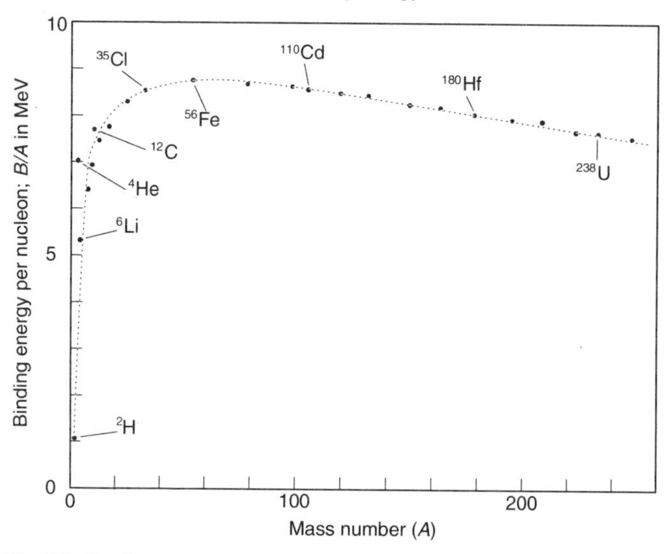

Fig. 1.2 **Binding energy per nucleon plotted against mass number.** A few representative nuclides are shown.

energy at which neutrons are emitted from a fission event, to about a fortieth of an electron-volt which corresponds to the energy of neutrons moving with the energy corresponding to the most probable velocity according to the **Maxwell-Boltzman distribution law**.

The best way to visualize how neutrons react with nuclei is to plot the nuclear cross-section against neutron energy but we first need to distingish between the three different outcomes of a neutron-nucleus interaction. Firstly, neutrons can

Absorption cross sections for low energy neutrons					
Use	Material	σ_a (barns)	Use	Material	σ_a (barns)
Fuel	Uranium-235 Uranium-238	687.0 2.75	Moderator	Water Heavy water	0.664 0.001
Reactor poisons	Xenon-135 Samarium-149	3.5×10^6 5.3×10^4		Carbon Beryllium	0.005 0.01
Control rods	Boron Cadmium	775.0 2450.0	Reactor vessel Fuel rods	Steel Zirconium	2.53 0.18

be scattered by a nucleus or bounced off it in what is called an *elastic collision.* Secondly, they can be *captured* by the nucleus to form a different isotope and eventually decay to a more stable state. Thirdly, they can cause the nucleus to *fission* in those rare elements which have fissionable nuclei, mainly uranium, plutonium and thorium. The last two kinds of interaction largely determine the size of nuclear fuel elements and the first, elastic collisions, are important for moderation and determine how fuel elements are spaced.

A neutron of the same energy can behave very differently depending on the particular isotope with which it collides. Fig. 1.3 shows the plots of neutron energy against fission and capture cross-sections for uranium-235 . Its fission cross-section broadly increases as neutron energy falls while its capture cross-section is over 100 times less than that of uranium-238 at low energies and about 10 times less than the latter's fission cross-section below 1 eV. Fig. 1.4 shows similar plots for uranium-238 but with very different results. It will be seen that below 1 MeV neutrons are incapable of causing fission in uranium-238 but, on the other hand, uranium-238 can *capture* neutrons very effectively right down to below 10 eV, with a number of very large *resonance peaks* above this value. Scattering cross-sections are important because heavy elements can bounce back neutrons which hit them and can therefore act as *reflectors* to prevent neutrons escaping and because light elements will absorb energy from a neutron and therefore slow it down, accepting kinetic energy themselves. Light elements can therefore be used as moderators provided they do not capture too many neutrons in the process.

The task of the designer of a nuclear power plant which uses thermal neutrons (as opposed to a fast reactor) is to get the high-energy neutrons out of the fuel as quickly as possible and into the moderator and so avoid their capture by uranium-238. Then he must let the neutrons stay in the moderator for enough time to be slowed down to less than 0.1 eV before being able to collide with the uranium-235 in the fuel again. This requires around a hundred collisions and a time of 100 microseconds in carbon-12 and about 20 collisions in the hydrogen nuclei of water. The table on p. 14 summarizes nuclear cross-sections for some of the substances encountered in the nuclear industry.

Resonance

We have mentioned that the peaks seen in the nuclear cross-section plots are called resonance peaks. Resonance is a concept familiar in physics and every-day mechanics. When a child is sitting on a swing small pushes will cause the swing to go higher and higher provided you time your pushes with the periodicity of the swing. At other timings not only must you push harder but the effect

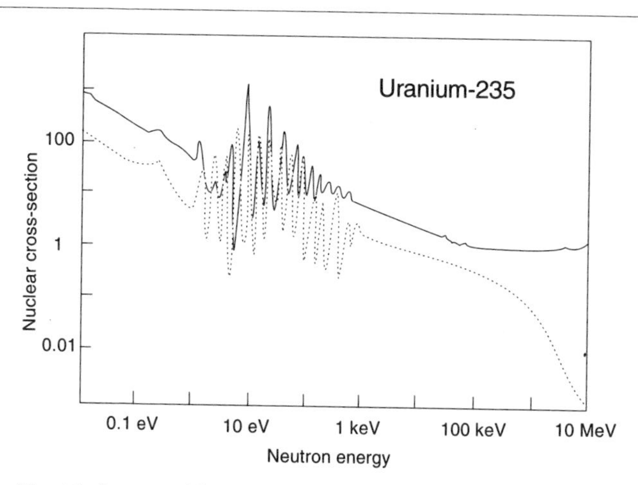

Fig. 1.3 **Capture (dotted line) and fission (solid line) cross-sections for uranium-235**. The greater fission cross-section below 1 eV should be noted. This region avoids both the many resonance peaks of uranium-235 but also those of uranium-238 (see below).

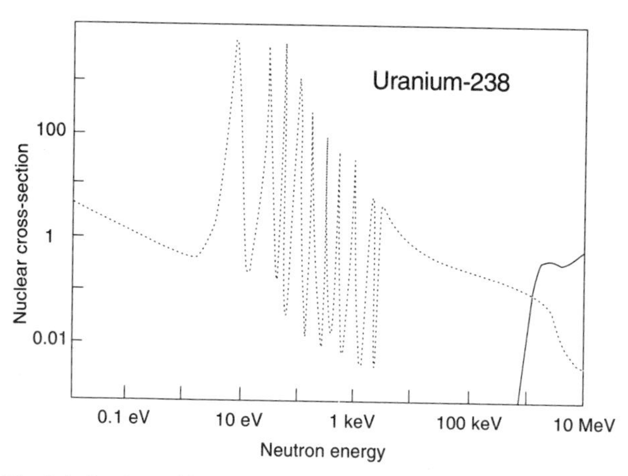

Fig. 1.4 **Capture (dotted line) and fission (solid line) cross-sections for uranium-238**. Note the rapid fall in the fission cross-section below 1 MeV.

of your mistimed pushes will usually reduce the height of the swing. In other words, the closer you time your pushes to the movement of the swing, the more energy is transferred from you to the swing. The same effect occurs in other vibrating systems. The nuclei in the fuel of a reactor have a range of vibrational energies, less than 100 eV which are clustered into peaks and which means in turn that any neutrons, only partially slowed during their passage through the moderator, will have a high probability of capture if they stray into the fuel.

Conversion to plutonium

An important factor in the efficient use of nuclear fuel in a reactor is the conversion of a fraction of the uranium-238 to plutonium-239, because during the life of the reactor, the fission of the newly formed plutonium can replace, in part, the uranium-235 that is necessarily burnt up. This ensures that a larger amount of energy can be extracted from a given amount of fuel although it cannot, in thermal reactors, replace all the uranium-235. In breeder reactors which use fast neutrons and therefore have no moderator, the total number of fissionable nuclei can increase during burnup. Such reactors not only convert but also *breed*.

Fission devices and power reactors

The art of making an atomic bomb is to ensure initially that, for a given mass and density of fissionable isotope, the number of effective neutrons produced is below one and no chain reaction occurs. Then if its mass or its density can be increased sufficiently suddenly, the nuclear fuel can be made to explode (see Chapter 7). There is, therefore, a *critical mass* below which the chain reaction will not occur.

Somewhat different criteria apply to nuclear reactors. It is essential to control the number of neutrons present, or neutron intensity, so that just one *effective* neutron is produced for every disintegration of uranium-235. In general, if the flux rises, more of a neutron absorbing substance, like boron, is moved into the reactor and if it falls, less.

As we have discussed, a neutron which collides with a nucleus causes one of three things to happen: the nucleus captures the neutron or the neutron bounces off, changing direction and losing some of its energy (see Fig. 1.3) or, in some elements, fission may occur. Natural uranium with only 0.72% uranium-235 cannot provide enough neutrons to sustain a chain reaction after the disintegration of an atom of uranium-235 if ordinary light water is used as a moderator. Although the water nuclei will slow the neutrons, too many of the latter will be absorbed in the process. It is therefore necessary to enrich the uranium with between 1 and 5% of uranium-235. Heavy water or deuterium oxide will also slow neutrons but its capture cross-section is about a six hundredth of that of light water so more neutrons survive and unenriched uranium can be used as a fuel. This is the basis of the successful Canadian design of power reactors called CANDU.

There are many ways in which the essential elements of a nuclear reactor can be put together. Choice will depend on the power required, the availablity of fuel, coolant and moderator and the purpose for which it is built, e.g. power, propulsion, research or isotope production, but there are only a comparatively small number of designs suitable for large-scale power stations. Another important distinction is between civilian power reactors and *military reactors*. In a military reactor the fuel is left in the reactor only long enough to obtain a reasonable yield of plutonium without the production of an excess of other fission products which are difficult to handle because of their radioactivity. In a civilian reactor such an approach would be uneconomic and a much greater buildup of long-lived fission products has to be tolerated.

Reactor preferences and national history

After the development of the atom bomb there were two types of fissionable material available: uranium-235 and plutonium-239. Another man-made isotope of uranium, uranium-233, is also fissionable. To produce uranium-235 the US government built three immense plants which made use of the differences between the rates of diffusion of uranium hexafluoride gases containing the two isotopes. Enriched uranium-235 was therefore expensive and its export carefully controlled at the time for political and military reasons. Plutonium can be made by a rather simpler process in which the common form of uranium is bombarded with neutrons in a nuclear reactor. The British, who did not have access to enriched uranium at the time but did want to make an atomic bomb, therefore chose to build a nuclear reactor using natural uranium, graphite moderator and carbon dioxide as the coolant. As these were military reactors, the fuel rods were removed after a short time and transferred to a reprocessing plant which separated the plutonium from the other elements present, a problem in chemistry much simpler than the physical separation of uranium isotopes. As a by-product, these reactors at Calder Hall in Cumberland used their coolant to make steam and drive generators from which electricity was sold to the national grid. They were therefore the first commercial power producers in the world. The Canadians also had no enriched uranium but neither did they want to make an atom bomb. They took a third route with natural uranium and deuterium oxide (*heavy water*) as moderator for building power stations.

This early history has largely influenced subsequent developments in different countries with argument still as to which approach is more efficient or safer 40 years later. Some of this indecision stems from the shear size of the individual projects and the length of time needed for their completion and evaluation. Further, they are so large and their technology so new that governments and committees have inevitably had to make the important decisions. The next chapter describes at a more practical level the problems raised by the need to design an economic power station.

Chapter 2

The Reactor Problem

The last chapter considered the basic structure of the atom and its nucleus and some of the properties of the nucleus which show how energy can be recovered from fission and fusion. This chapter turns to more practical problems in particular the definition of the more significant structures in nuclear reactors and the way in which they have to be arranged so that useful power can be extracted in a controlled way.

Neutrons

Uranium-235 disintegrates to produce various fission products, neutrons and energy:

$$^{235}\text{U} + \text{neutron} \Rightarrow {}^{A}_{Z}X_{N} + {}^{A'}_{Z'}X'_{N'} + x \text{ neutrons} + \text{E},$$

where X and X' are fission products, x is the number of neutrons and E = energy. The rules for balancing such an equation require that the number of protons and neutrons should be the same before and after a disintegration, so that $Z + Z' = 92$ and $A + A' + x = 235$, but this can hold for many pairs of products. The energy produced is roughly equivalent to 200 MeV or 3.2×10^{-11} joules per disintegration so that one gram of uranium-235 (2.56×10^{21} atoms) would release 8.19×10^{10} joules, equivalent to burning nearly 14 barrels of crude oil.

The 200 MeV is divided as follows:

MeV	167	5	7	6	11	5
Form	Kinetic energy of products	β-particles	Prompt γ-particles	delayed γ-particles	Neutrinos	Fast neutrons
Effect	Absorbed in fuel elements		Irradiate structure		Lost	To chain reaction

Prompt particles result from the neutrons emitted by the fission of uranium-235. *Delayed* particles come from the neutrons emitted by the fission products formed by the primary disintegration. *To chain reaction* indicates that these neutrons are available for continuing the chain reaction.

Fast neutrons initially have an average energy of about 2 MeV and, as we have seen, a much lower probability of inducing a further fission in another nucleus of uranium-235 than if they more closely matched the thermal energy of the fuel. For maximum effect their energy must be reduced to about 3 eV by passage through a *moderator*; 3 eV is in the range of the normal thermal processes which effect nuclei. Neutrons of this energy are called *thermal* or *slow* neutrons and reactors which mainly use thermal neutrons are *thermal reactors*. Other types of reactors, such as those used in *breeding*, the process of producing more fuel than was initially present, make use of high-energy or fast neutrons; hence the terms *fast breeder* and *fast reactor*.

The main elements in a thermal-fission reactor are shown schematically in Fig. 2.1. They are discussed below together with other topics that affect the design of reactors.

Moderators

A moderator reduces the energy of neutrons without capturing too many of them. In other words a neutron and a moderator nucleus should collide like a billiard ball, with the neutron changing direction and transferring energy. Such nuclei have low molecular weight and carbon, beryllium, water or heavy water are good moderators and can circulate or be placed around the fuel elements. As neutrons gradually lose energy they pass through energy levels at which they are likely to be captured by other elements, in particular the common isotope of uranium, present in the fuel. For this reason it is important to space the fuel elements so that most of the neutrons pass from the fast to the thermal state without re-entering the fuel.

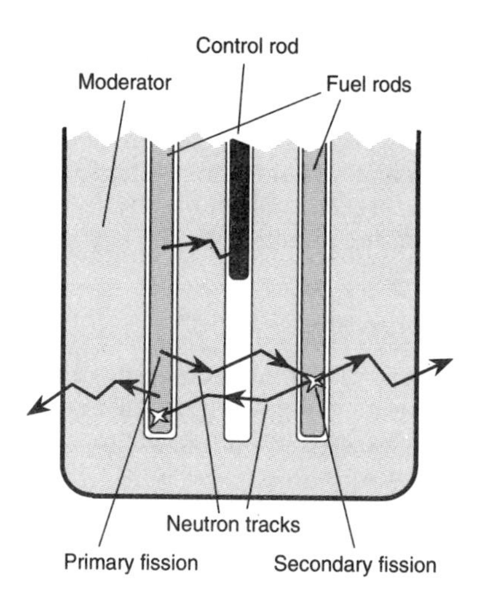

Fig. 2.1 **Schematic diagram of a fission reactor**. Neutrons are shown escaping, being absorbed and initiating fission.

Control

Other elements, such as boron and cadmium, have the ability to absorb neutrons with an efficiency intermediate between reactor poisons and moderators. Therefore the number of neutrons available to sustain the chain reaction can be controlled by mechanically withdrawing or inserting rods made of these substances. The neutron flux will fall as the nuclear fuel is burnt until the amount of uranium-235 approaches half its initial value and the reactor has to be refuelled, but during the life of the fuel the flux must be maintained by slowly withdrawing the control rods. The same control elements are usually also needed to shut down the reactor and protect it against wear and tear in the fuel elements and any other accidents.

Reactor poisons

Other elements, especially those present among the fission products, capture neutrons very effectively. Of these xenon-135 and samarium-149 capture neutrons about a million times more effectively than the light elements found in moderators. Very little of these elements can shut down a reactor or *poison* it (see table on p. 14). The gas xenon-135 is a decay product of iodine-135, a common fission product, and in a normally running reactor it is made and destroyed at the same rate. In a reactor that is starting up or shutting down this equilibrium takes time to re-establish and can cause control problems.

Reactivity

It is important to determine how responsive a reactor is to changes in its operating conditions and how easily it is to control. *Reactivity* is the change in the number of neutrons or neutron *flux* which follows a change in some factor like the position of a control rod. The flux of *prompt* neutrons, i.e. those coming from the primary disintegration, will change tenfold about a second after moving a fuel rod in a working reactor and it would be difficult to achieve stability with such extreme sensitivity. Fortunately there is a small proportion ($< 1\%$) of *delayed* neutrons coming from the fission products whose flux alters more slowly with time; typically with response times up to a minute. Fission reactors are therefore designed to operate with a prompt neutron flux just below that required to sustain a chain reaction, i.e. just below *criticality*, and to rely on the delayed neutrons to bring them up to this figure. In other words, if a chain reaction cannot be sustained in the absence of delayed neutrons, their presence prevents the reactor responding too rapidly to any change in operating conditions.

Temperature effects

Reactors must function safely over a range of temperatures, particularly during startup and rundown. If raising the temperature caused the neutron flux to increase and thus more heat to be generated, the reactor would be in danger of *thermal runaway*. However, the higher the temperature the less dense the moderator and therefore the less chance of a neutron losing energy by collision with a moderator nucleus. The neutrons will therefore have higher energies on average and a better chance of being captured by uranium-235 (*resonance capture*). Converting water to steam stops the chain reaction in water- or deuterium-moderated reactors for the same reason but it also causes the loss of almost all its cooling capacity.

Doppler broadening of the resonance absorption peaks of elements like uranium-238 is another factor that has to be taken into account. As neutrons lose energy they must at some time have the same energy as the resonance absorption peaks of the uranium-238. If such a neutron escapes from the moderator and enters the fuel it will be captured. Higher temperature broadens these absorption peaks and so increases the proportion of neutrons with matching energies. More neutrons will be captured and the *resonance escape probability* will fall (see the next section).

The reactor problem

These effects and others, like the proportion of neutrons which escape the core, contribute to the *reactor problem*, whose solution determines whether a design can sustain a chain reaction without the chance of a rapid increase in the number of neutrons and the risk of *meltdown*. They all depend not only on the fuel, the moderator and the coolant but also on the positions and shapes of the components. Calculating the effects of various arrangements is no easy task. Much calculation and experiment were needed to determine the disposition of the most suitable materials in the early years of the nuclear age and so solve the reactor problem. That is in ensuring that just one neutron can survive to continue the chain, no more and no less.

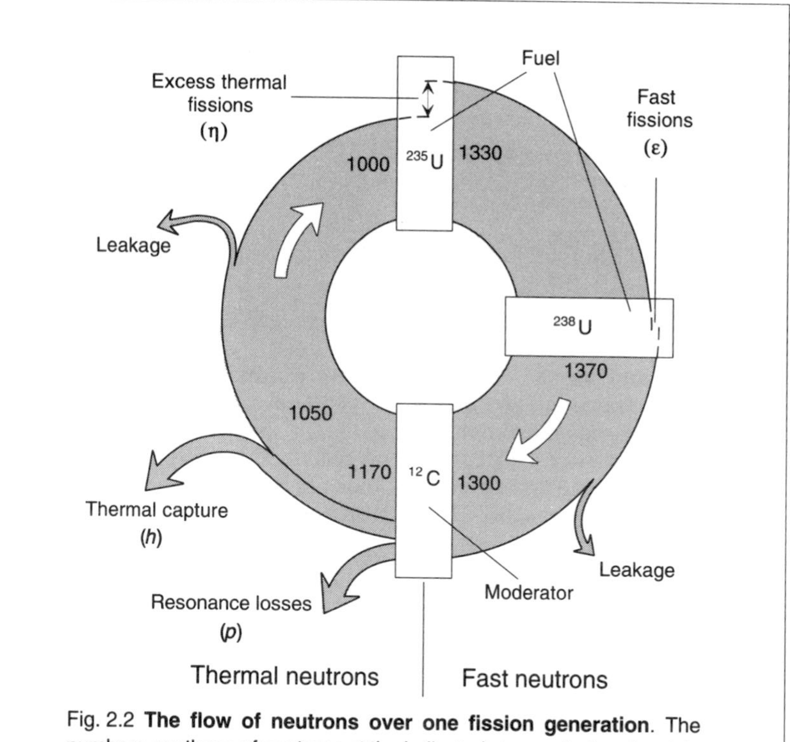

Fig. 2.2 **The flow of neutrons over one fission generation**. The numbers are those of neutrons at the indicated stages, fast on the right and thermal on the left. The symbols in brackets refer to the factors mentioned in the text that are associated with the various methods of neutron loss.

The factors governing a single generation of neutrons is illustrated in Fig. 2.2. When 1000 thermal neutrons arrive at the fuel containing uranium-235 (top of the figure), they will not produce overall anything like the 2500 fresh neutrons which the number of neutrons expelled from a single disintegration might have led us to expect. This is because not all incoming neutrons will meet a uranium-235 nucleus and many will be absorbed by non-fission reactions with both isotopes of uranium. A factor $\eta \times 1000$ fast neutrons will ultimately be produced where η is typically 1.33.

A small number of these fast neutrons can induce fission in uranium-238 which increases the total number of fast neutrons available by the *fast fission factor* of ε (≈ 1.03). Some of this increased total of 1370 fast neutrons is lost from the reactor by leakage and only 1300 pass into the moderator. As these neutrons lose their energy in the moderator a proportion is captured by the nearby uranium-238. This is the *resonance escape probability* (p) and about 0.9 of the neutrons escape and survive to the next stage. Lastly, some of the neutrons are absorbed by the moderator itself, so the fraction of the neutrons available for the next generation must be multiplied by the *thermal utilization factor* (f) of, again, 0.9. In addition there will be some further leakage of thermal neutrons.

A balance has to be made in the design between increasing the resonance escape probabilty by lengthening the mean neutron path in the moderator and allowing a further loss of neutrons as the thermal utilization factor falls. A mean path of 19 cm in graphite, which is the distance required for the 100 or so collisions needed to thermalize neutrons, is commonly used.

From all this the *neutron reproduction factor* (k) can be found for the simple case, neglecting any leakage, from the relation

$$k_\infty = \eta \varepsilon p f.$$

This is called the *four-factor formula*, the subscript in k_∞ indicating that the reactor is of infinite size and therefore without neutron losses; if $k < 1$ then the reactor is subcritical; if $k = 1$ it is critical and it is supercritical if $k > 1$.

Raising or lowering the output of a reactor is done by altering the neutron reproduction factor but any excess over one must be kept small. Reactivity has been discussed and with it the need to rely on the delayed neutrons, with their longer half-lives, to control the reactor. If the reproduction factor exceeds the sum of that for the prompt neutrons (taken as one) and the delayed neutrons (typically 0.007) then the number of neutrons will increase exponentially with a doubling time in milliseconds. This is the condition of *prompt criticality* and the reactor may be impossible to control thereafter.

Chapter 3

Energy from nuclear fission

Conventional power plants burn fossil fuels, such as coal, oil or natural gas. They make use of the *chemical* energy stored in these hydrocarbons to produce heat when burnt in the presence of oxygen from the air. Nuclear power, on the other hand, uses the energy stored in the atomic nucleus to provide heat which in turn boils water and drives the turbines of an electrical generator. Nuclear bombs, whether they are *fission* devices which derive energy from the breakup of uranium or plutonium nuclei, or *fusion* devices, which give energy from the forced combination of light atomic nuclei, such as deuterium, tritium or lithium, are, once initiated, essentially uncontrolled processes. Any nuclear mechanism able to give useful power must be controlled so that the energy is produced slowly enough to make steam and drive generators safely.

The attraction of nuclear power is that fossil fuels will eventually run out while uranium is comparatively plentiful and can even be made to 'breed' more fissionable material as plutonium-239. The raw materials for a fusion device are deuterium which occurs in virtually unlimited quantities in the sea and lithium which is about 20 parts per million in the Earth's crust. Unfortunately fission reactors, using uranium, are highly complex and produce highly radio-active material as a by-product which must be stored or processed. Fusion reactors which produce much less by-product radioactivity, have not yet been made to give more energy than is needed to run the reactor.

Reactor classes

We describe in this chapter the main features of five kinds of reactor systems, found in nuclear power stations in a number of countries. These include the two kinds of light-water moderated reactors, the *pressurized-water* and *boiling-water* designs, which have compact reactor cores but rely on a coolant which will drastically change its properties if it boils from water into steam. There is also an interesting Canadian design, using heavy water as both moderator and coolant, which has a good record for reliability but is not as fuel efficient as some more modern designs. Then there are the gas-cooled reactors which have massive graphite cores and a coolant which cannot boil. They are also of two kinds: *carbon dioxide-cooled*, many of which have been built mainly in the United Kingdom but also in France and elsewhere, and secondly a rather miscellaneous class of high-temperature reactors (*HTR*) which are *helium-cooled* and have not been used commercially despite the advantages of using a very non-reactive coolant, largely because of the difficulty of maintaining the structural integrity of the combined fuel and moderator at high temperatures. Another design, the Soviet RBMK reactors which are graphite moderated but water cooled are described in the chapter on nuclear safety (Chapter 5). Finally fast-breeder reactors are discussed.

In each case the overall design is described together with the more important parts of the arrangement of the fuel elements and control rods in the reactor core. The next chapter describes progress in providing power from nuclear fusion.

Light-water reactors

More reactors, using ordinary or light water as a moderator and coolant, have been built than any other type, particularly in the USA. They are of two kinds; pressurized-water reactors (PWR) and boiling-water reactors (BWR). In both the fuel is in the form of small cylindrical pellets about one centimetre in diameter and up to two centimetres long, made of enriched uranium oxide powder which has been heated to a high temperature so the particles form a homogeneous solid matrix of ceramic. To keep the fuel separate from the water, the pellets are encased in tubes or *fuel pins* about four metres long made of an alloy of zirconium, called zircaloy. A power station reactor may contain over 50 000 fuel pins, holding between 100 and 150 tonnes of uranium oxide. After about 4% of the uranium atoms (60% of the uranium-235) have been burnt up, the reactor can no longer maintain its power output and has to be shut down, the core flooded with borated water and the top of the pressure vessel unbolted and lifted off. It is then possible to remove the spent fuel assemblies and put them in a storage pond. Generally one third of the fuel assemblies are removed each time and the old and new assemblies rearranged to ensure optimum burnup.

Pressurized water reactors (PWR)

In a PWR the reactor core, the heat exchanger and the pipework connecting them must be able to contain the light-water moderator which is pressurized to more than 138 bar (2000 pounds per square inch). The inlet water, coming from the steam generators at 295°C, is heated to 325°C as it passes over the fuel rods and is pumped round a circuit including the steam generator. The latter boils water in another, entirely separate, secondary circuit at a slightly lower temperature and pressure to provide the steam for the turbines (Fig. 3.1). In a modern power station one reactor is connected to four heat exchangers, which provide steam for two turbo-generator sets. This is called a *four-loop* configuration. The radioactive primary circuit from the reactor core to the steam generators is therefore kept within a primary containment vessel which is itself surrounded by a reinforced concrete building, with the turbine steam circuit entirely separate.

PWRs are compact and were first used as the power unit for nuclear submarines, but the reactor vessel and the piping for the primary circuit must withstand very high pressures. Typically, the primary pressure vessel has a steel shell 200 mm thick and is 13 metres high and 5 metres in diameter. The heat exchangers have in the past caused considerable problems due to corrosion in the welds. The fuel assemblies are of square section (see Fig. 3.2) with the fuel pins held by spacer grids which are designed to allow the passage and mixing of the cooling water. Each assembly is usually 4 metres high and carries 264 fuel rods and 24 control rods.

Pressure is maintained in the primary circuit by a separate pressurizer connected to one of the loops and provided at its top with a *pressure-operated*

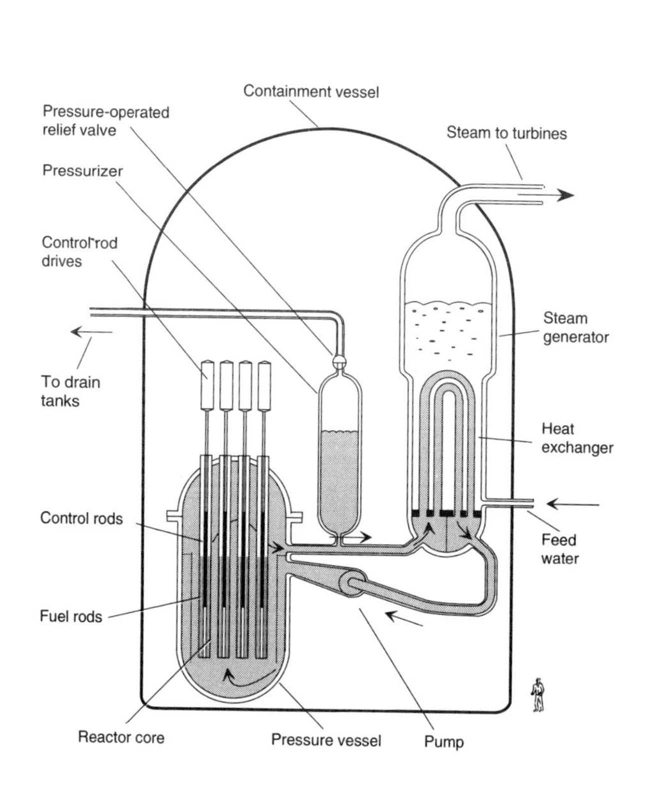

Fig. 3.1 **Schematic drawing of a pressurized-water reactor**. Only the main components within the containment are shown and only one of the steam generators is drawn. The arrows indicate the direction of circulation of the light-water moderator and coolant. The pressurizer contains immersion heating elements which produce a local supply of steam to maintain the preset pressure of the whole primary cooling circuit.

relief valve (PORV). The relief valve is connected to drain tanks and sumps designed to condense and store any radioactive steam which might be released through the relief valve. The pressurizer itself contains electric heating elements that are controlled by pressure sensors in the primary circulation and generate steam locally to maintain the pressure within a narrow range.

A considerable attraction of the PWR design is that it is both *inherently stable* and *inherently load following*. It is inherently stable because the reactivity of both the fuel and the moderator decreases as the temperature rises and increases as temperatures fall. This means that if, say, control rods are moved out of the core the power of the reactor will increase until the increased temperature reduces the reactivity and the reactor stabilizes again. It is inherently load following because opening the control valve supplying steam from a heat exchanger to the turbines will cause a fall in steam temperature. This in turn causes the temperature of the

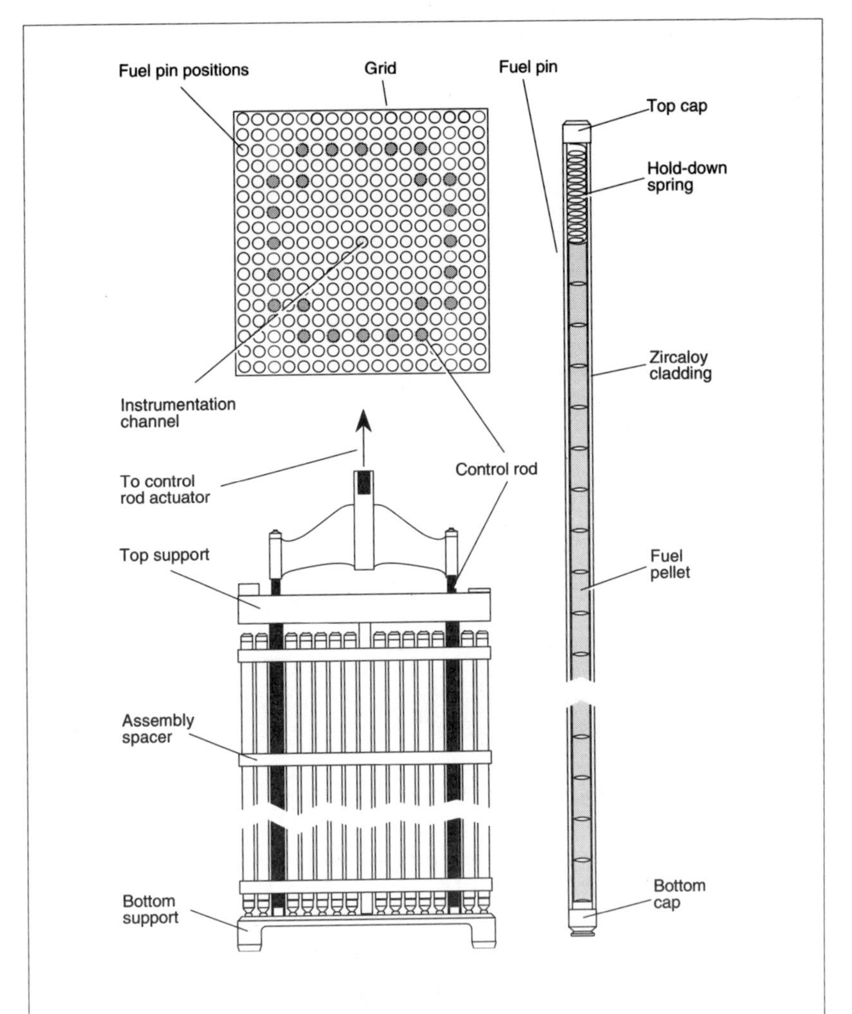

Fig. 3.2 **Fuel assembly of a pressurized water reactor**. The fuel pins are just under 10 mm in diameter and 3.8 m long and are held in the grid shown by five spacers which allow a free circulation of water around each pin. The 24 control rods are moved as a group by arms, like the pair shown, which radiate from the central spigot.

cooling water passing through the reactor core to fall and so increase the reactivity and the power of the reactor. Reducing the steam flow has the opposite effect.

The PWR is a very compact design in which one reactor supplies sufficient heat to provide about 1200 MW of electricity. It is therefore cheaper to build. The very compactness of the core with its rather low thermal capacity compared to a graphite-moderated reactor, coupled with a coolant which can exist in two phases (water and steam) make it faster to respond to changes in operating

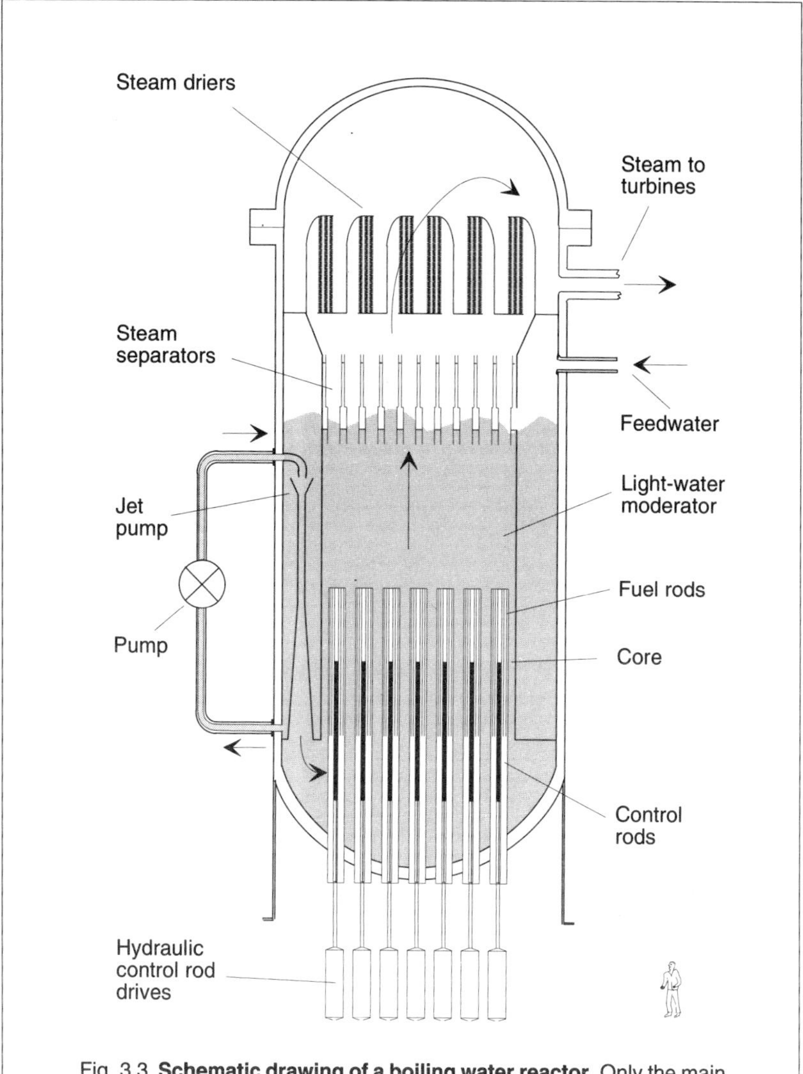

Fig. 3.3 **Schematic drawing of a boiling water reactor**. Only the main components within the primary pressure vessel are shown. The arrows show the circulation of the light-water moderator and the steam derived from it. Water is circulated within the system by a number of jet pumps, powered by external centrifugal pumps.

conditions, particularly any associated with system failure. It therefore needs a greater complexity in its built-in safety features. Nevertheless the considerable experience now available about its performance and the high reliability of more recent designs is widely thought to compensate for its increased complexity and, as we have noted, there are more PWRs than any other kind of power reactor.

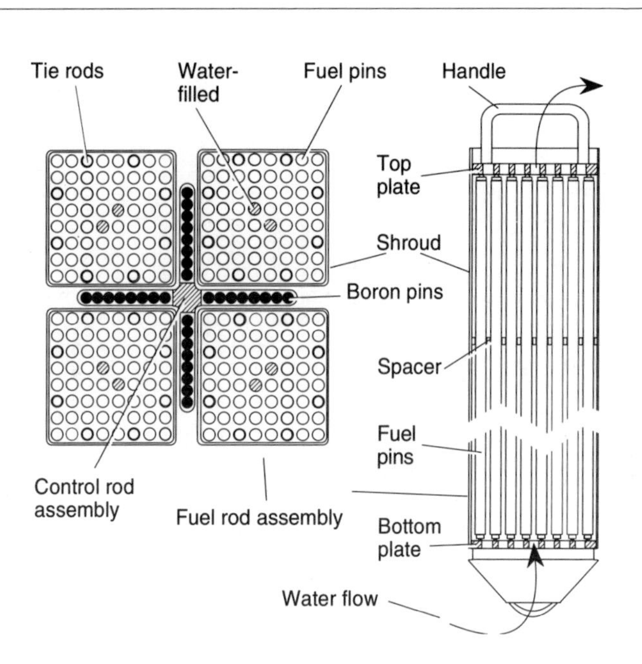

Fig. 3.4 **Layout of fuel and control rod assemblies in a boiling-water reactor**. Each fuel assembly contains eight tie rods which hold the top and bottom plates together (see *right*) and two water-filled tubes near the centre to even out the neutron flux. The spacer is of cellular construction to allow water flow. Each fuel assembly is enclosed in a zircaloy shroud with a valve at its base which allows the water flow to be set according to the position of the channel in the core.

Boiling-water reactors (BWR)

In a BWR the moderator is allowed to boil and the steam drives the turbines directly with the elimination of the secondary water circuit and steam generators (see Fig. 3.3). There is no need for a separate pressurizer and the nuclear core and the equipment for separating and drying the steam are all contained in a very large pressure vessel which, because of its size, is made of a thinner section of steel. The reactor is only designed therefore to operate at about 70 bar (1000 pounds per square inch), producing lower temperature steam with the efficiency of power production necessarily less. A further possible problem is that the steam fed to the turbines is radioactive and leakage through the rotating seals has to be carefully controlled.

The main difference between a BWR and PWR is related to the state of the water moderator and coolant as it moves over the fuel pins. Because boiling is allowed to occur in the former both moderator and heat transfer effectiveness will vary locally, causing *channelling*. It is important to control the degree of channelling particularly when the steam supply is increased to meet a higher demand with a consequent fall in pressure and increased boiling. This leads to a decrease in thermal neutrons but also an increase in fuel cladding temperature. Although, like PWRs, these reactors are inherently stable, they are not *inherently load following* because increasing the supply of steam causes reactivity to fall and this must be compensated by moving control rods out from the core.

Because all the steam separating and drying equipment is located above the core, the control-rod operating mechanisms must be placed below the pressure vessel which has the advantage that, when the top of the reactor is removed for refuelling, the control rod drives can be left in place unlike those in the PWR. The arrangement of the fuel is also rather different. The fuel pins are contained within rectangular carriers (see Fig. 3.4) with the control rod assembly arranged as a cruciform component which slides between four adjacent carriers and is actuated hydraulically from below. During refuelling the top of the reactor vessel is unbolted and removed, the steam dryers and separators removed and the vessel filled with water which can overflow into a surrounding pond. The central fuel assemblies are then lifted out and transferred to this pond and then moved underwater through channels to the primary storage area. Peripheral fuel assemblies are then relocated to more central positions and fresh assemblies placed at the periphery. The control rods remain in position during refuelling.

Refuelling requires at least eight days every 12 to 18 months and involves replacing a third of the fuel assemblies.

Heavy-water reactors

The CANDU design

As already mentioned, the technology for enriching uranium in uranium-235 was secret in the 1950s and its products not commercially available. The Canadian government therefore authorized the development of a reactor which made use of natural uranium with its 0.72% uranium-235, of which they had a plentiful supply. The only way of conserving the scarcer neutrons is to use deuterium oxide or heavy water as the moderator (see Chapter 2). Furthermore, the Canadian engineers were not then confident of making the very thick-walled pressure vessels demanded by the US designs. They therefore tested and made a form of reactor in which the primary circuit coolant, maintained at high pressure, was lead through a very large number of stainless steel pipes to horizontal reactor tubes made of zircaloy (see Fig. 3.6). They were arranged horizontally so that the pipework could be kept clear of supports for the low-pressure containment vessel and the control equipment. Each reactor tube cantains a number of fuel elements.

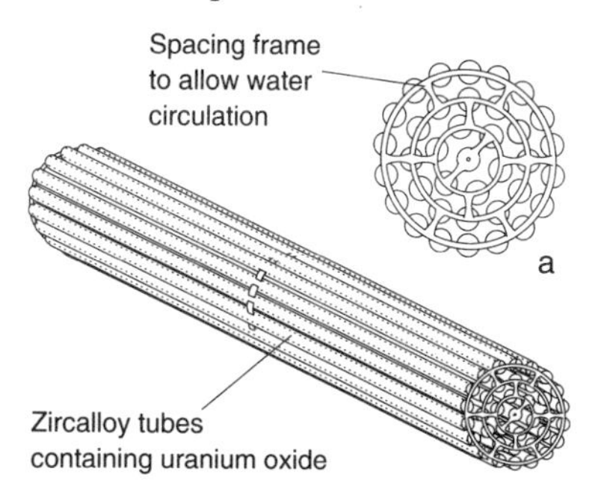

Fig. 3.5 **Fuel assembly for a CANDU type of reactor.** Several of these assemblies can be inserted into a high-pressure coolant tube. Inset a: enlarged view of end plate.

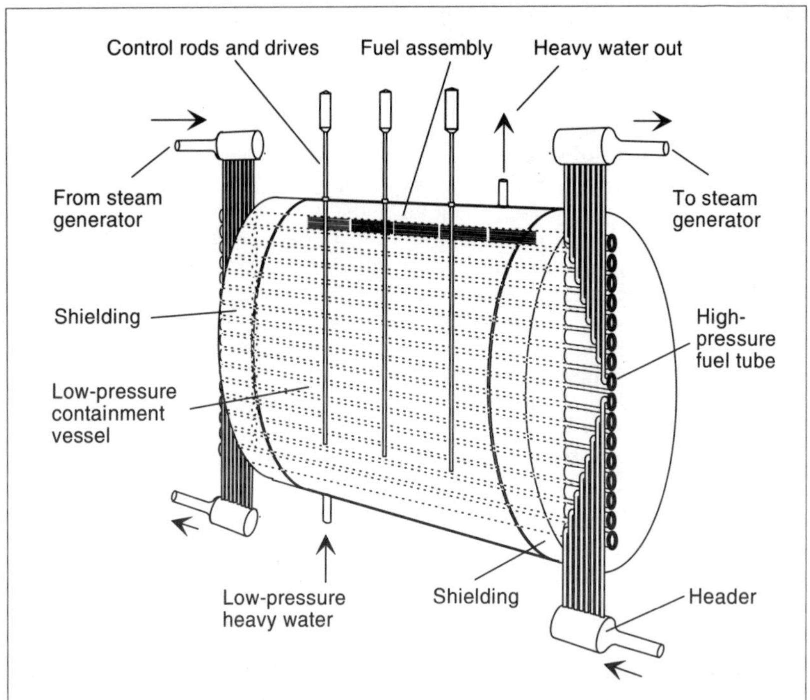

Fig. 3.6 **Simplified schematic drawing of a CANDU reactor**. Only 16 fuel channels out of about 350 are drawn and they are narrower than indicated, filling the volume of the containment vessel. Deuterium oxide (heavy water) coolant at a pressure of 100 bar (1450 pounds per square inch) and a working temperature of 310°C is pumped down each tube before being collected into headers and sent to a steam generator. The steam generator (not shown) boils light water to make the steam to drive the turbines. The heavy-water moderator, at a much lower pressure, is contained in the large horizontal reactor vessel and surrounds the high-pressure tubes. There are many more control rods than are shown in the figure. A number of fuel assemblies (see Fig. 3.5) are placed in each fuel channel.

Because both ends of the pipe carrying the fuel elements are accessible, machinery could be made for isolating a single tube under pressure and feeding in fresh fuel assemblies (see Fig. 3.5) at one end and removing spent assemblies at the other. The reactor need not be shut down for refuelling as in some other designs and a more even burnup is achieved for individual elements as they are progressively moved along the tube.

These are very considerable advantages but there are some disadvantages too. Roughly one tonne of heavy water is needed for every megawatt of electrical capacity of the power station and in the early days of the CANDU development there were frequent delays due to lack of this moderator. This is now rectified and the higher cost of heavy water is partly set off against the lower cost of the natural form. The very large number of joints in the pipework, particularly those between the stainless steel and the zircaloy at the boundaries of the core, was also a source of weakness in the design; in the early days, very many were found to be faulty. But the most serious disadvantage is the rather low temperature at which these reactors work leading to a lower thermal efficiency compared to the PWR design, despite the design of special turbines, optimized for the lower temperature steam.

Steam-generating heavy-water reactor (SGHWR).

A prototype of another design rather like the CANDU was built in the UK but using enriched uranium. The zircaloy pressure tubes containing the light-water coolant were vertical and connected to stainless steel tubes which completed the circuit to the steam drum, circulating pump and back to the reactor; the bulk moderator outside this system was contained at a low temperature and pressure in a tank, called a *calandria*, because of all the pipe work passing through it; most importantly the short-term power output of the reactor could be controlled by raising and lowering the moderator level, which was covered with helium in the calandria; long-term output changes were compensated by varying the boric acid concentration in the moderator. Although an interesting design, able to be scaled down for smaller power stations, a full-scale reactor was never built.

Gas-cooled reactors

Gas-cooled reactors are very different in design from those we have considered so far, all of which use light or heavy water as both moderator and heat transfer agent. In gas-cooled reactors the moderator is graphite, a relatively inert substance able to withstand high temperatures, and the coolant is a gas which transfers the heat to heat exchangers like those found in the PWR design. This gas can be either carbon dioxide or helium and, unlike water, maintains its uniformity and stability at high temperatures.

Magnox reactors

The first commercial gas-cooled reactor was the British Magnox power station at Calder Hall, Cumbria, UK, which has operated successfully for 35 years and

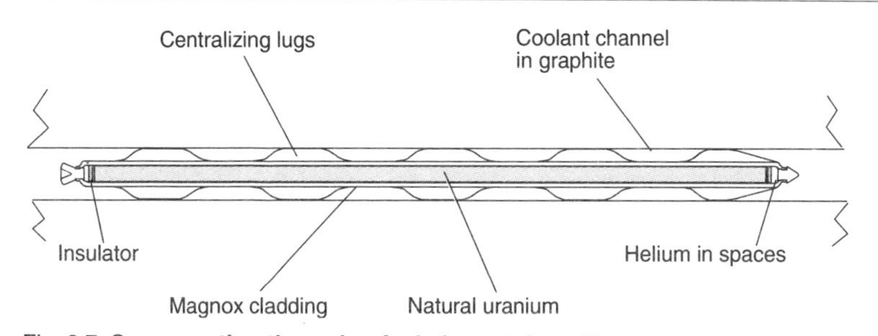

Fig. 3.7 **Cross-section through a fuel element for a Magnox-type gas-cooled reactor**. Two further sets of lugs at right angles to those drawn hold the element central in the coolant channel. Deep fins are also machined into the fuel element between the lugs to increase heat transfer. After it has been assembled and the end caps welded on, the cladding is hydraulically compressed on to the uranium rod to further increase thermal conductivity. Eight elements, placed on top of each other, make up a channel stack.

is likely to go on doing so. Other Magnox stations are now being decommissioned as more efficient designs are built. The natural uranium fuel is encased in finned tubes of magnesium alloy (see Fig. 3.7) which fit into the coolant channels that run through the blocks of graphite moderator. Carbon dioxide is pumped through the channels in the core to the heat exchangers in which ordinary water is boiled to provide the steam for the turbines. The graphite core in the later Magnox designs weighed about 4000 tonnes and contained 50000 fuel elements in over 6000 channels, containing 600 tonnes of uranium. Because of the comparatively high temperature and pressure, these components have to be contained in a large vessel. The original Magnox stations used thick-section steel, which was expensive and difficult to fabricate, but in the late 1950s French engineers showed that the primary pressure vessel could be made of reinforced concrete with only a thin internal steel skin to prevent seepage. The power stations at Oldby and Wylfa in the UK are of this type.

A problem, common to all gas-cooled and graphite-moderated reactors, is that carbon dioxide reacts with graphite at high temperatures to form carbon monoxide. This reaction both degrades the graphite and, because the carbon monoxide ionizes at these temperatures, causes a carbon coating or furring to be deposited on all surfaces, which reduces the thermal conductivity of the fuel elements. One way of reducing this effect is to add methane to the coolant gas.

Nine commercial Magnox power stations have operated in the UK with an increasing design efficiency and output, culminating in the Wylfa station on Anglesey which was commissioned in 1971. The French have also built six Magnox type stations, starting with designs and output similar to the earlier British examples and then developing a unique integral design of their own. There is also one each in Italy and Japan. Since these stations were built most countries have abandoned gas-cooled reactors and built light-water reactors of which there are now over 200 in the world outside the the USA.

Nevertheless, if the provision made for decommissioning Magnox reactors is realistic, they can be accounted a reasonable commercial success, sufficient in any event to encourage the building in the UK of a second generation, the advanced gas-cooled reactors, or AGR which is the only commercially successful design of high-temperature gas-cooled reactor. It uses enriched uranium oxide as the fuel and is cooled by carbon dioxide.

Advanced gas-cooled reactors (AGR)

The carbon dioxide in Magnox reactors is heated to 375°C which means that the steam temperature entering the turbines is considerably less than that used in modern turbine sets. The overall efficiency of such a plant cannot exceed about 30% and to achieve any improvement a higher gas temperature is essential. Higher gas temperatures mean higher fuel rod temperatures but above 600°C the physical properties of uranium metal deteriorate under neutron bombardment and the Magnox fuel pin cladding would melt. As a consequence a new design is needed with uranium oxide and stainless steel as the fuel and cladding respectively.

AGR therefore confine the fuel in stainless steel pins, grouped together in fuel elements within a graphite sleeve (see Fig. 3.8). The elements are held in a square grid of graphite blocks which are specially shaped to allow the blocks to

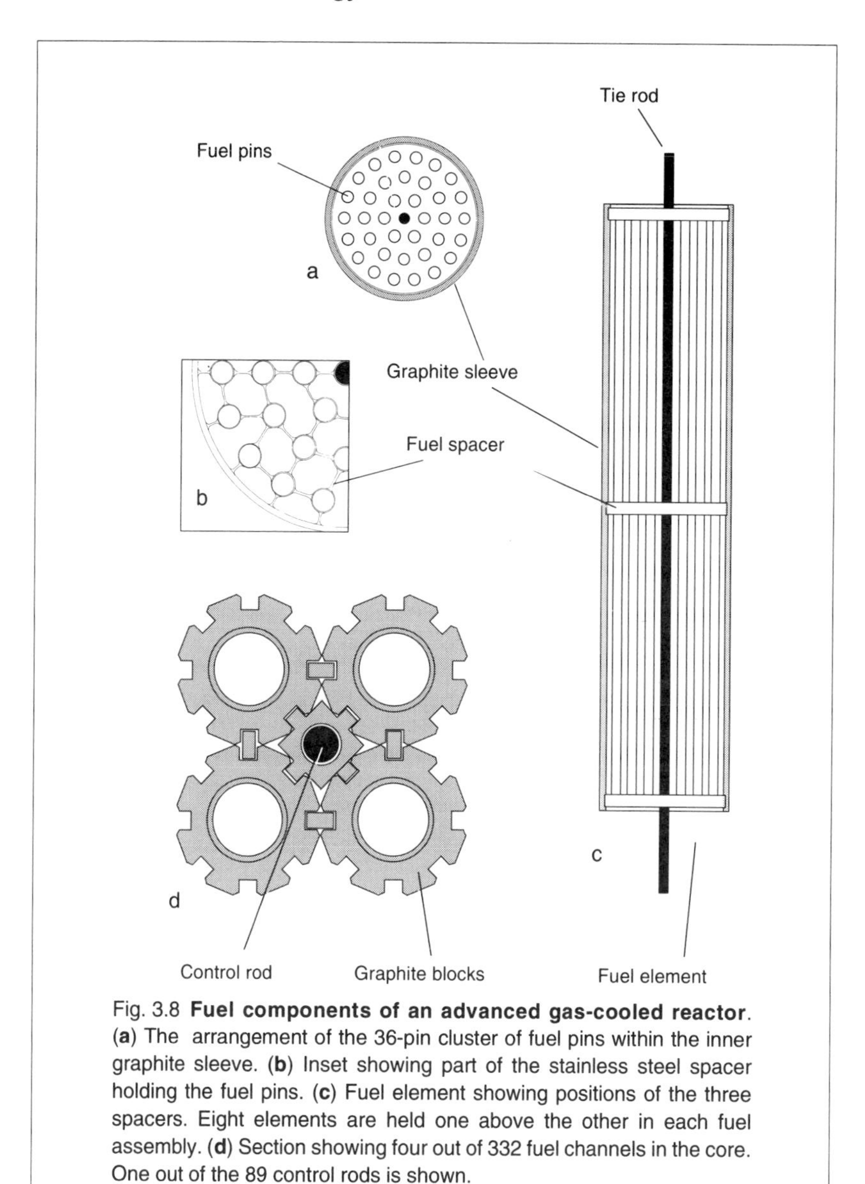

Fig. 3.8 **Fuel components of an advanced gas-cooled reactor**. (**a**) The arrangement of the 36-pin cluster of fuel pins within the inner graphite sleeve. (**b**) Inset showing part of the stainless steel spacer holding the fuel pins. (**c**) Fuel element showing positions of the three spacers. Eight elements are held one above the other in each fuel assembly. (**d**) Section showing four out of 332 fuel channels in the core. One out of the 89 control rods is shown.

be interlocked by graphite tongues, with space for control rods. The fuel is about 120 tonnes of a ceramic of uranium oxide, enriched between 1.4% and 2.6% with uranium-235, depending on its planned position in the core. The core weighs 1800 tonnes and is penetrated by 332 fuel channels (see Fig. 3.9) and a similar number of secondary channels, 89 of which take control rods. The remainder pass a downward flow of coolant gas, in which monitoring instruments can be placed.

The coolant carbon dioxide is heated to 645°C at a pressure of about 40 bar

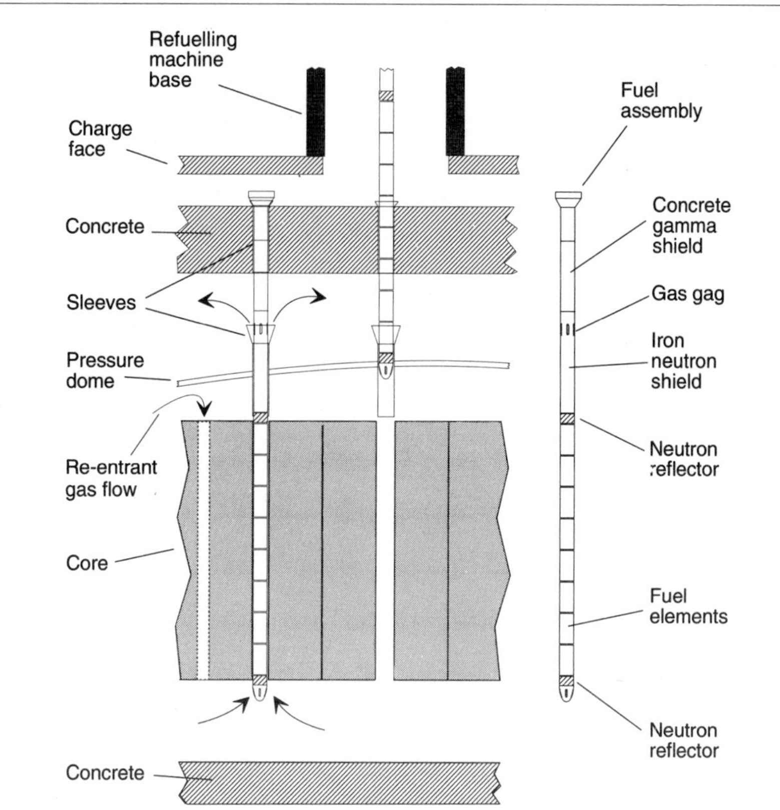

Fig. 3.9 **The main elements of a UK advanced gas-cooled reactor.** A fuel assembly is shown fully inserted (*left*), partly withdrawn (*centre*) and isolated (*right*). The arrows indicate the circulation of the carbon dioxide coolant. The charge face contains removable hexagonal plates which allow the refuelling machine access to an individual fuel assembly. The gas gag provides a flow control for the reactor coolant gas passing into the steam generators and there is a re-entrant gas flow downwards between the large graphite blocks (see Fig. 3.8). The upper part of the fuel assembly, above the neutron reflector, is called the plug unit and is designed to shield personnel working above the reactor. Fuel assemblies are some 25 m long. The drawing is not to scale.

(576 pounds per square inch) and it is pumped through the system by eight gas circulators, any one of which can be withdrawn for maintainance without shutting down the reactor. A pair of reactors at each site power two 660 megawatt turbo-generator sets with a thermal efficiency of over 40%, comparable with the best conventional power stations and much better than the earlier Magnox designs.

The more recent British gas-cooled reactors have achieved over 90% availability, although there were initial problems with the control rods and refuelling was not possible at full power.

High-temperature reactors (HTR)

If stainless steel fuel cladding could be avoided even higher gas temperatures would be possible because both ceramic fuel and the graphite moderator are theoretically stable up to about 1200°C. High-temperature reactors therefore differ from the AGR design by having an all-ceramic core in which the fuel is dispersed within the graphite and by being cooled by helium. With the gas temperature at about 1000°C the coolant could, in principle, drive a turbine directly and eliminate the costly steam generators needed in other designs. Experimental and large-scale prototypes have been built but the ready availability of light-water reactors and the decline in orders for nuclear reactors has slowed their further development.

An experimental helium-cooled power reactor was built at Fort St. Vrain in Colorado, USA, but it is now shut down and the design and its proposed follow-ons are now unlikely. The fuel was compacted into the centre of small spherical particles just under a millimetre in diameter, in which the uranium compound is coated with several layers of carbon and silicon carbide to protect and contain the fission products. The fuel itself can be an oxycarbide of highly enriched uranium-235, thorium-232 or plutonium-239. A fuel rod 10 cm long is made by binding the spherical particles with graphite, and these are inserted into a hexagonal graphite block about 75 cm across and 35 cm high (see Fig. 3.10). Each block had some 150 channels of 3 cm diameter for the helium coolant and for the insertion of control rods. The reactor core is made up of several thousand of these blocks. Such a design of fuel block is called *prismatic* and distinguishes it from the next type to be discussed.

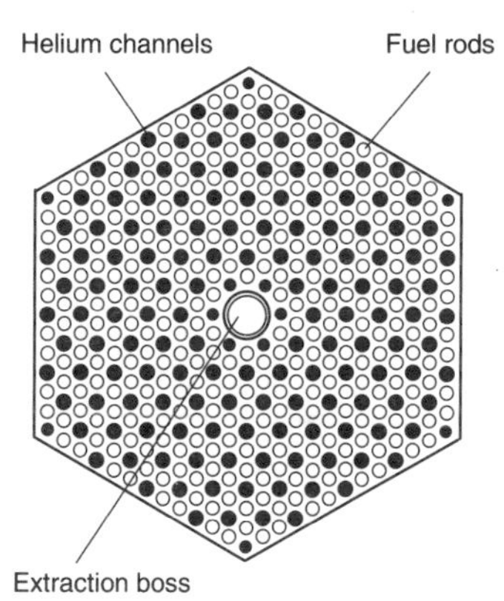

Fig. 3.10 **High-temperature reactor fuel block**. These blocks are closely packed together but are spaced vertically by a few centimetres.

An important type of helium-cooled reactor has been built in Germany and designed by the Arbeitsgemeinschaft Versuchs Reaktoren (*AVR*). It is a high-temperature reactor of a unique *pebble-bed* design in which the fuel of enriched uranium with added thorium is contained in graphite spheres. They are 60 mm in diameter and are loosely packed into a graphite container and neutron reflector which also houses the normal control rods that are therefore placed round the periphery of the core. In addition, emergency control rods can be forced through the pebble bed from its top. Helium is pumped downwards to prevent the spheres rising and the reactor and steam generators are integrated into a prestressed concrete core, with the steam driving conventional turbine sets. The pebble-bed design has the unique property that a proportion of the 700 000 fuel spheres can be removed for inspection each day and returned if satisfactory to the top of the pebble bed with additional spheres added as necessary. Very high helium temperatures up to 950°C can be achieved.

The AVR also has a large negative coefficient of reactivity and the consequent stability allows its power output to be controlled simply by varying the speed of the gas circulators and therefore the helium flow through the core.

It is intended that further development of this design will eventually include the direct operation of gas turbines to power the generators.

Breeder reactors

It was once thought that reserves of uranium worldwide were much less than they have been found to be, and that the number of nuclear power stations would be much larger than they now are. This led to the design of reactors which would both provide power efficiently and produce more fuel than they started with. This can be achieved by using some of the neutrons released by the nuclei of uranium-235 and plutonium-239 to sustain the chain reaction and others to irradiate uranium-238 and convert it into plutonium as in military reactors according to the following decay series:

neutron + uranium-238 ⇒ uranium-239 ⇒ neptunium-239 ⇒ plutonium-239

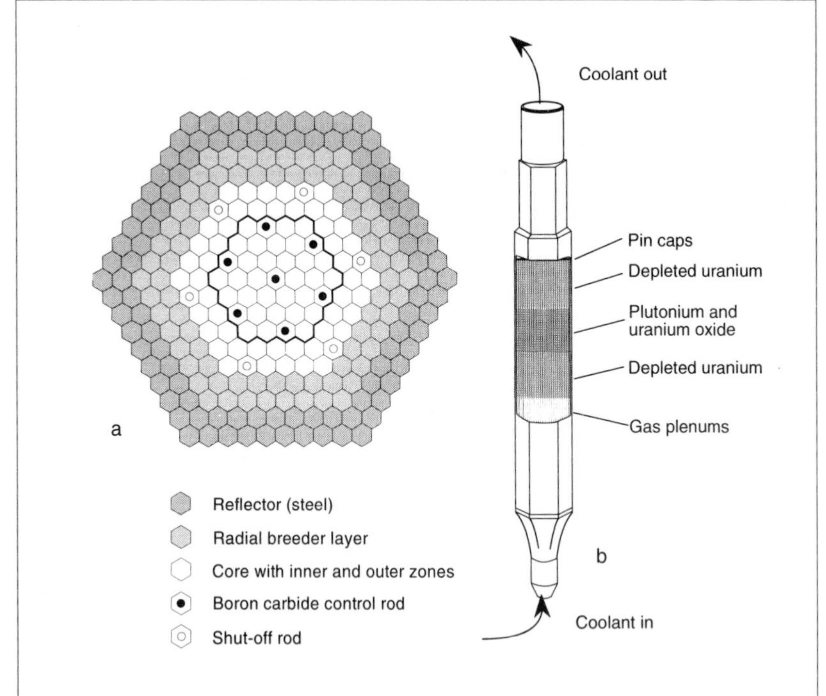

Fig. 3.11 **(a) Cross-section through a fast-reactor core. (b) Much foreshortened drawing of a fuel assembly from the core zone.** The cutaway window in (b) shows some of about 300 fuel pins packed into the assembly. Each pin has a central fissile region with fertile regions above and below and a gas plenum for storing gaseous fission products. Each pin is about 6 mm in diameter and nearly 3 metres long and slightly separated from its neighbours to allow the circulation of the liquid sodium coolant.

Each stage has a comparatively short half-life until plutonium is formed. Fissions from plutonium will then help to keep the reactor critical and contribute to the power generated or can, after purification, be used to refuel another reactor. In practice as much as 70% of the total uranium-238 and uranium-235 atoms can be fissioned in a fast reactor to provide up to 50% more energy than in a thermal reactor.

However, as we have seen, uranium-238 only fissions efficiently by bombardment with fast neutrons of about the energy at which they are first released from uranium-235 in a conventional thermal reactor. This can be done (see Fig. 3.11) by making the core small with minimal moderation so that enough high- energy neutrons can be captured by uranium-238 to produce plutonium and maintain the chain reaction. Such compact cores are difficult to cool. Most designs use liquid sodium to transfer the heat through a heat exchanger to a second sodium circuit situated within the pressure vessel and thence to a steam generator. In this way irradiated sodium will not be released into the boiling water if tubes fail in the steam generator.

Two layouts can be used: in the *pool* type the primary sodium circuit and the heat exchanger are kept within the containment vessel; in a *loop reactor* the primary sodium circuit and heat exchanger is taken outside. An advantage of using sodium is that it can be heated to 560°C without pressurization and so provide steam at over 500°C without the need for heavy and expensive pressure vessels.

Experimental and prototype fast-breeder reactors have been constructed at Dounreay, Caithness, UK and similar prototypes have been built in France and the USSR. The French have also recently completed the SuperPhénix fast-breeder power station at Creys-Malville.

The design problem is to produce a safe fast-breeder reactor which is economically competitive with thermal reactors even when the lower fuel costs are fully allowed for. At present, the price of natural uranium and enrichment is not sufficiently high to justify the extra cost. This may change. In the immediate future, further development in Europe is being co-ordinated collaboratively by the European Fast Breeder project.

Chapter 4

Energy from Fusion

As already mentioned the use of nuclear fusion rather than fission to generate power has considerable advantages. There is much less radioactivity produced, the nuclear reactions are simple and the raw materials, deuterium and lithium, very abundant. It has been calculated that if only 10% of the deuterium at present in the sea could be recovered, this would provide nearly five million times as much energy as there is in all the remaining fossil fuels on the Earth. The problem is that the technology is very difficult and still may be impossible.

The nuclei of any of the lighter elements can be made to fuse and liberate energy but it is likely that only the very lightest elements, deuterium, tritium, lithium and helium will be useful in any future fusion reactor. The problem is the positive charge which all nuclei carry and which will repel any nucleus which comes close enough to have a chance of fusing with it. The only way of overcoming this is to give the nuclei such enormous energies that they are literally forced together. The reactions are known to work because they are the basis of thermonuclear devices (hydrogen bombs) in which the energy comes from placing small fission bombs in a shell round a core of light elements and exploding them simultaneously (see p. 68). This creates the necessary temperature and pressure shock waves to *ignite* the charge.

Similar reactions also occur in nature in the interior of stars and these are obviously self-sustaining – the Sun goes on burning and has not blown up, although in fact the fusion reaction most common in stars is different from those which are being attempted in the laboratory.

Hot fusion

Hot fusion occurs when the temperature is high enough for the thermal energy in the nuclei to overcome the repulsion due to the positive charges, increasing the chances that a nuclear collision will occur at the expense of the occasions when the nuclei just bounce off each other (see panel on p. 39). In addition the density of the ions must be high enough to materially increase the chances of these collisions. The problem is that the temperature required, hundreds of millions of degrees, causes any element to be converted into a *plasma*, that is a gas containing only ions, in which all molecular bonds are broken and all the electrons are stripped away from the nuclei. The plasma becomes electrically charged and its movement creates magnetic fields; fields which interact with themselves and need precise control. Plasma also emits very high energy photons or X-rays. Furthermore, the plasma will vapourize any containment vessel with which it comes into contact!

There are, however, advantages additional to the ready supplies of fuel. At any one time the amount of fuel in a fusion reactor will only be enough for a few seconds running and the reactor will shut down almost immediately if it

fusion reactions

The physics of fusion reactions is simple. If two light nuclei can be made to fuse together then the fused nucleus always has less mass than the originals. The difference in mass is converted into energy.

Although any nucleus lighter than nickel could be used, only the very lightest have a *fusion cross-section* sufficiently large that a controlled reaction has a reasonable chance of working. The following reactions both provide considerable energy and components with a workable fusion cross-section. The symbol ● stands for a proton and ○ for a neutron.

Deuterium alone

D + D ⇒ proton + tritium ●○ + ●○ ⇒ ● + ●○○

D + D ⇒ neutron + ^{3}He ●○ + ●○ ⇒ ○ + ●●○

Deuterium and tritium

D + T ⇒ neutron + ^{4}He ●○ + ●○○ ⇒ ○ + ●●○○

Deuterium and ^{3}He

D + ^{3}He ⇒ proton + ^{4}He ●○ + ●●○ ⇒ ● + ●●○○

Calculating the energy output

If we take the first of the reactions listed above then the mass of the reactants (m) is

$$2 \times m_{deuterium} = 4.0298 \text{ u}$$

The mass of the products is

$$m_{proton} = 1.00782 \text{ u}$$
$$m_{tritium} = \underline{3.01605 \text{ u}}$$
$$4.02387 \text{ u}$$

The mass difference (4.0298 – 4.02387) u = 0.00593 u. Since 1u = 931.502 MeV/c^2 and 1 MeV = 1.6×10^{-13} J, then 0.00593 u equals 8.838×10^{-13} J.

develops a fault. Indeed, less than a tenth of a gram of gas is introduced at any one time in current experiments. Although tritium, which is radioactive, will be a fuel in the first fusion reactors, it has a half-life of only 12.3 years compared to the thousands of years of many fission-reactor products. But at the same time the containment vessel will become highly radioactive under the bombardment of the neutrons which the reaction should produce.

There are currently under development two rather different solutions to these problems. The first involves the confinement of plasma by a magnetic field and the second, called *inertially confined fusion* or *laser fusion*, uses an inwards explosion to confine and heat a small fuel charge.

There have been many attempts at magnetic confinement, but at present the *tokamak* arrangement seems to have much the best chance of early success (see Fig. 4.1). Tokamaks were first proposed and built in the USSR in the late 1950s. They consist of a vessel shaped like the American doughnut, a *torus*, which forms the secondary 'winding' of a large transformer. A magnetic field formed by coils wound round the torus forces the plasma to move round the ring (see **toroidal field**). The rotating plasma acts as the secondary winding of the transformer which induces an additional slowly-pulsed magnetic field at right angles to the first. This confines the plasma as it rotates and stops it drifting on to the surfaces of the containment vessel. See **tokamak**.

The Joint European Torus (JET) at Culham in the UK is an experimental fusion reactor which has attained temperatures, densities and energy retention times which individually exceed those required for a useful power source (see panel on p. 41). The maximum temperature attained has been 28 keV, equivalent to 320 million degrees. The maximum density is 4×10^{20} ions per cubic metre and the best energy retention time has been 1.8 seconds. Only very recently has it proved possible to achieve all these values simultaneously. Some idea of the progress made is shown by the following. Multiplying the temperature, ion density and retention time together gives a number called the *triple fusion product*, with units keV seconds per cubic metre, which can be used as a measure of progress. Twenty years ago the highest figure attained by any fusion device was 2×10^{17}; in 1990 JET achieved 2.5×10^{20} and in late 1991, 9.0×10^{20}.

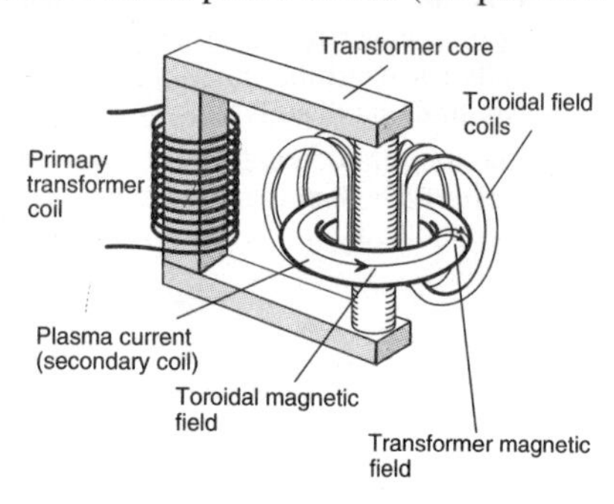

Fig. 4.1 **Schematic drawing of a tokamak reactor**.

One of the most difficult problems is to prevent material from the inside surfaces of the torus vaporizing and contaminating the plasma, where the additional electrons absorb and then reradiate energy, causing a disastrous fall in temperature. Improvements have been made by coating the inside of the torus with a thin layer of beryllium and replacing carbon by beryllium tiles in certain critical areas, but a large-scale design modification is now being undertaken which will divert and clean up the plasma during a power pulse. It is this plasma 'poisoning' or 'carbon bloom' which is seen as the last major obstacle to the manufacture of a successful fusion reactor.

Plasmas of hydrogen and deuterium have been made and significant neutron yields from the fusion of deuterium nuclei measured, but in late 1991 tritium was introduced and a very significant output of heat achieved. A peak output of 14.3 MW was sustained for two seconds in these experiments, corresponding to peak rate 6×10^{17} per second of mainly 14 MeV neutrons. This is a very important step forward although, of course, the amount of energy produced is still much less than that required to sustain the reaction. In order to keep tritium contamination low before the installation of the new divertor modification, the number of runs with tritium has had to be severely limited for the present.

It has never been part of the plan for JET to achieve *ignition*, the self-sustaining condition in which the energy from the production of alpha particles (helium nuclei) will sustain the reaction for several seconds. Instead the International Thermonuclear Experimental Reactor (ITER), a joint program involving the EEC, Japan, the USA and the USSR is now being designed to incorporate improvements suggested by JET's history and to study in particular the lithium blanket which will surround the torus, trap neutrons and provide the energy for future power production. ITER is intended to demonstrate the commercial feasibility of fusion power and should be constructed early in the next century.

Joint European Torus (JET)

A simplified cross-section of this experimental fusion reactor is shown in Fig. 4.3. The fusion chamber is a torus of major radius 2.96 m with a D-shaped cross-section of 4.2 m by 2.5 m. The toroidal coils which surround it carry a peak power of about 400 megawatt. The transformer core is a series of eight open iron rectangles, 11.5 m high, equally spaced round the torus. The primary coils surround the inner vertical members of the core. There are also outer circumferential coils to stabilize the plasma.

The long transformer pulses supply most of the heating to the system as the plasma acts like a single turn secondary, short circuited on itself, but as the plasma heats up its resistance falls, so that increasing power becomes progressively less effective. Additional radiofrequency heating supplies 32 megawatt of energy to specific places in the plasma in order to improve performance. There is also equipment for injecting neutral gas (deuterium or tritium) at very high velocities into the centre of the plasma.

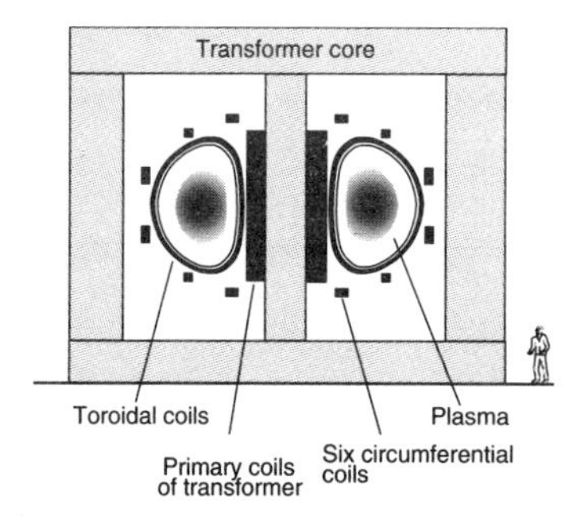

Fig. 4.3 **Simplified cross-section of the Joint European Torus.** Only the electrical and magnetic components are shown.

JET makes enormous demands on the electricity supply as it requires about 700 megawatt during pulses which is well in excess of the power which can be supplied by the generating authority. Additional power is supplied by two 9 m diameter flywheels, each weighing 775 tonnes. They are run up to speed between pulses and at the beginning of a pulse the motors are switched to generating mode and the rotational energy converted into electrical energy as the flywheel slows down. Each flywheel generator can deliver a pulse of 400 megawatt.

Production of useful power

Not only must a fusion reactor produce more power than is consumed in igniting it, the extra power must give a good commercial return. It does not seem likely that this will be achieved in the next 20 years but already some design features are becoming clearer. The next development stages will have larger fusion chambers than those used in JET to reduce the proportion of surfaces which cool and contaminate the plasma. They will need a lithium blanket outside the torus to trap the neutrons generated in the plasma, heat the water which drives the steam turbines and produce fresh tritium (see Fig. 4.2). Two reactions produce tritium from different isotopes of lithium: a neutron will interact with a ^{6}Li atom to give tritium and ^{4}He or it will react with ^{7}Li to give tritium, ^{4}He and an additional neutron. They will probably use superconducting magnetics to reduce the electrical power needed.

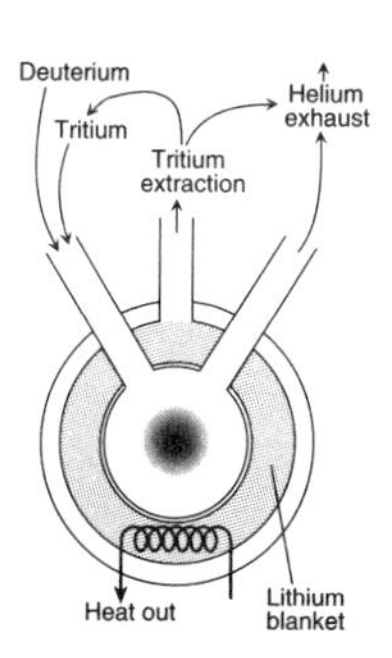

Fig. 4.2 **The main constituents of a working fusion reactor.**

Laser fusion

Laser fusion or *inertially confined fusion* also attempts to produce useful power from the fusion of deuterium and tritium. The subject is both helped and hindered by its close similarity to the current development of fusion or thermonuclear bombs. It has, for example, been long known that exploding small atomic bombs round a spherical capsule containing deuterium and tritium will cause the latter to fuse and explode. Further, the process of developing smaller fusion bombs has resulted in the military scale approaching that required for a useful power reactor. At the same time much detailed information has been classified and has had to await its independent development in e.g. Japan before becoming generally available.

If the beam from a high-energy laser or a linear accelerator is made to heat the surface of a hollow sphere containing deuterium and tritium gas surrounded by a shell of solid frozen deuterium and tritium, the resulting inward explosion compresses and heats the gas so that its nuclei fuse. The outside is formed from an *ablator*, a material designed to vaporize explosively when the sphere is irradiated (See fig. 4.4). The material exploding away from the surface causes an equal and opposite reaction imploding on to the solid fuel which compresses it to a radius of about 1⁄30 of the original. In so doing it should heat the gas so ignition takes place, the *spark* condition.

Considerable progress has been made in the design of lasers of sufficient power to obtain a spark in the minute amount of highly compressed and heated gas. Using high-powered lasers delivering about 15 kJ for 1 nanosecond, temperatures of over 100 million degrees at the centre of the sphere, with a neutron yield of 2×10^{13} and about 1% conversion of laser energy into fusion energy have been measured.

One problem is that the highest compression of the plasma will only be achieved when the plasma at the *stagnation point*, the point of maximum pressure at the centre, is uniformly spherical. For this the uniformity of irradiation has to be improved. One solution in *directly driven* designs, in which several laser beams are spaced round the sphere and irradiate different sectors, is to divide each beam into multiple beamlets which are spaced to give a much more uniform pattern. *Indirectly driven* designs are those in which a hollow target is placed round the fusion sphere and multiple laser beams irradiate the target through apertures; the target gives off X-rays which in turn irradiate the sphere. Such a design is inherently more uniform but pays a price for the conversion of the laser radiation to X-rays. The NOVA project at the Livermore Laboratory in the USA works on this principle using a laser capable of pro-

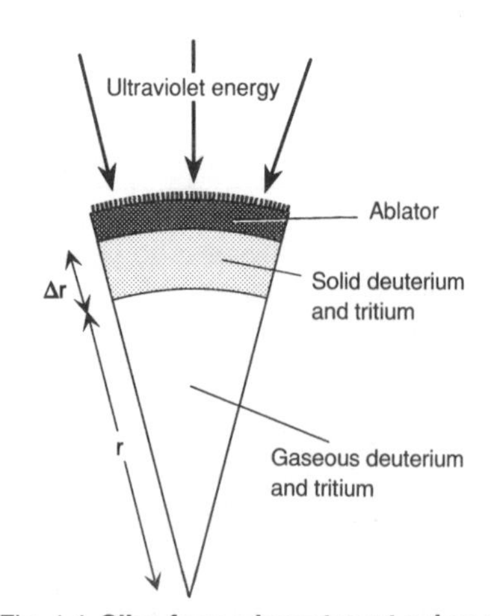

Fig. 4.4 **Slice from a laser target sphere.** The aspect ratio is $r/\Delta r$ and the overall radius is about 3 mm.

viding 70 kJ for 2.5 nanoseconds.

So far high powered neodymium glass lasers which operate in the near UV or green regions of the spectrum are used. Shorter wavelengths are desirable because they will be more efficient in both directly and indirectly driven designs. Krypton fluoride gas lasers are now being developed with the requisite power at 249 nm, in the far UV. Other improvements are concerned with the *aspect ratio* of the fuel, that is the radius of the compressible gas divided by the thickness of the shell of solid fuel (see Fig. 4.4). So far the best results from the Gekko 12 project at the University of Osaka, Japan have been with high aspect ratios and solid fuel layers of only 10 μm. A workable reactor will need a thicker layer of solid fuel which makes the problems of compressing the gas more difficult.

The present status of laser fusion is comparable to that of magnetic fusion although tokamaks like the Joint European Torus have achieved higher temperatures and pressures. The main problem with lasers still seems to be a practical engineering design in which a succession of fusion spheres can be irradiated, the heat extracted and the tritium recycled as seems possible with the lithium blanket proposed for next generation tokamaks. No doubt engineering solutions will be possible, once the basic physics is proved but they will be formidable.

Cold fusion

There have been several attempts to circumvent the need for extremely high temperatures and pressures in fusion reactions. One of these is to use muons which enter the Earth's atmosphere as cosmic rays and have the same electric charge as electrons, but are 207 times heavier. They can replace electrons to produced muonic atoms and because of the principle of the **conservation of momentum** their radius of rotation will be 207 times smaller than that of an electron. The effect of this is that a muonic atom will be able to approach another atom much more closely and thus increase the chances of nuclear fusion occurring. As would be expected such a fusion would displace the muon which would then be free to catalyse another fusion.

There has been considerable research into these reactions, particularly in the USSR, where complex molecules of deuterium and tritium have been found to share both muons and electrons which at least theoretically might result in a few hundred fusions during the microsecond half-life of the muon, but so far only a few neutrons generated by such fusions have been found to occur. Muons are formed by the rapid decay of the π mesons that are formed in large accelerators. Unless thay can be made some other way it does not seem likely that this could be an economical source of power.

Test-tube fusion

In the spring of 1989 it was announced by Fleischmann and Pons in Salt Lake City, USA, that nuclear fusion had occurred in a simple laboratory experiment using readily available materials and components. It involved the electrolysis of heavy water (deuterium oxide) in which a palladium rod formed the central cathode and a coil of platinum wire the anode (see Fig. 4.5). When a current was passed through the cell the heavy water ionized and the deuterium ions moved to the cathode. Palladium is used because it is able to adsorb large amounts of deuterium (and hydrogen) which passes into the crystal structure of the metal where it can build up to a very high density. Indeed, this property of palladium has been used for the safe storage of hydrogen. After deuterium ions had been loaded into the palladium over a period of weeks or months, it was found that there was a sudden increase in heat production in the electrolysis cell which sometimes continued for several hours. This heat was in excess of that calculated to have been caused by the current used in the electrolysis. There was also a small number of neutrons, measured as gamma rays emitted when neutrons collided with a molecule of water in the surrounding bath. This excess energy was thought to be the result of thermonuclear fusion.

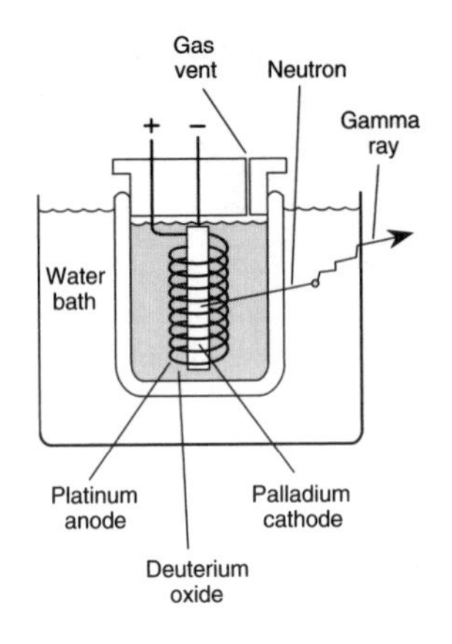

Fig. 4.5 **Fusion in an electrolysis bath.** The waterbath temperature will indicate heat produced. Any neutrons released from the palladium will be slowed down in the water before capture with the emission of a gamma ray.

Although it was not immediately apparent at the initial announcement, the excess heat in one of their experiments was 0.08 watts or 4.6% of the amount of energy put into the electrolysis vessel. One problem with this value is that it is critically dependent on the fate of the gases evolved during the electrolysis. If some passed into the air above the electrolysis cell and there recombined and returned the heat of recombination to the cell, the actual excess energy could be very different or non-existent. Even more important, if fusion occurred, even the 0.08 watt measured would have been accompanied by about a million times more neutrons than were in fact found. It should have been very dangerous to be in the room with an unshielded cell. Some time after the initial publications the gamma-ray spectra eventually published showed that the neutrons seen had the wrong energy peak for those produced by the fusion of deuterium nuclei. They were probably from the background cosmic rays.

Few of these problems were obvious at the initial news conference, and many laboratories throughout the world immediately attempted to duplicate these potentially important results. There have been reports of similar amounts or even more energy produced and of some neutron production but rarely both from the same laboratory. It was postulated that a completely new kind of fusion process had been discovered which does not involve neutron production or that perhaps all the fusion products remained within the palladium. Reports of very careful measurements for the 'test-tube fusion' process have shown that nuclear fusion was very unlikely and in particular there is no evidence for neutron, helium or gamma rays either emanating from the palladium or somehow trapped within it.

The consensus view seems to be that 'test-tube fusion' does not occur, although a few laboratories continue to work on it. The rewards of success would be enormous and perhaps significant insights may yet be found. The scientific community would agree that most of the controversy could have been avoided if proper publication after refereeing by experts had been chosen in the first place.

Summary and comment

There can be no doubt that fusion reactors are likely to be large and difficult to design with commercial success perhaps 50 years away. But there is equal certainty that fossil fuels will eventually become prohibitively expensive to recover and finally become exhausted. It has not been demonstrated that renewable energy sources can provide anything more than a fraction of the world's likely requirements. The alternative is between, on the one hand, continuing to build fission power stations for the foreseeable future and, on the other, investing sufficiently in research to ensure that fusion power is available by the year 2050.

Chapter 5

Nuclear Safety

In nuclear safety questions it is usual to distinguish the part which is inherent in the type of reactor, the *design criteria*, and the part which stems from the particular way in which a given design has been implemented, the *engineered safeguards*. Essentially, design criteria are concerned with normal operation and the engineered safeguards are there to cope with accidents and malfunctions. Certain designs may be more forgiving of construction and operator mistakes and suggestions have been made which could result in safer reactors. These suggestions have the additional merit of indicating the perceived weaknesses in some current designs.

Loss of coolant accidents

The most important question that has to be asked about the safety of a design is: how does it cope with a *loss of coolant accident* or LOCA? Loss of coolant is caused by a rupture of the vessel containing the pressurized water or gas or, more probably, the bursting of the main pipes entering and leaving the pressure vessel. In pressurized-water designs (either PWR or BWR), loss of pressure will cause the water to immediately turn to steam with two consequences. First, and this is sometimes forgotten, the water loses its moderating ability and the chain reaction *stops* but, second, the fuel pins are no longer being cooled. The temperature at the centre of a fuel pin is estimated to be about 2400° C in normal operation and, although the chain reaction has stopped, the highly radioactive fission products continue to produce heat, initially at about 7% of the normal thermal output, so that within less than two minutes the fuel cladding will begin to fail.

The situation is rather different in a gas-cooled reactor. The chain reaction will continue because the gas has little moderating capacity but the graphite reactor core and cladding have a much higher heat capacity than a PWR core. Provided that the reactor is immediately shut down by inserting control rods, the temperature will increase to that required for fuel rod failure in a gas-cooled reactor in about 60 minutes. Even this temperature will not damage the graphite core, damage only starting another 10 hours later even without forced cooling.

The primary engineered safeguard against a loss of coolant accident is the emergency core-cooling system (ECCS) which in water-cooled reactors consists in pumping in cold replacement water which contains borate (*borated water*) or gadolinium nitrate. The boron and gadolinium nuclei capture neutrons and prevent the chain reaction restarting, and the water can be cooled by circulation through emergency heat exchangers. In gas-cooled reactors emergency cooling circuits, separate from the main steam generating circuit, are provided so that carbon dioxide or helium can be pumped through the core and cooled outside the reactor vessel.

Other engineered safeguards in most types of reactor consist in enclosing the fuel rods in suitable cladding (zirconium alloy or stainless steel) and, for all types of reactor, there is the pressure vessel which contains the core and its attendant heat exchangers. These are meant to contain pressure surges between 10% and 15% above the normal operating pressure; they would not contain anything other than the kind of hydrogen explosion experienced at Three Mile Island, Pennsylvania, USA. Safety valves set to open at a predetermined pressure are provided and any steam discharge can be passed through water to condense it. Finally, there is the massive concrete building which acts as a radiation shield and is supposed to contain radioactive steam or gas liberated in an accident.

Tests on small reactors

How does one know whether in the real world and with a real LOCA these engineered safeguards (and others) will be sufficient? The answer of course is that one does not. The design is proposed and passed by the national regulatory authorities mainly on the basis of a sophisticated computer simulation of the likely events. It is not possible to make a controlled test of a full-sized power reactor although tests on small reactors were made in the early 1970s. The latter results were not very reassuring. In one, the fuel bundles were allowed to heat up to 966°C for two seconds using the fission heat in the fuel. In the other, normal fuel rods containing aluminium dioxide instead of uranium dioxide were heated electrically to 1150°C before exposing them to steam. In both tests there was extensive damage to the zirconium alloy fuel cladding, the rods becoming twisted and swollen and in some cases punctured. The steam reacted with the cladding and with the fuel or aluminium dioxide. Further, the coolant channels between the fuel rods were so blocked as to reduce the flow by almost a half. None of these results was predicted from the computer simulations and the programs and design criteria had to be considerably rewritten.

Nuclear accidents

Important lessons have also been learnt from nuclear accidents, culminating in the core meltdown at Three Mile Island and the disaster at Chernobyl. There have been a number of other, less dramatic, accidents, nearly all of which have been the result of operator error. Typical of these was the 1952 Chalk River, Ontario accident involving a prototype of the CANDU type of reactor. The operator reduced the water cooling instead of the heavy-water moderator supply. When this was discovered immediate insertion of the control rods was demanded but the wrong switch was pulled. Eventually the core melted and caused a hydrogen explosion which destroyed the core. No one was injured and very little radioactivity released.

Significant radioactivity was released in the rather different kind of accident which happened in 1957 at Windscale, (now Sellafield) in the UK. This was a carbon-dioxide-cooled graphite-moderated reactor used for military purposes and running at a low temperature. Neutron bombardment causes a molecular rearrangment of the graphite which alters its crystalline structure and leads to the storage of energy and the distortion of the blocks of moderator which lose some of their effect. The storage of energy, called *Wigner energy*, only occurs at low temperatures because high-temperature reactors like the Magnox and AGR designs, in effect, anneal the graphite continuously, causing it to return to its initial state. In low-temperature designs however, it is necessary to shut the

sequence of events at Three Mile Island

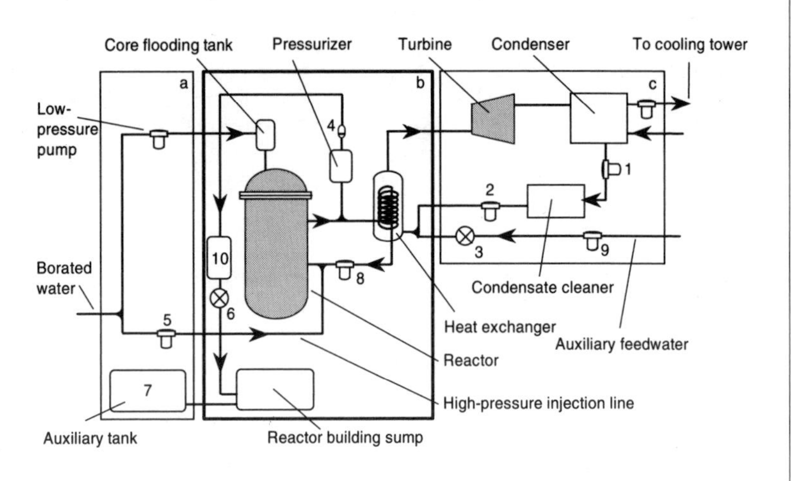

Time	Action
0	Condensate pump (1) shuts off
0	Feedwater pump (2) trips; turbines trip because no water in sump
under 2 sec	Reactor scrams (rapid insertion of control rods)
within 6 sec	Pressure builds in heat exchanger; relief valve (4) opens
about 12 sec	Reactor at 2355 psi; reactor cooling system starts
	Auxiliary feedwater pumps run but no flow because valve (3) not opened after maintainance
	Reactor pressure drops to 2205 psi but valve (4) stuck open
1 min	Pressurizer water level apparently rises; heat exchanger dries out
2 min	Emergency core cooling system (5) activates
4.5 m	One high-pressure pump (5) turned off because pressurizer water level appears, wrongly, to be too high
8 m	Valves (3) opened and auxiliary feedwater flows
10.5 m	Second high-pressure pump (5) turned off
after 15 m	Relief valve (4) still stuck open, pressure builds in drain tank (10) so rupture disk (6) collapses; sump pump sends radioactive water to auxiliary tanks (7)
20 – 75 m	1015 psi; 290°C; relief valve open; pumps (5), (9) and sump pump running
75 – 110m	Both coolant pumps (8) turned off because of excessive vibration; core temperature rises; off scale after 14 m; no natural circulation through core
140 m	Relief valve (4) closed off, but pressure rises to 2150 psi and it opens again
after 230 m	Hydrogen explosions indicated; pressure falls to 500 psi; core probably uncovered with partial meltdown and radioactivity release
After about 10 hrs hydrogen bubbles in the reactor noted; removed by pumping the gases in the pressure vessel through a newly-installed hydrogen combining system, which runs for 3 weeks.	

reactor down and raise the temperature slowly to return the graphite to its earlier form. If this is not done carefully the carbon can catch fire and this is what happened. Unfortunately the fire went undetected for several days and a great deal of radioactivity escaped to the atmosphere.

Three Mile Island 1979

Three Mile Island was much more serious and merits rather detailed description, not only because it, too, involved operator errors but also because we can use it to illustrate the overall complexity of a nuclear power station. It has also perhaps been the most meticulously investigated accident in history.

The figure in the panel on p. 48 is a schematic diagram depicting the major elements in the design of the power station. It consisted of three buildings: the main reactor building (b), the turbine hall (c), much like that of a conventional power station, and an auxiliary building (a) which contained holding tanks and pumps. From the sequence of events outlined in the panel it can be seen that a comparatively minor incident, the tripping of the condensate pump, started a series of events which resulted in partial meltdown. At least three other equipment malfunctions contributed: a valve in the auxiliary feedwater line had been left closed after maintainence; the safety valve above the pressurizer stuck open, and the water-level indicator in the pressurizer was faulty. Because the safety valve stuck, the high-pressure injection pumps could not increase the coolant pressure sufficiently to stop it boiling. The steam so generated passed to the drainage tank within the reactor building, building up enough pressure to rupture the seal which allowed radioactive coolant to find its way first into the sump and then into the tanks in the auxiliary building. These in turn allowed xenon and iodine to escape into the atmosphere. None of these malfunctions would have had an immediate effect during normal operation and no one would have predicted that they could all have occurred together.

There were no direct injuries but the event proved calamitous to the US nuclear industry. Not only was public confidence in nuclear power shattered but the electricity supply company lost its investment of hundreds of millions of dollars in a day. The result has been the complete standstill of new construction in the US industry.

Chernobyl 1986

The RBMK type of reactor involved in this disaster was different from any of those discussed earlier although the Hanford-N military reactor in the USA is of somewhat similar design. Chernobyl was a graphite-moderated water-cooled reactor in which the zirconium-niobium fuel rods were held in groups of 18 in an assembly which can be removed for refuelling through tubes from the top. The cooling water circulated through 1680 tubes which passed into the reactor containment vessel then through the graphite before passing out again to the steam separator and boiler which powers the turbines. Most of the tubes contained fuel assemblies, but some of them contained ‘fixed’ neutron absorbers, which can be adjusted or removed by the refuelling machine, and others the movable control rods. Helium and nitrogen were used to prevent graphite oxidation and the containment building does not have to withstand high pressures. The fuel was uranium oxide enriched to 2% with uranium-235.

The design has advantages: like CANDU reactors, it replaces a thick-walled pressure vessel by many smaller tubes but at the cost of many welded joints; in addition the reactor does not need to be shut down for refuelling. Further, the operators have considerable power to trim the reactor by moving fixed absorbers in order to maintain high and even neutron fluxes. Monitoring equipment was provided throughout the core for this purpose.

Design problems

One disadvantage is that the containment vessel is breached by the cooling tubes with the possibility of leakage round each tube. This may not be serious under normal conditions because of the low pressures but would become more so during an accident.

Three other design problems have been identified, all or some of which may have contributed to the accident. When the reactor runs near equilibrium with many of the fixed absorbers withdrawn and the fuel being replaced, a reduction of the density of the cooling water, either due to increasing temperature or steam forming in the tubes, will under certain conditions cause the effective flux of thermal neutrons at the fuel rods to *increase*, boosting fission. The reactor has, in other words, a positive void coefficient of reactivity at this time. The designers knew of this and there was a rule that the reactor should not be run for a prolonged period at less than 20% power.

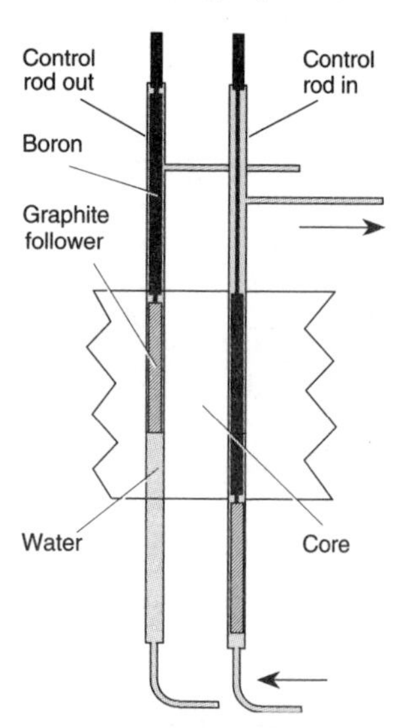

Fig. 5.2 **Schematic drawing of the Chernobyl control rods.** The arrows show the direction of coolant flow.

A second problem stemmed from the designers' quest for the highest possible thermal efficiency. They needed therefore to extract heat from the graphite moderator which can be done most efficiently by having a large number of cooling tubes running through a core which is taller in relation to the lattice grid, that is the distance between the fuel rods, than in most other kinds of reactor. With a core which is shorter relative to the lattice grid, inserting a control rod near the top has a larger effect on all the neutrons in the system. This is not so in the RBMK design particularly when a buildup of xenon-135 at the centre of the core is depleting the immediately surrounding neutron flux.

Lastly, a feature of the control rods, partly consequent on the tall core design, may have contributed to the accident. Boron control rods which move and simply displace water in the cooling channels of the core will not be very effective because both water and boron absorb neutrons. Attaching graphite followers, which do not absorb neutrons, to the control rods and arranging that they move with them will restore the boron's effectiveness. But because of height restrictions the graphite followers could not be made long enough to fill the entire space when the control rods were fully withdrawn (see Fig. 5.2). This meant that, if a fully withdrawn rod was reinserted, 2.5 m of neutron-absorbing water would be replaced by non-neutron-absorbing graphite, causing a surge of neutrons. This is the condition known as *positive scram.*

The events

It had been decided that before the next planned shutdown, a test would be made to find out whether the remaining kinetic energy in the turbine and generator could be used to power the cooling pumps for the 40 seconds required to start the emergency generators. Similar experiments had been done successfully on two previous occasions but some electrical changes had been made in the system. Half power was quickly attained in the early hours of 26 April 1986 and one generator shut down with the intention of starting the test when the power had been further reduced. In order to prevent the emergency cooling system from interfering with the test, it was switched out in violation of safety regulations. There was some difficulty in stabilizing the reactor but in any event a request was received from the grid controller to keep one generator running until late in the afternoon. This was done but there appears to have been an excessive buildup of the reactor poison, xenon-135, in the reactor core and not enough time to stabilize the situation by increasing power again before running the test. To compensate for this the control rods were almost completely withdrawn carrying their graphite followers well above the bottom of the core. Eventually the supervisors decided to proceed with the test, but at a power output of only 200 Mw, much lower than that originally planned.

When the test started there were four pumps connected to the grid and four to the remaining generator. After 35 seconds the voltage from the slowing generator fell sufficiently for its safety cutout to trip and disconnect the power to the four pumps. It is almost certain that the voltage did not hold up for the full 40 seconds because the test began with the generator already running at much reduced power. Immediately the other four pumps started to vibrate probably due to cavitation as the high temperature water from the steam condensate accelerated through the pumps. Cavitation would cause the pumps to fail and for only froth to arrive at the reactor core. These pumps were then also switched off.

At this point a catastrophic surge in heat output from 6% to 50% occurred in 10 seconds. A *scram* was ordered but the fully withdrawn control rods could not be replaced in time and it is probable that the positive scram condition occurred. Indeed it is also likely that the core was already too damaged to accept the rods. The fuel rods burst and the fuel mixed with the water which, turning to steam, lost any remaining cooling ability. A thermal explosion resulted, rupturing the containment vessel and the reactor building, and dispersing hot radioactive material around the power station site causing more than 30 fires, most of which were under control after four hours. Unsuccessful attempts were made to flood the reactor compartment and stop the graphite core fire but it was not until 10 days later, and after placing 5000 tonnes of lead and dolomite on the reactor by military helicopter, that the fire was brought under control.

At least 31 of the plant personnel and firefighters died either during the accident or soon after from burns and radiation. In the first 10 days after the accident about 2×10^{18} becquerels (5×10^{7} curies) were released into the atmosphere in the form of fission products including caesium-137 and iodine-131. Large areas of western Russia, Scandinavia and northwest Europe were contaminated.

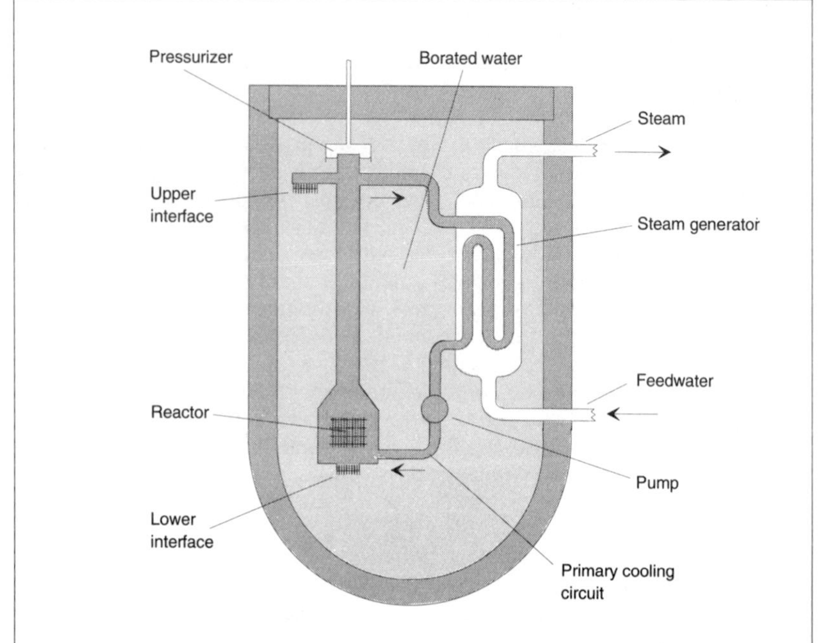

Fig. 5.3 **Process Inherent Ultimately Safe reactor concept**. The upper and lower interfaces separate the hot coolant from the surrounding borated water. Pump failure or other disturbance to the primary coolant circuit will cause the borated water to enter the core.

Future reactor designs

There are at least 560 nuclear power stations in operation today and there has been one catastrophic and another serious power station accident during the last 30 years. But because of the long-term effects that such an accident may ultimately have and because of the opposition of sections of the public, this has inevitably led to further consideration of how to make nuclear power stations safer. One approach has been to implement more sophisticated assessments of the probabilty of any part of a power station breaking down or of being misused as a result of operator error. This approach is called *probabilistic risk assessment* (PRA) and the first power station to be fully designed using this technique has been the Sizewell B pressurized-water power station in the UK (see panel, p. 229). This power station incorporates many design improvements and extra components as the result of this approach and is due to go on stream in 1994.

Another approach has been to prepare outline designs for major improvements to light-water reactors. A common feature is a large reduction in the size of the reactor so that each unit produces about 800 MWe (megawatts of electrical output). Smaller reactors can be assembled in the manufacturing plant, usually ensuring a higher quality of workmanship, and they are easier to cool and contain in the event of an accident. The new designs also rely on a high degree of passive control. Instead of pumps injecting coolant into the reactor, coolant flows from a pond built into the top of the reactor building. It is also possible to allow both the coolant within the pressure vessel to circulate and the reactor chamber to be cooled by convection. These and related measures should allow a shutdown reactor to cool safely without human intervention.

A more radical design has been proposed in Sweden but has never been built. It is called *Process Inherent Ultimately Safe* (PIUS) and is a reactor designed to shut down without human intervention and even after sabotage. The idea is to immerse the reactor and heat exchanger in cold borated water, all inside a concrete pressure vessel (see Fig. 5.3). The reactor coolant is connected to the borated water at two positions at both of which the hot circulating water overlays the cold borated water. Both systems are pressurized from a common source and provided the pump runs and the coolant circulation is intact, the borated water and reactor coolant do not mix. Any disturbance, however, will cause the borated water to enter the reactor and shut it down after which the core will be cooled by passive convection by the large bulk of borated water.

There are potential problems in reactor maintainence and refuelling, and it is thought that it might be too sensitive to surges in the pumped circulation. In addition the concrete pressure vessel would probably only allow rather low water pressures and therefore steam temperatures at the turbines. Further, the nuclear industry has gained considerable experience of current designs and is reluctant to attempt something radically different. Power stations built with a larger number of small reactors appears to be the most likely solution although the high-temperature pebble-bed design which is being explored in Germany and is described on p. 35 has several interesting safety features.

Most of the potential improvements relate to light-water reactors and they are perceived to have been the most economically successful of all current designs, not least because of the rather poor earlier commercial record of some gas-cooled reactors, particularly in the UK. The newer British Advanced Gas-cooled Reactors may change this perception because they have greater thermal efficiency than the Magnox designs and have achieved very high availabilities.

Can nuclear reactors explode?

In Chapter 7 we describe fission and fusion bombs and it is worth noting here that nuclear explosives need over 90% enriched uranium-235 or plutonium and that very high degrees of accuracy in the timing of detonation are essential. Neither of these requirements can be met in nuclear reactors. The danger in the latter is of a thermal not a nuclear explosion, however devastating a thermal explosion might be.

Chapter 6

Uranium Fuel Cycle

The cost of fuel should include not only the difficulty of its extraction, processing and transport but also the cost of disposing of its spent products. Even natural gas which burns to water and carbon dioxide extracts some price if the effect of carbon dioxide in the atmosphere is included. Other fuels have more tangible disposal costs such as the removal of sulphur from the flue gases of oil- and coal-fired power stations and the disposal of ash from coal. Nuclear fuels are unique in that their radioactivity makes their mining and purification dangerous but, above all, the products of nuclear fission are more radioactive and more toxic than the minerals from which they came. These fission products will have to be stored for thousands of years at a cost, presently unknown because such long-term storage has not yet been attempted. The stages from mineral to ultimate storage make up the uranium fuel cycle which is described in this chapter (see Fig. 6.1).

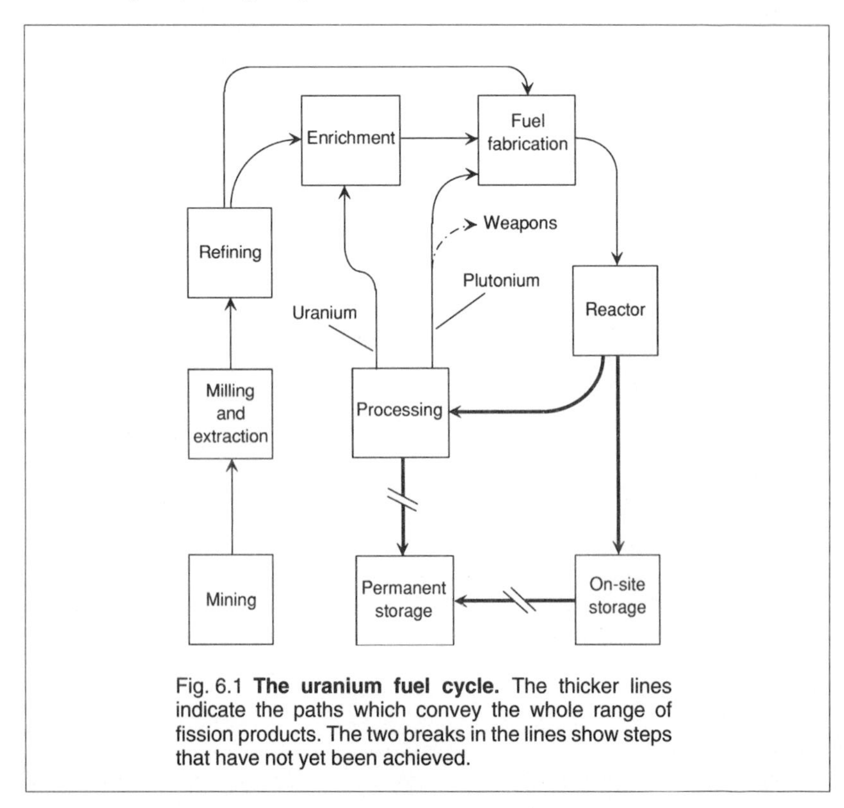

Fig. 6.1 **The uranium fuel cycle.** The thicker lines indicate the paths which convey the whole range of fission products. The two breaks in the lines show steps that have not yet been achieved.

uranium

A hard grey metal with a number of isotopes. Symbol U, at.no. 92, r.a.m. 238.03, rel.d. 18.68, mp 1150°C. Uranium-235 is the only naturally occurring readily fissile isotope and exists as 0.72% of natural uranium. Uranium-233 is a manmade isotope formed in breeder reactors. Uranium and all its compounds are radioactive. The primary mineral of uranium is *uraninite*, UO_2, which has the highest content of U of any uranium mineral. *Thorianite*, ThO_2, also contains UO_2. Uraninite contains lead, thorium, and the metals of the lanthanum and yttrium groups. It occurs as brownish to black cubic crystals and is an accessory mineral in granitic rocks and in metallic veins. When massive, and apparently amorphous, known as *pitchblende*.

There are many other complex uranium minerals. Many occur in pneumatolytic and hydrothermal deposits and in the alteration products associated with them. Numerous oxidized and hydrated salts occur, esp. phosphates, silicates and arsenates of uranium and other metals. Many have a spectacular appearance and are brightly coloured. Important minerals are *autunite*, $Ca(UO_2)_2(PO_4)_2.12H_2O$, of yellow colour, *torbernite*, $Cu(UO_2)_2(PO_4)_2.12H_2O$, a beautiful rich green colour and *carnotite*, $K_2(UO_2)_2(VO_4)_2.3H_2O$, a canary-yellow mineral. These and *coffinite*, $U(SiO_4)_{1-x}(OH)_{4x}$, are all worked as uranium ores.

The abundance of uranium in the Earth's crust is 2.3 ppm. Uranium deposits occur in most classes of rock, but dominantly in the Precambrian basement or in sediments immediately overlying the basement. Vein-type deposits occur in Proterozoic rocks, sometimes remobilized later, and in rocks associated with the Hercynian and Alpine orogenies. In conglomerates, there are substantial amounts of the world's uranium reserves, e.g. associated with gold in the Precambrian Witwatersrand deposits. Sandstone may be impregnated by carnotite and other minerals, as in the Colorado-Wyoming area of the USA.

Mining and extraction

Uranium is a relatively common element in the Earth's crust because the large size of its common ionic form, U^{4+}, has caused it to float to the surface of the molten rock in the mantle. It also occurs in other ionic forms and in many minerals, often at high concentrations (see panel, p. 55). Most of these more concentrated deposits have already been mined but there are still large quantities of ore in the 0.1% to 1% range of concentration.

The commercial product is called *yellowcake*, which is a powder of about 70% U_3O_8. It is made by dissolving the crushed ore in either mineral acid or sodium carbonate depending on the acidity of the mineral and then precipitating and heating it. Yellowcake is often produced at the mine, and miners and process workers are exposed to radioactivity not only because of the uranium and other elements like radium and thorium found with it but also from radon. This is a radioactive gas with a half-life of 3.82 days. It decays to form other radioactive products which will be ingested with any dust that may be about. An additional long-lasting hazard is that the depleted ore will still be radioactive and is often left at the mine site in the tailings and sometimes forgotten.

Yellowcake has then to be refined and converted to the metal or into the gas, uranium hexafluoride (UF_6), which is the usual feedstock for enrichment. This is done by dissolving it in nitric acid and then extracting the resulting liquor with an organic solvent called tributyl phosphate in odourless kerosene (TBP/OK) which forms chemical complexes with the uranyl nitrate but not with other elements like boron and cadmium which have high neutron capture cross-sections. A number of subsequent chemical steps result in highly pure uranium tetrafluoride, which can be either reduced to give uranium metal or further fluorinated to provide uranium hexafluoride or *hex*.

Enrichment

This is the process in which the proportion of the isotope uranium-235 is increased from its natural abundance of 0.72% to that required for the particular reactor or other device. Enrichments of between 2% and 5% are commonly needed in a commercial power reactor although much higher figures in excess of 50% may be required for fast-breeder reactors and bombs. Because isotopes cannot be readily separated chemically, separation has to rely on physical differences, which in the past has meant the atomic masses. Many separation processes have been used or are being developed: currently the most important are diffusion, centrifugation and differential ionization of isotope by laser. The original uranium-235 enrichment for the first fission bombs was, in essence, the use of many mass spectrometers in which powerful magnets deflected the lighter uranium-235 to a greater extent than the heavier uranium-238. Although still used as a method of separating pure isotopes for the laboratory, it was expensive for mass production and soon abandoned.

Gaseous diffusion

Uranium hexafluoride is a highly reactive substance which is gaseous above 56°C and has the advantage that fluorine has a low molecular weight (19) and only one isotope. Because lighter molecules move faster, more of a light gas will pass through a barrier made of submicroscopic holes than a heavy one. In fact the rate of diffusion of a gas is inversely proportional to its molecular weight; so a gaseous diffusion enrichment plant depends on the small difference between the two forms of UF_6 which have molecular weights of 352 and 349. The only way, therefore, to get good yields of uranium-235 is to build the plant of many stages and to cascade the less enriched material from one stage back to the previous stage (see panel on p. 57 and Fig. 6.2).

The optimum arrangment of a complete plant requires that a large number of diffusion stages operate in parallel near to the position where the natural uranium hexafluoride is pumped in, but that the number in parallel should decrease slowly towards the enriched output end and more abruptly towards the depleted tail region. See **cascade**.

Three such plants were built in the USA during the late 1940s. Each needs some 2000 stages to produce 90% enriched uranium-235 and were enormously expensive to build, run and maintain. In the 1970s they consumed 4% of the total electricity made in the USA and produced 17 million *separative work units* (SWU) of uranium-235 each year.

Centrifuge separation

A second way of separating the isotope is to use hundreds of centrifuges. In the rotor of each centrifuge the forces acting on the gas set up gradients which are

isotope separation by diffusion

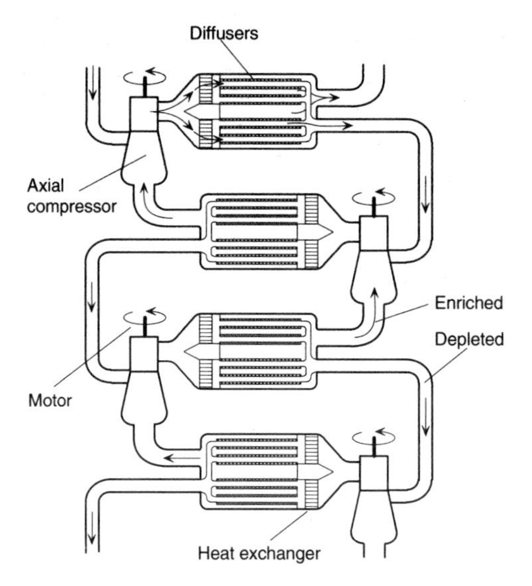

Fig. 6.2 **Stages of a diffusion separation plant for uranium hexafluoride.**

The figure shows four of the many hundreds of stages of a gaseous diffusion plant. As drawn, the gas becomes more enriched for uranium-235 as it proceeds upwards and more depleted as it proceeds downwards towards the *tails*. The capacity of such a plant is defined in terms of **separative work units** (*SWU*) which in turn depend on the throughput of natural uranium and the concentration of tails which the plant produces. Typically, a diffusion plant producing 0.2% of uranium-235 in the tails will make 200 g of 3.2% enriched uranium-235 per SWU.

The large US diffusion plants each produced about 6 million SWU per year and other plants have been built elsewhere.

The diffusers are ceramic cylinders and contain millions of minute channels with an average diameter of about 10×10^{-9} m. Diffusion is assisted by pumps with valves in the depleted flow circuits to equalize their gaseous resistance with that through the diffusers. Heat exchangers remove the surplus heat introduced from compressing and pumping the gas.

slightly different for each isotope, with more of the heavier at the periphery and more of the lighter near the centre. Gas is fed down a stationary central column and into the rotor, while the heavier and lighter gases are removed from the bottom and the top of the rotor respectively. This is done by arms which are specially shaped to induce an axial flow of the gas, up at the periphery and down at the centre (see Fig. 6.3). Modern materials and the design of suitable magnetic bearings, stabilizers and induction motors have made it possible for the centrifuges to run unattended for years.

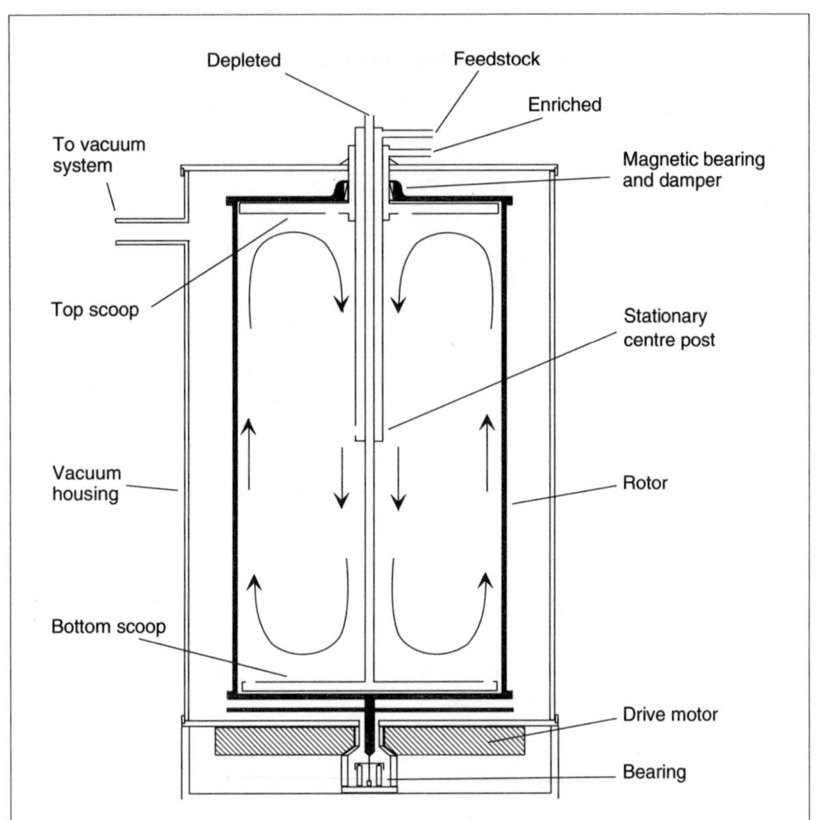

Fig. 6.3 **Schematic cross-section of gas centrifuge rotor**. In the drawing the revolving parts are filled in in black and the stationary components are outlined. Gas is fed to the centre of the rotor and is extracted at the top and bottom by scoops which are shaped to promote the axial flow. The rotor casing is at a low pressure to reduce frictional losses and heating. The peripheral velocity is 400 metres per second.

The degree of separation achieved in each centrifuge is higher than in a diffusion stage in the earlier process because separation depends on the difference in mass of the two forms of uranium hexafluoride and not on their ratio as in the diffusion process, but the amount of material which can be processed at each stage is very small. A centrifuge plant with about half the capacity of the the three US diffusion plants together would need two million centrifuges; it is not a cheap option. The US government has operated such a facility and a European consortium, called Urenco, has built one each in Germany, the Netherlands and the UK. The current output of the three Urenco plants is about two million SWUs a year which can be augmented at any time, adding centrifuges in parallel to increase output or in series for greater enrichment.

Other methods involving gas flow

There has been considerable research into other cheaper methods of achieving the kinds of enrichment needed for commercial power reactors. In particular a

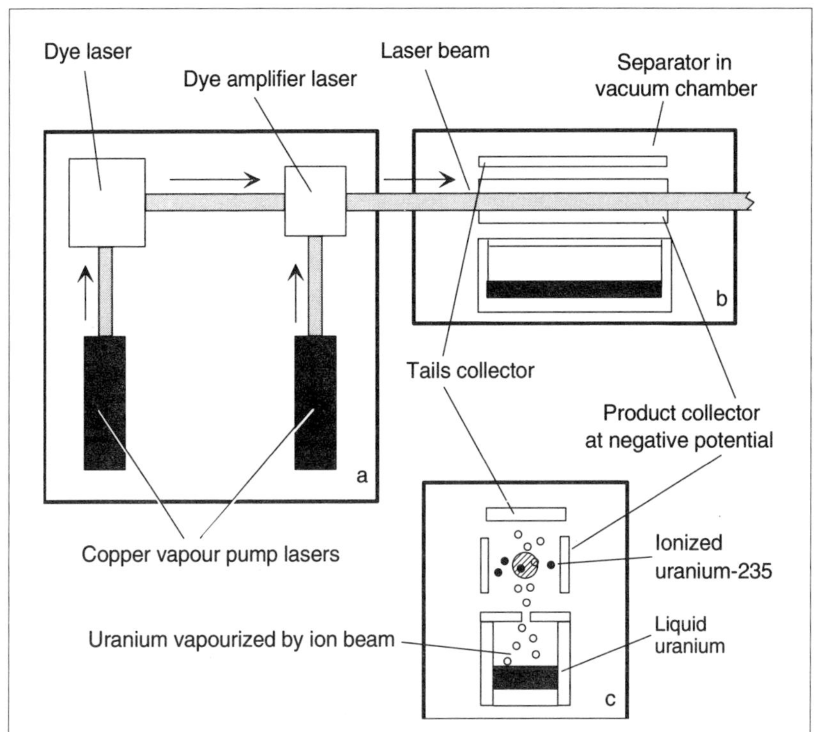

Fig. 6.4 **Schematic drawing of a laser enrichment facility**. The multiple lasers at (**a**) deliver coloured light at frequencies exactly tuned to the energy levels of uranium-235 which has been vapourized in the separator (**b**) by electron-beam heating. Only the uranium-235 becomes ionized and the positive charges move to the collector plates, where they are collected as molten metal. Drawing (**c**) is a cross-sectional view of the separator and shows a possible arrangement of the main components.

method called the *Becker nozzle process* uses a mixture of 5% uranium hexafluoride and 95% hydrogen which is forced out as a jet into a swirling chamber where the lighter fraction is separated by a special aerofoil-like surface.

Laser enrichment

Beams of light from a dye laser can be tuned to provide energy at the specific wavelengths which will ionize the vapour of uranium-235 but not uranium-238 (see Fig. 6.4). Once ionized the isotope can be collected by applying a large negative potential to a plate parallel to the vapour stream. The theoretical separation factors are much higher than in either of the gaseous processes. The main problems appear to be the yield, controlled by the laser power, and the containment and collection of the molten uranium. Considerable effort is being expended on this technology with the US government aiming to replace its ageing diffusion plants by laser enrichment rather than by centrifugation because of the high separation factors possible. Experimental devices have been made in the US and France and, despite the fact that the method can produce uranium-235 for weapons, much technical information is now available.

Spent reactor fuel

Reprocessing

Spent fuel rods have to be placed in a storage pond near the reactor for several months to reduce their radioactivity and dissipate their high initial heat production but thereafter two courses are open to the reactor operator: the spent fuel can be moved to a more permament site for indefinite storage or it can be sent to a reprocessing plant. Here the fuel is broken up, the cladding stripped, and the remaining fuel and fission products dissolved in nitric acid from which the uranium and plutonium are purified.

In the USA no spent fuel from commercial power stations is processed; it is simply stored. The amounts are large: of the order of 500 tonnes of such waste is produced each day divided among some 70 power plant complexes. In the UK a different policy has in the past ensured that gas-cooled plants have their spent fuel removed to a reprocessing plant and then reused as newly fabricated fuel rods containing enriched uranium or plutonium. Commercial considerations based on the relative cost of fresh and reprocessed fuel have, more recently, caused the UK nuclear generating utilities to reconsider this policy and to design suitable dry storage facilities.

Spent fuel arriving from reactors is contained in fuel pins which will have become distorted because of irradiation and some of them may leak. They will also contain fission products which may be powerful neutron poisons or may chemically attack the fuel cladding. Fairly simple chemical extraction methods would quickly achieve an acceptable degree of purification to allow recladding and reuse but a number of other criteria dictate that very high degrees of purification are essential. The processed fuel must be sufficiently free of radioactive impurities that it can be handled with the same procedures as the pure enriched uranium or plutonium used initially. Secondly, unlike normal chemical processing, the plant must be designed and operated so that no fissile material can build up into a critical mass at any position in the plant. Although there is never the possibility of a nuclear explosion (see p. 65), a critical mass would produce considerable heat and radioactivity which *criticality control* procedures are designed to avoid. In addition very careful auditing procedures must be in place to ensure that the strict regulations imposed by international safeguards on the movement of enriched nuclear fuel are observed. Recoveries of over 99.8% for uranium and plutonium are routinely achieved.

The panel on p. 61 and Fig. 6.5 describe the essential features of a reprocessing plant designed for the remote handling of uranium oxide spent fuel. Note how the fuel canisters are kept in storage ponds before being hauled up to the shearing *cave*, where they are cut into short lengths before dropping into the dissolver cells. Plutonium and uranium are recovered for further processing and the remaining fission products stored pending some long-term solution to the problem of storage.

Long-term storage

Spent fuel contains a large number of fission products with varying half-lives and toxicity. The proportion of each product depends on how long the fuel is kept in the reactor, the nature of the moderator, and the operating conditions. A

uranium oxide reprocessing plant

Fig. 6.5 shows a highly schematic drawing of a reprocessing plant designed for fuel which has been irradiated with thermal neutrons. Such plants are designed to handle the fuel rods which come from either pressurized-water, boiling-water or gas-cooled reactors and are equipped with special shear assemblies to suit these types of fuel.

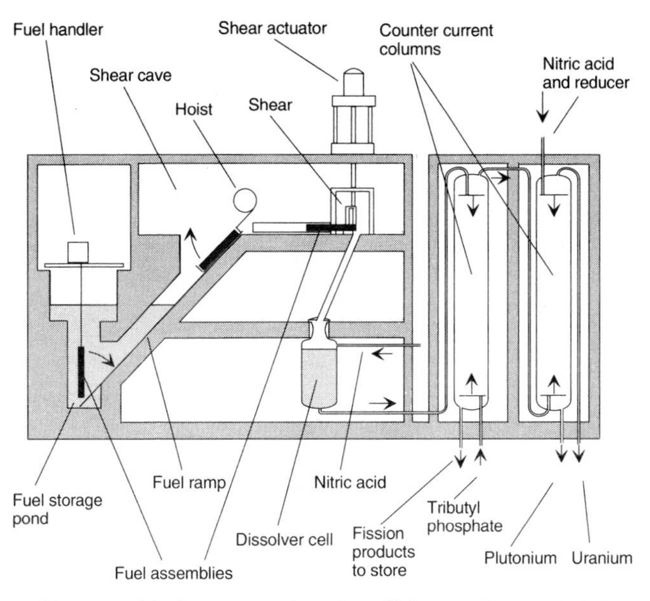

Fig. 6.5 **Diagram of fuel reprocessing plant**. Valves and pumps not shown.

Spent fuel is brought in heavily shielded flasks to the fuel handling area where the flasks are opened by remote control and the fuel canisters placed in the storage pond. When required a canister is hoisted up the ramp to the shear cave and moved hydraulically to the shear. This cuts the whole assembly into small pieces which fall down a chute into the digester where nitric acid dissolves uranium and its fission products but not the cladding. This solution is transferred to holding tanks (not shown) before passing to the purification plant.

Purification requires two countercurrent columns: in the first the nitric acid solution passes slowly down the column and an organic compound, tributyl phosphate diluted with paraffin (kerosene) passes up. Uranium and plutonium pass into the tributyl phosphate and are pumped to the base of the next column, while other fission products still in nitric acid solution are removed for neutralization, nitric acid recovery and storage.

In the second column the plutonium and uranium in the organic phase move up to meet nitric acid and a reducing agent moving down. The latter changes plutonium nitrate to a trivalent form, which is insoluble in the organic material but soluble in the nitric acid. The plutonium salts therefore appear at the base of the column and the unchanged uranium salts at the top. Further purification, chemical modification, solvent and nitric acid recovery results in uranium and plutonium ready for enrichment or fabrication into fresh fuel elements. This Purex purification process has been operating satisfactorily for over 40 years.

measure of the radioactivity produced is the heat generated; this is plotted against time in Fig. 6.6, where it is also compared with the relative toxicity. It will be seen from this figure that, although both the radioactivity and toxicity fall between a thousand and ten-thousandfold in the first 100 years after the fuel is removed from the reactor, nevertheless some remains for over a million years with a late buildup of toxicity due to actinides after a hundred thousand years. We are concerned here with *high-level wastes* which include unprocessed spent fuel, the waste from the first stage of the extraction process in a reprocessing plant, and concentrated waste from subsequent purification steps. *Low-level wastes* are all other radioactive materials, such as those produced during medical procedures. These can be stored in drums and buried. High-level wastes on the other hand, must be sufficiently secure that they cannot leak into the environment even after thousands of years. A number of disposal strategies have been proposed and they can be divided into three main classes. They all have advantages and disadvantages.

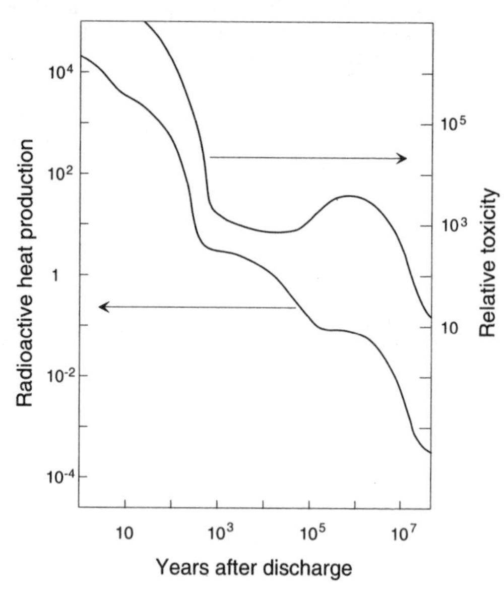

Fig. 6.6 **Reactor product decay**. The toxicity (upper curve) and heat production (lower curve) in the years after fuel is removed from reactors. *Logarithmic scales.*

Forming insoluble glass-like material

This option takes high-level wastes and melts them with glass powder to form insoluble glassy blocks. Certain borosilicate glasses have been found to be very resistant to internal irradiation and seem to be chemically and thermally resistant. In France a continuous casting process for glass and fission products has been developed called the AVM process, and it seems likely that this will form the basis for a widely used method. Glass either as small lens-shaped blocks or as larger ingots could be conveniently stored on the surface until the initial heat production declines and later placed in deep storage.

Deep-mine disposal

Deep-mine disposal is suitable for storing high-level wastes that need no further processing and two kinds of geological formation are favoured. The first is in salt mines, because salt is so soluble in water that its very existence means that no water has ever reached it. Salt is also plastic when heated as by the remaining fission heat of the waste. It will therefore flow round the steel drums and after the drums eventually corrode form a plastic mixture with the waste that is able to withstand geological movement. The second preferred formation is granite, a crystalline rock, occurring in geologically stable areas and very impervious to water if the strata are sound. It has advantages over salt in that the rock would be dry and non-corrosive and the waste in the mine could be recovered if necessary. Deep storage sites such as salt and granite will however remain difficult to monitor once sealed and test bores may be necessary to check for seepage over the centuries.

Changing reactor design

Another approach is to change reactor designs to increase neutron energies so that more of the longer-lived actinides get converted into shorter half-life elements. This would really be a fast reactor with all its complexities and is not at present an economical solution, especially as it is only a palliative measure for reducing long-term toxicity.

Worn-out power stations and decommissioning

Although only marginally part of the fuel cycle the fate of old power stations is best considered here. All power stations wear out and their owners have to decide what to do with them. Nuclear power stations have, in addition to the normal wear and tear on buildings and equipment, the accelerated ageing due to neutron bombardment. However well a reactor is designed, some neutrons escape and, over the lifetime of a power station, irradiate the pressure vessel and pipework. The steels used contain iron, but also cobalt and nickel, and after neutron irradiation they convert in part to the isotopes, iron-55, nickel-63, nickel-59 and cobalt-60. This causes the steel to become weaker and more brittle, and its strength must be regularly monitored in a succession of testpieces, irradiated for the purpose. At some time, predicted on the basis of accelerated testing to be from 20 to 40 years, the reactor vessel and its ancillary equipment will become unsafe and its owners will be faced with a decision on the station's future. They have several choices which will be influenced by commercial considerations and the public's requirements.

Decommissioning

Decommissioning is a general requirement, not only applicable to large power plants but also to experimental reactors, fuel processing and enrichment plants whenever they have come to the end of their useful lives either for safety or economic reasons. The following three stages are recognized officially.

1. Final shutdown and surveillance, involving the removal of all fuel and taking appropriate measures towards the eventual disposal of accumulated waste and plant decommissioning.

2. Restricted site release, involving the dismantling of those plant and buildings external to the reactor biological shield and the completion of the processing of accumulated operational waste.

3. Unrestricted site use, involving the complete dismantlement and removal from the site of all materials, equipment and structures with radioactivity levels above those acceptable for unrestricted use of the site.

These stages are expected to continue over a long period: defuelling of the first Magnox power station is not expected to be finished before mid-1992 and absence of adequate storage facilities means that stage 2 will not be complete until after 2005; it is expected that stage 3 will not start for a hundred years when the radioactivity of the remaining structures should need little in the way of special measures. This delay has obvious attraction for nuclear operators as the danger, difficulty and expense of immediate dismantling are great.

So far a very few small reactors have been dismantled. In one US instance, it has been at a greater cost than that of the original reactor and to minimize radiation dosage the steel components had to be cut up expensively under water.

Other alternatives include converting the now rather old turbines and generators for use in a conventional power station. This may not be economically possible for those many power stations where one of the attractions of nuclear power was the distance of the site from sources of fossil fuel.

Renewal

It might be economical to remove the old reactor vessel and store it in a pond for perhaps 20 years to allow the radioactivity of the iron-55 with a half-life of 2.6 years to fall to a more acceptable level. During this time a new reactor and pressure vessel could be installed. The problem might well be that 40 years on, designs will have changed and there is no guarantee that the old concrete shell or power station would be suitable or economically viable.

Mothballing, burial and entombment

If the power station has its fuel and other valuable components removed in stages 1 and 2 of the decommissioning process, then the more radioactive components can be left behind the biological shield with all pipe work sealed, a single entrance left for a periodical inspection to see that the roof is not leaking and no one has entered. Such *mothballing* is quite an attractive option, particularly because there is an implied maximum duration for such a process.

One option which is being pursued in Canada where they hope to reach stage 3 after only 30 years is to dig a pit and line it with concrete directly under the floor of the reactor. All the radioactive components can then be placed in the pit and filled over with concrete. This is certainly better than another alternative, *entombment*, in which a concrete structure is built over the whole edifice as it exists on the surface. Such a structure has to be sufficiently well designed and constructed to ensure that residual radioactivity cannot leak out, and people and water cannot get in. Its major advantage is the possibility of monitoring the site over a very extended period, but at an aesthetic cost of a slowly increasing number of large concrete structures in the landscape.

The cost

The actual cost may not be as low as was once, in the 1970s, predicted. It was thought then that this would add about 1% to the price of the electricity generated over the lifetime of the stations and this is generally thought by the industry to be still true. More recently some estimates have indicated a higher figure, not least because society is demanding higher standards of implementation. The uncertainty was sufficient to ensure that the UK nuclear utilities were not included in the sell-off of the British electricity industry.

Chapter 7

Fission and Fusion Bombs

So far we have been mainly concerned with the controlled release of energy which results when a nucleus of certain isotopes of uranium and plutonium capture a neutron. In the table in Chapter 2 (see p. 18), we noted that about 200 MeV of energy was produced from each nuclear fission and that this was found mainly in the kinetic energy of the fission fragments and neutrons and in the gamma rays. Some was also released later from the new nuclei produced in the *fission products*.

Roughly the same amount of energy will come from the nuclei of the fissionable isotopes, uranium-235 and -233 and plutonium-239, or from the fertile uranium-238 isotope, which cannot sustain a chain reaction because it only produces, on average, one neutron from each fission. Moreover, the amount of energy emitted is nearly independent of the kind of fission products made and the energy of the neutrons which initiate the fission. We can therefore calculate the instantaneous release of energy from a quantity of fissionable material detonated in an atomic bomb.

The energy in a bomb

Some of the energy is in the fission products and will not contribute to the immediate explosion; the usual figure for the available energy is 180 MeV. The number of nuclei in 12 g of carbon-12 is 6.02×10^{23} (Avogadro's number) and there is the same number in 235 grams of uranium-235. If all these undergo fission then 10.83×10^{25} MeV will be produced, which is equivalent to the production of 4.44×10^{12} calories. This unit is convenient because one kiloton (about one thousand tonnes) of the high explosive trinitrotoluene (TNT) is taken, by convention, to produce 10^{12} calories. Therefore 57 grams or 2 ounces of uranium-235 can provide as much energy as a thousand tonnes of TNT. The table gives the same yields in other units.

The complete fission of 57 grams of uranium-235 yields:	
1.45×10^{23}	Fission nuclei
10^{12}	Calories
2.6×10^{25}	MeV
4.18×10^{19}	Ergs
4.18×10^{12}	Joules
1.16×10^{6}	Kilowatt-hours

The useful fissionable isotopes, uranium-235 and -233 and plutonium-239 are relatively stable substances which yield between two and three neutrons during fission, but not all of these will survive to initiate a further fission. In particular,

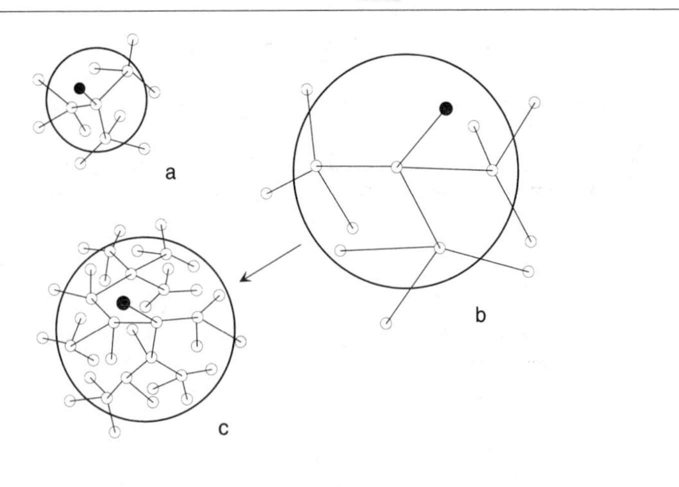

Fig. 7.1 **The effect of volume and density on neutron loss**. The closed and open circles indicate the original and subsequent fissions respectively. Each fission is assumed to give three neutrons. (**a**) has the same density of uranium-235 as (**c**). (**b**) has a lower density and a longer mean path length for neutrons.

some will escape from the mass of material and if, on average, less than one remains, heat will be produced but no explosion. Such a device will be *sub-critical*. The management of these escaping neutrons thus holds the key to the safe handling of a bomb before detonation and its successful detonation later.

Neutrons will escape either because they are close to the surface of the fissionable material or because the density of the material is so low that most neutrons fail to collide with a nucleus. Fig. 7.1 illustrates these different conditions. In (a) the density of fissionable nuclei is high but most of the neutrons escape because the surface area is large relative to the initial volume. If the volume is increased (c), keeping the density the same, then more of the neutrons will be captured by a nucleus and the device will become *supercritical* and explode. Alternatively, if the same amount of fissionable material as in (c) is kept at a lower density (b), enough neutrons will escape without colliding with a nucleus and no explosion will occur until after compression to the volume illustrated by (c). The problem is to bring the subcritical components together so fast that the energy from fission, now increasing exponentially, does not blow the super-critical mass to pieces before the desired explosion is achieved.

Two methods of detonation have been used. Fig. 7.2 illustrates the more primitive *gun* method in which two hemispheres of uranium-235, each separately subcritical, are brought together by firing an explosive charge. The barrel of the 'gun' can be lined with uranium-238 which both reflects neutrons and contributes extra neutrons from its own fission so that the same subcritical state will be achieved with a lower charge. The barrel will also act as a *tamper* like those used with conventional explosives to contain the explosion for a short time and prevent the fissionable material flying apart before most of the chain reaction is completed. A bomb of this type was used at Hiroshima.

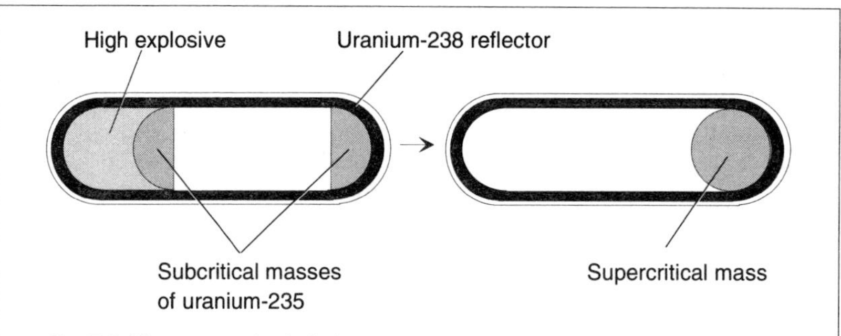

Fig. 7.2 **The gun method of triggering a fission bomb.** *Left*, the primed device. *Right*, the moment of impact of the subcritical masses, just before the explosion.

The second, density increasing, method is illustrated in Fig. 7.3, and depends on precisely shaped high-explosive charges of lens-like shape placed round a sphere of plutonium. The charges are triggered by high-speed switches, called *krytrons*, which on a suitable signal detonate all the charges simultaneously, producing a spherical shock wave that compresses the plutonium. The procedure appears to be to store energy from a battery in a number of special condensers at a high voltage and then release this energy through the krytron to detonate the high explosive. This is the type of bomb dropped on Nagasaki. An initial flux of neutrons is required in both designs and comes from the bombardment of beryllium by an alpha-particle emitter like polonium-210,

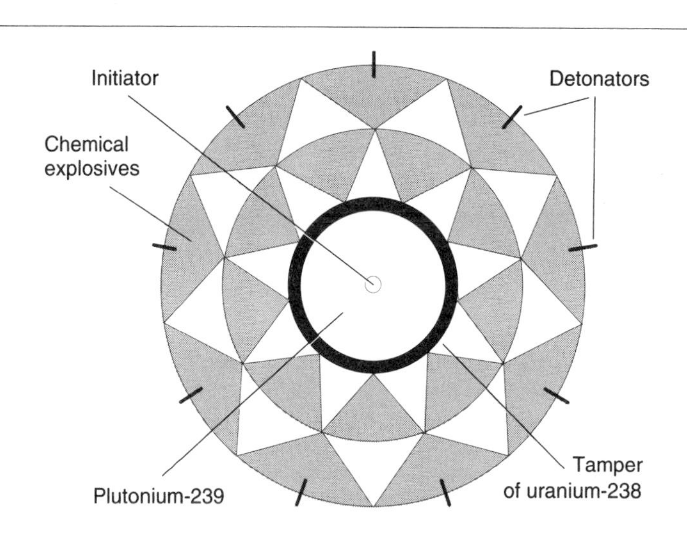

Fig. 7.3 **Schematic drawing of an implosion bomb using plutonium-239.** Simultaneous detonation of the shaped chemical explosives causes the uranium-238 tamper ring to implode and compress the plutonium into a supercritical state. At the same time an initiating burst of neutrons ensures that a high proportion of the plutonium explodes before all the components fly apart.

contained in a small sphere made of concentric layers of these materials and called the *urchin.*

It seems that the gun method is now obsolete, possibly because it depends on accelerating the two hemispheres towards each other and this might occur inadvertently during an aircraft accident or other mishandling. Moreover, pure fission bombs of any type are not currently made by the superpowers who now depend on a combination of fission and fusion to obtain greater destruction from similar sized devices. These hybrid bombs will be described in outline later.

Explosion time scale

Any fission explosion initiated by one or a few neutrons will go through a number of neutron generations. These last for the average time it takes for a neutron to travel from its parent nucleus to a second fissionable nucleus, which depends on the velocity of the neutron and the distance each travels before capture. For fast neutrons the time is about one hundred millionth of a second, 10^{-8} s. Bomb makers call this a *shake*.

The number of generations to produce a given number of nuclear fissions and amount of energy can be easily calculated if we know the proportion of fission neutrons available for continuing the chain reaction. The total number of fissions needed to provide the same energy as 0.1 kiloton of TNT is 1.45×10^{22} (from the table on p. 65) and this would require 51 generations. A 100 kiloton explosion needs 58 generations, so most of the energy is released in the extremely short period of 7×10^{-8} seconds or seven shakes, the whole procedure taking about 0.6 of a microsecond. But before this time has elapsed the amount of heat produced will have caused the nuclear explosive, its constraining tamper and the weapons casing to explode. As they fly apart the uranium or plutonium will become subcritical again and a proportion of the charge unused.

It is considerations of time therefore that make the design of nuclear triggers and the exact arrangement of the various components of a bomb so important in making an effective device. Uranium enrichment or plutonium production is not the only obstacle to nuclear proliferation.

Fusion bombs

Because thermal energy in the form of extremely high temperatures is needed to overcome the Coulomb barrier to nuclear fusion in bombs as in potential fusion power stations, this kind of fusion is called *thermonuclear fusion* and the bombs *thermonuclear bombs* and sometimes hydrogen bombs. In fact these bombs are hybrid devices, relying on nuclear fission to achieve the high temperature needed to ignite the fusion fuel and even a third stage of uranium-235 and uranium-238 fission to further increase the yield.

The main fusion explosive is lithium deuteride, which is a greyish powder at room temperature and easy to handle; it is made with the isotope lithium-6. Neutrons from the initial fission explosion cause the lithium-6 nuclei to convert to tritium and helium according to the following reaction:

$$^{6}\text{Li} + \text{neutron} \Rightarrow {}^{3}\text{H} + {}^{4}\text{He},$$

with the tritium now having enough energy to fuse with the deuterium and then

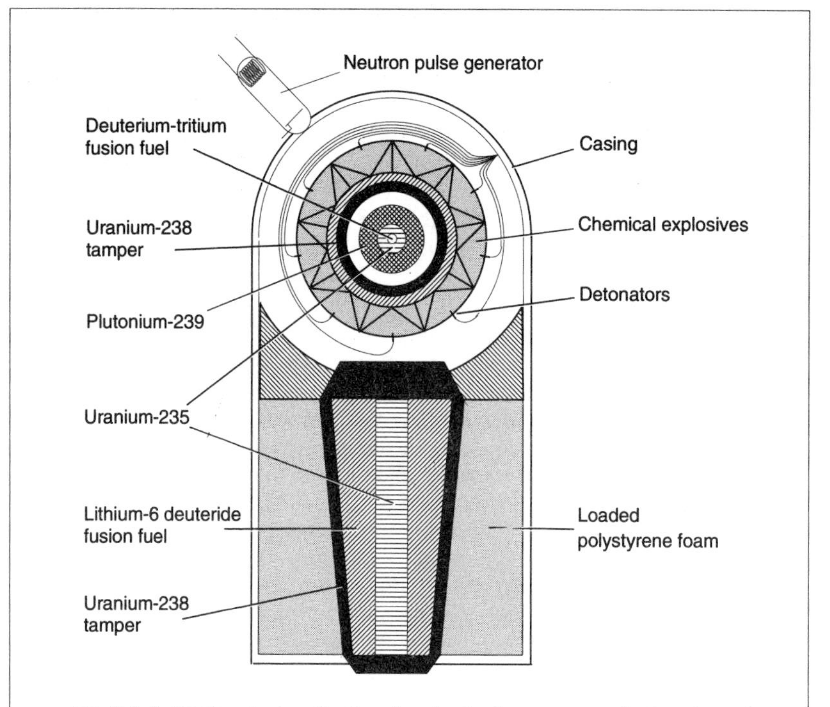

Fig. 7.4 **A fission-fusion-fission bomb**. In the upper part, the shaped explosives cause the plutonium-239 and uranium-235 to explode with the small amount of deuterium and tritium producing extra neutrons. X-rays and gamma rays from this explosion vapourize the loaded polystyrene foam around the lower cylindrical component which in turn compresses and heats the tamper and lithium-6 deuteride to the latter's fusion temperature. The diagram and description comes from H. Morland, *The Secret That Exploded* (Random House, New York, 1981).

to generate sufficient heat, neutrons and alpha particles to sustain the reaction until the weapon finally blows apart.

The main technical problem of such bombs is to arrange that the *primary* fission explosion does not drive apart the *secondary* fusion charge before it can be triggered and exploded. No official drawing or description of a fusion bomb has been published by the US authorities, but the layout illustrated in Fig. 7.4 is believed to be essentially correct. It was pieced together from many published and other non-classified sources by Howard Morland in 1978-9 and published only after US government attempts to censor it were abandoned in the courts.

The weapon can be considered as two coupled components: the first is a spherical component (the *primary*) very like an implosion fission bomb, in which the shaped charges, probably of 1,3,5,-triamino-2,4,6,-trinitrobenzene cause the tamper, a sphere of uranium-238 to contract and accelerate into a plutonium and uranium-235 spherical charge; the second is a tapered cylindrical component (the *secondary*) again with a tamper but now mainly containing the lithium-6 deuteride fusion fuel and a core of uranium-235. The secondary is surrounded by a heavy polystyrene foam which appears to contain pentane and probably finely-divided metal to absorb X-rays and gamma rays.

It is the coupling of the primary and secondary which is the technical innovation and may be the 'secret'. A shaped plug of neutron-absorbing uranium-238 shades the fusion fuel of the secondary from direct irradiation. Further, the energetic products of the primary explosion are neutrons, gamma and X-rays, of which only the last two are photons and travel at the speed of light. They have therefore time to be reflected from the inside casing and ignite the plastic foam into a plasma causing a second chemical explosion which forces the conical uranium-238 tamper to implode. This, together with the fission explosion, both heats and compresses the lithium deuteride to the ignition temperature of the thermonuclear reaction. The additional uranium-235 and the uranium-238 tamper will then fission and augment the explosion.

Other features are worth noting. Gamma and X-rays are only reflected when at near glancing incidence; it may therefore be that the hemispherical shape at the upper part of the drawing conceals different reflectors internally. As in the plutonium fission bomb, the shaped chemical explosive charges are set off simultaneously by the krytron triggers and in addition a *neutron-pulse generator* injects a priming pulse of neutrons into the fission core. This core also contains a pocket of tritium and deuterium gas under pressure whose fusion can significantly alter the yield of the whole weapon. This is the 'dial-a-yield' feature by which it is possible to inject known volumes of the gases during final arming to control the explosive yield.

Such a bomb is therefore a fission-fusion-fission device and the advantage to the bomb-user, but not to anyone else, is that the maximum power can be increased by 10 to 100 times over that of a plain fission bomb and all contained in a warhead about 50 cm in diameter and 200 cm long. A fission bomb with sufficient uranium-235 or plutonium to provide the same energy, would certainly be considerably larger and blow apart before much of the charge could explode.

Fission-fusion-fission bombs with their uranium-238 tampers and uranium-235 triggers and enhancers are very 'dirty' bombs because of the fission products dispersed from these components. Relatively 'clean' bombs can be made without uranium-238 at the expense of decreasing the yield. Small versions of the latter can be used as tactical weapons that give an intense burst of neutrons from the fusion reaction which will damage people but not property. One possible way of making such a weapon is to rely on lasers to ignite a deuterium-tritium charge in the same way as the laser-driven fusion power technique but with a larger initial target sphere.

Fission product 'signatures'

One long-lasting effect of an airborne explosion of a bomb or of an accident at a nuclear power station is the discharge of fission products into the atmosphere which can be detected by monitoring stations in many countries. These monitoring services can distinguish between different isotopes, the proportions of which will depend on the amount and kind of the fissionable materials used and on the average energies of the neutrons which cause the fissions. It is therefore possible to identify the nature of a bomb or of a class of power station from the 'signature' detected in the atmosphere, a result of considerable importance for the international control of weapons proliferation.

Destruction by nuclear weapons

We shall be considering the nature of the biological damage by radiation in later chapters and there are also many books on the destructive power of nuclear weapons. We will, therefore, not consider the question in detail now. Some things are worth summarizing however. The radius of the circle of destruction round the *epicentre* of any explosion varies roughly with the cube root of the yield. So a 10 megaton thermonuclear weapon will destroy virtually everything in a 20 km circle compared to a 10 kiloton fission weapon with a destructive diameter of 2 km. Similarly the lethal dose of neutrons and gamma rays will extend over about the same radius. The effect of the fission products depends on whether the bomb is exploded close to or high above the Earth. In the latter the vapourized cloud of products will be dispersed by high-altitude winds and will remain airborne for many months, during which many short-lived isotopes will decay away leaving the major products, strontium-90 and caesium-137, to fall to Earth over a long time. A ground-level explosion, on the other hand, will cause all the fission products to be intermingled and absorbed on to the dust and debris formed in the resulting dust storm. The dust will be carried downwind and can deliver a lethal dose of many different kinds of fission products to populations living within a few hundred miles of the epicentre of the explosion.

Chapter 8

Radiation detection

Humans can only detect a very limited range of radiation, the visible spectrum and, as heat, the infrared and parts of the microwave region. It is partly this invisibility of most of the electromagnetic spectrum and all nuclear radiations, coupled with the dangerous nature of some of them, which makes the whole subject of radiation and radioactivity a cause of great concern to many people. The detection and measurement of radiation in a way that can be understood is therefore of great importance.

Nearly all devices which detect radiation convert electromagnetic waves or nuclear particles into electrical signals which can be heard as clicks, shown as a meter reading or recorded in some way. Others depend simply on the heating effect of microwave and far infrared radiation. The particular design will depend on whether we want to detect or to measure the radiation and whether we want to distinguish between parts of the electromagnetic spectrum and the different kinds of nuclear radiations. Detectors and their associated circuitry can be made to record the incident energy distribution. They can also be designed to detect very low levels or alternatively the exact time and direction of arrival of an energetic particle. In general the ability to detect nuclear radiation depends on the thickness of the absorbing material; a thin layer is sufficient to detect alpha particles and low energy electrons (beta particles) but high-energy gamma particles will require centimetres before most of them are stopped and converted to some other kind of signal.

Infrared detection

For wavelengths from approx. 100 μm to 1 mm (the far infrared) the usual means of detecting radiation is through its heating effect. The measurement of the radiant power therefore becomes a measurement of temperature, which in turn becomes a measurement of some associated physical or electrical property. It is usual for the absorber to be in the form of a solid and for the measurement to be made either of a change in resistance, when the detector is called a *bolometer*, or of the change in a thermoelectric potential, when it is called a *thermocouple detector* or *thermocouple*.

For a bolometer using a superconducting material as detector (e.g. tin near its transition temperature 3.7 K) the minimum detectable power is less than 10^{-13} W. At a wavelength of 100 μm this corresponds to less than 5×10^7 photons per second. Thermocouples have a sensitive element typically 1 mm × 1 mm, made by evaporating thin overlapping films of two different metals on to an insulating backing. The region of overlap is blackened to ensure the absorption of the radiation and so radiation falling on it will cause the temperature of the junction to rise above that of other junctions in the circuit, resulting in a difference in electrical potential at the output terminals. The metals antimony and bismuth are often used because they give the highest potential difference per degree Celsius. Minimum detectable power is about 10^{-10} W. If an area larger than 1 mm x 1 mm is needed several thermocouples can be connected together in series to form a *thermopile*. The potential differences of the junctions are added together, each being proportional to the radiation falling on it.

Visible radiation detection

Photoemission of electrons from solid materials becomes possible at wavelengths shorter than about 1.2 μm, with a quantum energy greater than 1 eV. It then becomes possible to detect single photons, giving much greater sensitivity than that available for longer wavelengths. Photons falling on a specially prepared surface, the *photocathode*, cause several *photoelectrons* to be emitted. Photoelectrons are then amplified by a combination of photoemission and an electron multiplier in a *photomultiplier* (see Fig. 8.3), in which the emitted photoelectrons are attracted to the first dynode of a chain, which is at a higher potential, where further amplification occurs. The process is repeated up the chain to the anode or collector where about a million electrons are received for every photon striking the photocathode. The electrical signal generated is then further amplified to drive counters or a recorder.

Alternatively, the spatial arrangement of incoming photons can be preserved and the photoelectrons accelerated on to a screen, as in a television camera tube, in order to give a greatly enhanced image of the light that initially fell on the photocathode. The device is then known as an *image intensifier*. Because photoelectrons leave the photocathode randomly even in the absence of light, the typical photomultiplier detector is less sensitive at detecting a constant intensity than is the human eye, which under the most favourable circumstances can see a constant object, such as a faint star, if a few hundred photons per second are received. The eye is able also to see a weak single flash of light if it contains more than a few hundred photons, but in this case the photomultiplier is more sensitive.

Photography and xerography

Photography is a familiar method for detecting visible and X-ray wavelengths. It requires high intensities but gives a permanent record and can easily add together the energy received over a defined period during an exposure. The primary process is the photoelectric ejection of an electron from a bromine or iodine atom. Since the minimum energy for this to happen is about 1 eV, emulsions usually cannot be made sensitive to wavelengths longer than about 1.2 or 1.3 μm. The ejected photoelectrons move about in the conduction levels of the crystal until they are held in trapping centres. They then attract some of the interstitial silver ions which are always present, to form a speck of silver which acts as a latent image. Subsequently the developing agent gives electrons to the latent image which then attract further silver atoms. The radiant energy required to produce a silver grain depends on the wavelength of the light and the sensitivity of the emulsion, and also on the development process. The minimum energy density needed to form a visible image is about 3×10^{-12} J per mm^2 (10^7 photons per mm^2).

Xerography (*dry writing*) starts with a layer of positive charges deposited on one surface of a film of semiconductor material, sometimes called the *photo-*

receptor. An image of the object (usually a flat opaque object such as a page of typescript) is formed on the charged surface by the incident photons creating electron-hole pairs (see p. 76) inside the photoreceptor. The electrons then drift towards the upper surface under the influence of the electric field in the photoreceptor, and neutralize only those surface charges in the regions where light has been received. In the third stage coloured (usually black) powder particles are placed near to the surface so that they are attracted to the charged areas and adhere there to produce a visible image. The powder particles are transferred to paper by electrostatic attraction and are fixed by heat treatment. Amorphous selenium is used for the photoreceptor.

Photographic emulsions have, at present, a higher resolution than the xerographic process and can be made sensitive to colour more easily. Special emulsions can be made to give false colour images from infrared energy. For X-rays sensitivity can be much increased by coating both sides of the support with thick emulsion and by the use of fluorescent intensifying screens.

Counters and monitors

Ionization chambers

An *ionization chamber* is a simple device (see Fig. 8.1) consisting of a volume of gas sealed between two plates about 1 cm apart and held at a potential difference of about 100 volts. The gas can be e.g. argon, nitrogen, methane or dry air depending on the particular radiation sensitivity needed. Any radiation reaching the space between the plates will collide many times with the atoms of the gas, giving up energy to the atom and ionizing it. The positive ions then move towards the low-voltage plate and the negative electrons to the high-voltage plate, causing an electrical signal which is amplified and heard as pulses or collected together (integrated) and read on a meter. It requires about 34 eV to ionize an atom of air so a particle expending 1 MeV of energy could cause up to about 30000 ionizations. Despite this, an ionization chamber is essentially a low-sensitivity device with the signal requiring considerable further amplification.

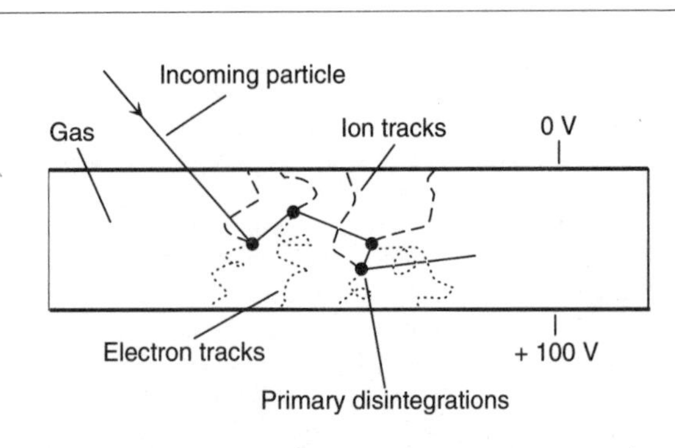

Fig. 8.1 **Ionization chamber**. This can be made quite large, say, 12 cm × 12 cm. An incoming particle successively ionizes atoms of the contained gas and the resulting electrons and ions move to the positive and negative plate, respectively.

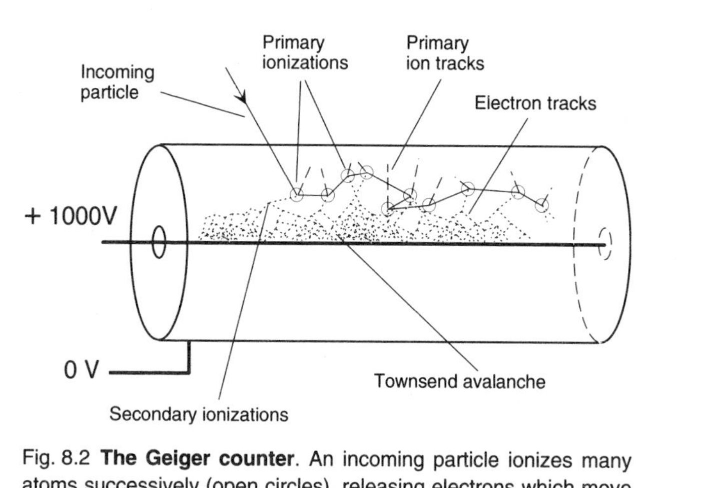

Fig. 8.2 **The Geiger counter**. An incoming particle ionizes many atoms successively (open circles), releasing electrons which move at high speed towards the anode ionizing other atoms as they go. These in turn ionize still more atoms so that eventually, near the anode, the whole tube participates and up to 10^{10} electrons collect at the anode.

The response of an ionization chamber to different kinds of radiation can be adjusted by covering it with absorbing material so that, for example, sensitivity more closely matches that of living tissue. Another modification is to use a gas with a high neutron scattering or capture cross-section like boron trifluoride. Ionization will then follow the rapid recoil of the gas nuclei from the collision or from the energetic particles released after a collision. It is even possible to coat the inside of the chamber with highly enriched uranium-235 which will fission in response to thermal neutrons with the fission products inducing intense ionization and characteristic electrical output signals.

Geiger-Müller tube

This development of the ionization chamber which is very commonly used is often called a Geiger counter (see Fig. 8.2). It is essentially the same device but in the shape of sealed metal cylinder at earth potential and a central wire as the anode. It operates, however, at a much higher voltage so that each electron produced from an initial collision is accelerated to a velocity high enough to cause a whole series of secondary ionizations (perhaps ten thousand) on collision with other gas molecules. The secondary ionizations then initiate a further series and so on to produce a vast burst of electrons near the anode, called a *Townsend avalanche*. The resulting electrical pulses are all of the same magnitude whatever the type of incoming radiation except that they are incapable of measuring gamma radiation above about 1 MeV. The Geiger counter is a simple reliable instrument, incapable of distinguishing the type of incoming radiation, but capable of giving an audible click or visual warning on a meter of the presence of radioactivity. The potential difference varies from a few hundred to over a thousand volts.

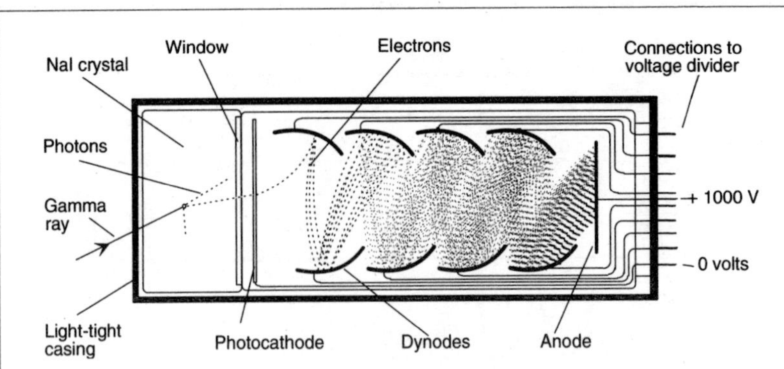

Fig. 8.3 **Scintillation detector for gamma rays**. A gamma ray excites an iodine atom in the NaI crystal, causing its electrons to move out from the outermost valence band to a new conduction band. As the electrons fall back to their more stable valence state they emit photons, one of which is shown striking the photocathode of the photomultiplier. This causes several electrons to be emitted which are accelerated to the first dynode, where the multiplication process is repeated until finally at the anode about 10^6 electrons are collected.

Scintillation detectors

Scintillation detectors detect and measure the flashes of light which certain atoms emit when they are irradiated by gamma and X-rays. They usually consist of a sodium iodide crystal to which a small amount of thallium has been added to alter the colour of the flashes to a wavelength which makes their detection easier. These are called NaI(Tl) crystals. The iodine has a high density and is much more efficient at absorbing high-energy gamma and X-rays than the gas in a Geiger counter. Usually a block of NaI(Tl) is enclosed in a metal tube with a glass window at one end through which the weak flashes of light (scintillations) are detected and counted by a photomultiplier (see Fig. 8.3).

Such an apparatus can be made extremely sensitive particularly if the photomultiplier is cooled to liquid nitrogen temperatures. This reduces the thermal electrons, often called *random noise*, which are emitted by the photocathode. Other types of scintillation counters will detect beta particles that cause scintillations either in a special organic liquid scintillator or more simply, in water by Cerenkov radiation. Random noise is reduced by *coincidence counting* in which two matched photomultipliers look at the counting vial and only pulses which arrive simultaneously at both photomultipliers are recorded. Random events in an individual photomultiplier are ignored. Another refinement, used in *scintillation spectrometers,* is to arrange the circuitry after amplification so that beta particles of different energies can be counted separately. This will allow, say, carbon-14 and tritium to be measured at the same time in biological experiments.

Solid-state devices

These are detectors in which the properties of a conventional solid-state diode are used in a special way. Such a diode is made of two pieces of silicon or

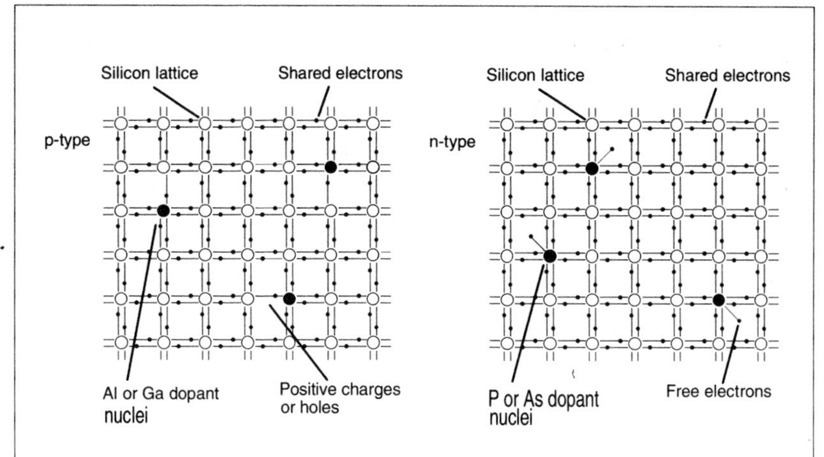

Fig. 8.4 **Semiconductor doped to give holes and free electrons**. An electron from any neighbouring atom can fill a hole, so the holes appear to move around independently of the crystal lattice. Similarly, electrons can move into the orbits of any neighbour. In this way electrons can diffuse through the crystal and fill the holes in the depletion layer between two pieces of differently doped semiconductor, forming a junction diode.

germanium which have been *doped* by adding small amounts of other elements. Both silicon and germanium are group four elements with four electrons in their outer *valence band* of electrons. These elements are able to form sheets of atoms in their crystal structure by sharing these four electrons between all adjacent atoms (see Fig. 8.4). If a small amount of an element with five outer electrons is added when the crystals are made, the added atoms will each provide an electron which is able to move through the crystal lattice and be replaced in its earlier position by another electron. On the other hand, a group three atom will enter the lattice but be short of an electron and leave a *hole* that is similarly free to move around. A semiconductor with an excess of negative charge carriers (electrons) is called an *n-type semiconductor* and one with an excess of positive charge carriers (holes) a *p-type semiconductor*. The semiconductor itself is electrically neutral.

If material with excess electrons is bonded to one with excess holes by a diffusion process the result is called a *p–n junction diode* in which the electrons and holes from the p and n regions diffuse and neutralize each other in what is called the *depletion region*. The atoms left behind become ionized and create an electric charge which prevents the further migration of electrons and holes. If we apply a voltage that neutralizes this charge then current will flow across the diode. This is the normal way in which a solid-state or semiconductor diode operates in, say, a television set and such a diode is said to be *forward biassed*. If, on the other hand, we apply an opposite voltage, the diode is *reverse biassed* and a negligible current will flow.

It was discovered a long time ago that radiation, striking the depletion region in a reversed biassed diode, creates electron-hole pairs and accompanying electrical pulses. Further, the amplitude of the pulse corresponded to the energy of the

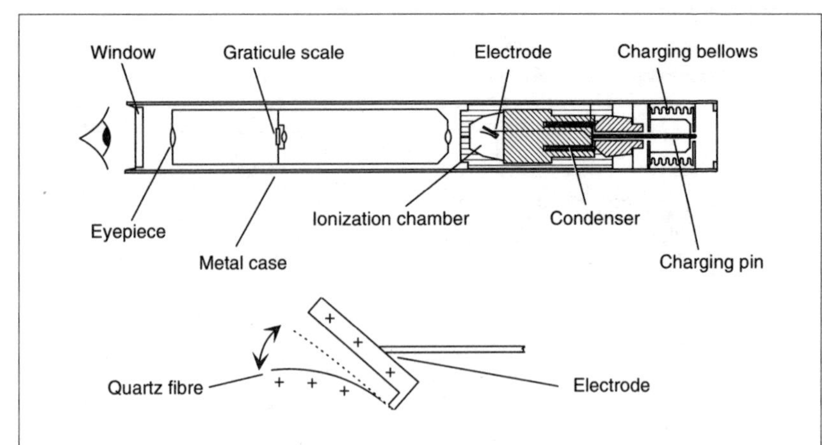

Fig. 8.5 **Schematic diagram of a quartz fibre electroscope**. The hatched components are made of transparent plastic so that the quartz fibre and scale on the graticule can be seen by pointing the device at a light source. Typically the condenser will hold the charge for about a week, after which the ionized air is replaced with the charging bellows. An external electrical source, with its own light, is then used to raise the potential of the central electrode and condenser until the quartz fibre tip is brought to zero on the scale. The electroscope is about 14 mm in diameter by 12 cm long.

radiation. Nowadays several thousand volts of reverse bias are used to improve the efficiency of the device and to increase the size of the depletion region. The detector is often made of p-type material into which lithium is first diffused and then allowed to drift into the p-type material under reverse bias and raised temperature. A large depletion region is the result but liquid nitrogen temperatures are then needed to prevent the lithium drifting back. Such detectors are said to be *lithium drifted* and aluminium or gallium is used as the positive dopant.

Semiconductor detection has several advantages: the material is solid and has a greater stopping power than the gas in other detectors; it is much more sensitive to low-energy particles, typically by one or two orders of magnitude; and, as already mentioned, it can be made from silicon or germanium depending on particular requirements. Silicon ($Z = 14$) is mostly used for the detection of alpha and beta particles and low-energy X-rays, while germanium ($Z = 32$) is better for higher energy X-rays and gamma rays.

Personal radiation monitors

These detectors are designed to be worn at all times and are therefore light and small. One kind is the *pocket fibre electroscope* or QFE which is very like the original electroscope used for detecting radiation by Rutherford. It is a miniature ionization chamber in which the central anode has a metal-coated quartz fibre attached to it (see inset in Fig 8.5). If a potential of about 100 volts is applied between the anode and cathode (the metal casing) the resulting charge causes the quartz fibre to move away from its supporting wire and act as a pointer moving over a miniature scale to the fully charged position. The

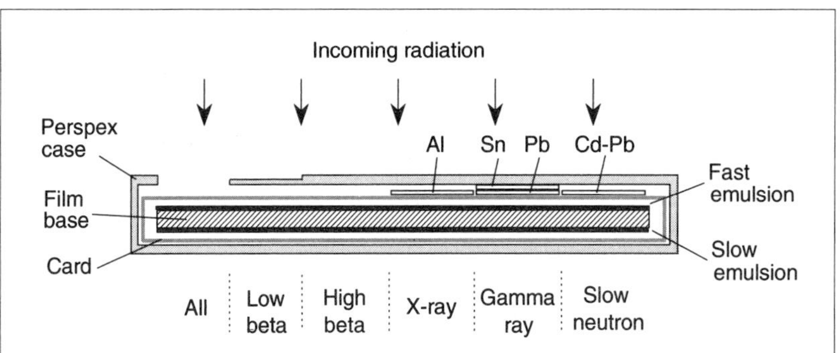

Fig. 8.6 **Film badge for personnel in the nuclear industry**. Photographic film in a light-tight envelope is placed in a perspex container which has a number of 'windows' (upper surface). Different thicknesses of plastic and metal plates attenuate different energy radiation and the blackening under the windows gives a permanent record of the energy and amount of radiation received. Each badge is about 3 cm × 4 cm. Simpler badges may be used in neutron-free areas.

potential is then disconnected. Radiation passing through the casing and ionizing the air in the tube will cause the charge to fall, at a dose-dependent rate, until the pointer returns to zero. The position of the pointer can be seen by looking through an eyepiece when the electroscope is lifted to the light. The device is calibrated so that an immediate reading is obtained for the dose received since last charging and ranges are available with maxima from 2 to 5000 millisieverts.

The second is the *film badge* which has the advantage of providing a permanent record but cannot be read immediately. It consists of a small piece of photographic film placed in a light-tight card envelope snapped into a hinged plastic container. The container is subdivided along its front face so that different kinds of radiation are preferentially attenuated (see Fig. 8.6). The film itself is coated on one side with a sensitive emulsion and on the other with an insensitive one. The pattern of exposure obtained on both sides of the film will therefore give a record of the amount and kind of radiation received over the period that the badge has been worn. The film in its envelope is then processed and the results recorded in the radiation worker's personal record sheet.

Film badges are worn by anyone working in a laboratory, hospital and other workplace who uses or works with radioactive materials. In situations where larger doses might be encountered the badge will be supplemented by monitors, like the personal electroscope, which give immediate warning. In addition, in areas where there is a risk of inhalation or ingestion of radioactive gases or dusts, there has to be provision for whole body monitoring and monitoring of air, clothes and workbenches. The personnel who monitor and collect the data are called *health physicists* and are an important part of any place at which radioactive materials are handled.

Chapter 9

Photons and Electrons

In the previous chapters we have been mainly concerned with the particulate radiation produced by nuclear reactors, especially the effect of neutrons on matter. Two forms of electromagnetic radiation have also been important, gamma radiation and those characteristic X-rays which are emitted when nuclei and atoms are in an excited state. This chapter, on the other hand, is concerned with the type of radiation which we all experience in its less energetic form as light or make use of as radiowaves.

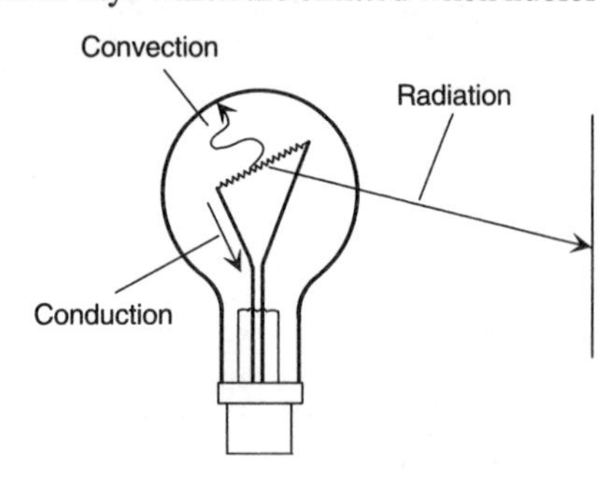

Fig. 9.1 **Energy flowing from a lamp.**

The ordinary electric filament lamp shining in a room (see Fig. 9.1) is a simple source of radiation. The electrical current passing through the filament heats it to a high temperature and the filament then loses energy in three ways: the metal supports of the filament *conduct* energy in the form of heat to the lamp holder; the inert gas in the glass envelope will be warmed and this gas will in turn heat the glass and base by *convection*; energy is also lost by *radiation* which includes the light which we see. This passes through the gas in the envelope, the glass and the air of the room until it reaches an object such as a wall.

If you reduce the voltage supplied to the lamp and therefore the current passing through the filament you will notice that the *colour* of the light becomes redder and the filament dimmer until eventually no light is seen, but you will still be able to detect heat on your hand if it is placed near the lamp. This is because the wavelengths of the radiation emitted are changing as the filament cools.

Wavelength or frequency of oscillation is one of the two fundamental ways of describing radiation; a range of wavelengths is called a spectrum. In fact the spectrum which comes from an ordinary electric lamp running at its designed voltage is like that of the visible light coming from the Sun except for having a greater proportion of its energy at the red end of the spectrum. The spectrum of radiation which the eye can detect is, however, only a very small part of what is called the *electromagnetic spectrum* which ranges in wavelength from radio waves of more than a hundred thousand kilometers to gamma rays of wavelengths of the order of size of an atom (see Fig. 9.1).

The electromagnetic spectrum.

Different parts of the spectrum have special names but in fact their properties

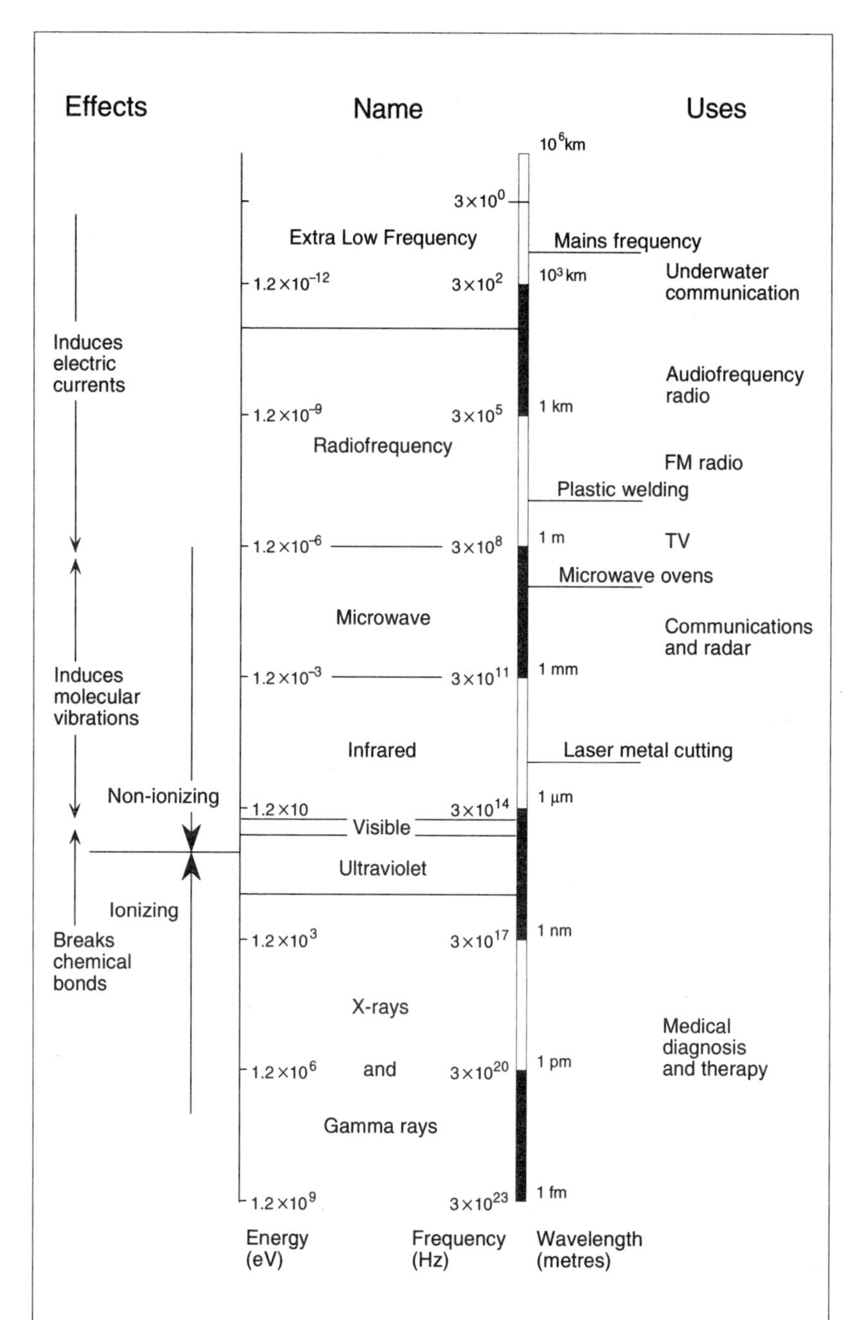

Fig. 9.1 **The electromagnetic spectrum**. Although names are given to individual parts of the spectrum there are no clear-cut distinctions between the different regions. The boundary between ionizing and non-ionizing radiation is particularly important.

vary gradually and there are no sharp distinctions between neighbouring regions. The visible region with which we are most familiar changes gradually into infrared at longer wavelengths and ultraviolet at shorter. Some people are able to see slightly longer and shorter wavelengths than others. Radiation of whatever wavelength moves through a vacuum at the same velocity (the speed of light) but the way it interacts with any matter which it encounters depends on its wavelength. Light of different wavelengths, for example, travels through a transparent medium like glass at different velocities, which enables a prism to spread white light into its coloured components. This kind of interaction is still most easily explained in terms of the wave-like nature of light that was the basis of classical physics until 1900.

Other phenomena, in particular the distribution of energy coming from a substance heated to a high temperature as shown in Fig. 9.4 is only explained by supposing that light is propagated as separate particles. Light and electromagnetic radiation behave therefore both as a train of waves and a train of particles.

How are the two related? Planck (see panel on p. 83) discovered that the energy of a photon was equal to the frequency of the radiation multiplied by a constant (Planck's constant) which means that the long radiowaves which have very low frequencies have very low-energy photons, while the photons of gamma rays are enormously energetic (see Fig. 9.1).

Electricity and magnetism

This kind of radiation is called elctromagnetic because of the relation discovered in 1864 by James Clerk Maxwell, who showed that any time-varying electrical charge moving along a conductor or through space has, and must have, an identical time-varying magnetic charge travelling with it but at right angles (see Fig. 9.2). It also accounts for the induction of electric currents in transformers, the propagation of radio waves and the generation of an electric current by the movement of a magnetic field.

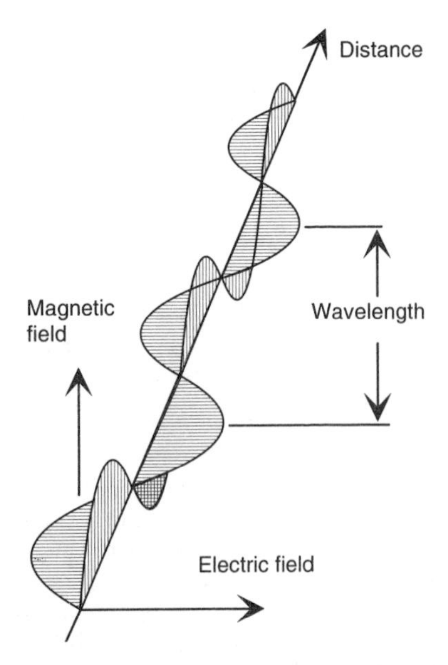

Fig. 9.2 **Relation between electrical and magnetic charges**.

Radiation and matter

The different ways in which radiation reacts with the material it encounters can be explained in everyday terms rather like those used by physicists before the acceptance of quantum theory.

Long radio waves are able to pass through gases, liquids and solids with little loss or attenuation and indeed find a principal use today in communicating with submarines under the sea. Shorter wavelength

quantum theory

Classical work on thermodynamics and radiation had predicted that the amount of radiation emitted by a *black body* should continue to increase as the wavelength was reduced. Experimental evidence showed that instead it fell after reaching a maximum which depended on the temperature of the body as in Fig. 9.4. To explain this discrepancy, Max Planck (1858-1947) had to abandon classical dynamical principles and formulate the quantum theory (1900), which assumed that energy changes occurred as abrupt instalments or quanta. Although at that time, it was only a mathematical interpretation which explained the experimental results, it came to revolutionize physics and the way we look at the world after Albert Einstein applied the theory to electromagnetic radiation (1905). Einstein was able to explain how the energies of photons emitted by atoms depended on the frequency of the light which fell on them, establishing that electromagnetic radiation behaved as both a wave and a particle. In 1913 Niels Bohr successfully applied quantum theory to the problems of subatomic physics.

A black body is defined as not only radiating energy at all frequencies but also absorbing all frequencies of radiation incident upon it. Such objects do not exist in nature but real objects like the Sun behave roughly like a black body of a temperature of 6000 K. Planck discovered a simple relation between the amount of energy radiated by a black body and the wavelength of the radiation. The energy equals the frequency of the radiation multiplied by a constant, Planck's constant, or put another way it equals the speed of light multiplied by the constant and divided by the wavelength. This is the formula

$$E = hf = \frac{hc}{\lambda},$$

where h, Planck's constant is 6.63×10^{-34} J s (J stands for Joule, the standard unit of energy, and s for second), f is the frequency of the radiation in Hertz (Hz) or cycles per second, c is the speed of light which is 3×10^8 metres per second, and λ is the wavelength. We can use this formula to work out the energy of light in the middle of the visible spectrum. This has a wavelength of 0.5 μm, so that

$$E = \frac{(6.63 \times 10^{-34}\,\mathrm{J}\,s) \times (3 \times 10^8\, m/s)}{0.5 \times 10^{-6}\, m} = 3.98 \times 10^{-19}\,\mathrm{J}$$

It is more convenient in this kind of measurement to use a unit of energy, the electron-volt (eV), which is not so large as a Joule. 1 electron-volt = 1.602×10^{-19} Joules. The energy of a photon of this wavelength is therefore 2.48 eV.

radiowaves like those used for television and communications are less able to pass through land and buildings, and at higher frequencies are confined to line-of-site applications. Microwaves are stopped or absorbed by metals which become heated as a result but can penetrate a few centimeters into other materials while transferring their energy by causing the molecules in their path to vibrate and heat up as in the effect of microwave cooking on water-containing tissue.

Radiation in the infrared, visible and ultraviolet region of the spectrum are often absorbed by surfaces which are therefore opaque to that region of the spectrum. Sometimes, however, a part of the radiation will be re-emitted and this gives the material a colour when illuminated by white light. The rest is converted into heat at the surface. Again ordinary glass is transparent to visible light but opaque to shorter wavelengths of ultraviolet, but glass can be coloured during manufacture so that it only allows through a part of the visible spectrum.

X-rays and gamma rays are absorbed by heavy elements such as lead but pass through materials made of lighter elements to varying extents. This is the basis for the use of X-rays in medicine. Both X-rays and gamma rays interact with the atoms present in a highly complex way depending on their energy.

Photons and quanta

Radiowaves and microwaves

A more complete explanation of these phenomena comes from Einstein's discovery of the properties of photons, the particles of light energy (see panel on p. 83). Photons passing through matter will eventually interact with an electron. What then happens depends on the energy of the photon but also on the nature of the electrons which surround the atomic nucleus. Photons such as those found in radio waves, microwaves and part of the infrared region of the spectrum have insufficient energy to alter the electronic structure of atoms. They lose their energy by inducing a current in the free electrons in a conductor or by converting it into vibrational energy manifested as heat. The heating effect can be very severe and can cause burns to skin and damage to biological molecules. Photons with greater energy and therefore with frequencies extending through the visible towards the X-ray and gamma-ray end of the spectrum are able to react with the electrons surrounding an atom in different ways.

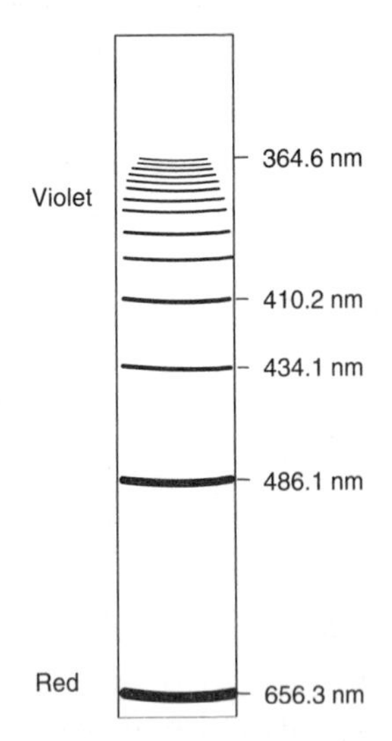

Fig. 9.3 **The Balmer series.** The line spectrum emitted by the excitation of hydrogen atoms.

Visible and ultraviolet

To understand this interaction we must look at the different states in which an electron can exist in the space round the nucleus. But first it is interesting to note some results that were not understood at the time and which were obtained for the simplest atom, hydrogen, with only one electron outside its nucleus.

Johann Jacob Balmer, a Swiss mathematics teacher reported his unexpected results in 1885 at the age of 60 and long before anyone realized their true significance. He shone a beam of white light through hydrogen gas and recorded the transmitted spectrum. His results (see Fig. 9.3) were very remarkable: most of the light passed unimpeded but it was absorbed at certain fixed wavelengths to give dark lines with a definite relationship to each other and getting closer together as the wavelength approached some limiting value. These wavelengths are now known as the Balmer series and in the next 25 years a number of similar series were discovered for hydrogen in both the infrared and ultraviolet regions of the spectrum. Similar results can be obtained from the other elements, complicated by their larger number of electrons.

In 1905 Einstein explained Balmer's results in his quantum photoelectric theory by assuming that the electrons surrounding the atomic nucleus can only exist at definite energy levels, no intermediate positions being allowed. Transfer from a lower to a higher level depends on an incoming photon having exactly the right amount of energy to pass to the electron and force it to make the jump. Only these photons will be absorbed from the beam of light and each dark band corresponds to the difference between allowable energy states. The displaced electrons immediately fall back to a more stable state giving off a photon of the same energy again, but these will be emitted in all directions.

X-rays and gamma rays

Most photons have insufficient energy to displace electrons from the tightly bound internal K- and L-shells of heavy elements but such displaced electrons are able to emit very high-energy photons in the form of X-rays as they return to a more stable state. Such X-rays are called *characteristic X-rays* and have a line spectrum typical of the atom which emits them in the same way as less energetic photons are emitted by the outer electrons surrounding a nucleus. *Bremstrahlung* or *continuous X-rays* are produced by the deceleration of high-energy electrons as they pass close to a nucleus. Such electrons are produced by accelerating electrons from a cathode to a high-voltage anode; these are the X-rays used in medicine. The energy of the electrons in an X-ray machine is determined by the accelerating voltage, i.e. the potential difference between the cathode and anode and such machines are described in megavoltage terms. See **Duane and Hunt's law**.

Most nuclear reactions, including those involving the ejection of alpha and beta particles, leave behind a nucleus in an excited, high-energy state which almost immediately decays to a more stable level by emitting a gamma-ray photon. Depending on the difference in energy level between these two states the gamma photon may have an energy between a 100 keV and 10 MeV and corresponding wavelengths between 10^{-11} and 10^{-13} metres.

The difference between X-rays and gamma rays depends therefore not on their energy, although in general gamma rays have more, but on how they are made.

Temperature and colour

We can now return to the colour of the light emitted by a filament and its dependence on its temperature. This determines not only the proportion of electrons occupying more energetic positions but also the involvement of those in shells nearer the nucleus, so increasing the available energy gaps and, as the electrons continually return to more stable states, allow the emission of photons of greater energy. Consistent with this, an enclosed electric arc lamp can heat the gas between its electrodes to temperatures which would melt any filament and the hot gas will emit photons in the short wavelength ultraviolet with spectral lines characteristic of a particular gas such as mercury or neon.

Even the very high temperatures found in an arc lamp fail to provide sufficient energy to cause the emission of many high-energy quanta in the gamma- and X-ray part of the spectrum. They do not, therefore, make a large contribution to the overall energy produced by a hot object. Equally there are many low-frequency quanta produced but these have so little energy that they do not make much of a contribution either. This explains the bell-shaped curve of a black-body radiator (see Fig. 9.4) which so perplexed physicists before Planck.

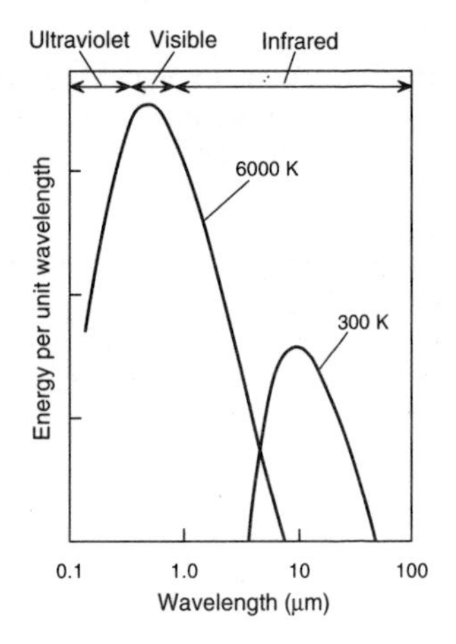

Fig. 9.4 **Black-body radiation**. The 6000 K peak corresponds roughly to the energy emitted by the Sun.

Ionization

Quantum photoelectric theory predicts that photons with energies less than those between any electronic energy level can never alter the electronic structure of an atom. They can induce currents in the free electrons in conducting metals and can raise the temperatures of atoms by contributing vibrational energy but they will never cause a photon to be emitted from an atom. In addition, the limiting value of wavelength towards which the Balmer series as well as all the other hydrogen series approach, corresponds to the wavelength of the photons with energy sufficient to strip the electron from the hydrogen atom and ionize it. This in turn explains why only the more energetic kinds of electromagnetic radiation are *ionizing*. The electromagnetic spectrum in Fig. 9.1 shows the boundary between ionizing and non-ionizing radiation with only the shorter wavelength part of the ultraviolet, X-ray and gamma-ray spectrum having sufficient energy. Because photons have first to cause an atom to eject an electron, they are called *indirectly ionizing radiations*. Charged particles which deposit their energy within the atom are called *directly ionizing radiations.*

The quantum interaction between photons and electrons is a fundamental attribute of matter, describing not only the familiar properties of colour and the emission of light on heating but also, in further developments not discussed here, how the electrons themselves also have a frequency component in their description and how the structure of the atom is maintained by them. Further, all other fundamental particles are also best described in terms of mass, frequency and energy.

Ionizing radiation and high-energy particles which can cause alterations to the properties of atoms, are of the greatest importance in any study of the biological effects of radiation. Before considering this important subject we must discuss in the next chapter the properties of the living matter with which they all interact.

Chapter 10

Cells, Tissues and Cancer

This and the following chapter consider the problems which are most immediately associated in many people's minds with nuclear energy and radiation: their biological effects, in particular the incidence of cancer which appears to be the main immediate result of low doses of ionizing radiation. The first chapter is concerned with basic biology; in effect the nature of the system which can be damaged. The next chapter will give a short account of the specific problems associated with nuclear and electromagnetic radiation.

The biology of the cell

Nearly all living organisms are made up of individual cells, either one or many of them; most of the exceptions being viruses and their bacterial counterparts, the *bacteriophage* or *phage* for short. Viruses and phage are in essence pieces of genetic material, protected from the environment by some sort of coat but quite unable to multiply except in their host cells. Cells on the other hand go through a cycle in which they grow, taking in food from outside themselves, and then divide to form two *daughter* cells more or less identical with each other and with their *mother* cell.

Some of these cells are organisms in their own right; protozoa, such as amoebae, and bacteria exist as single cells in an environment which provides their essential nutrients, and allows them to grow and divide until they exhaust that food. Higher organisms, as plants and animals are called, are multicellular; they start as a single cell, the fertilized egg, and grow and divide but the daughter cells can differ from each other, eventually giving rise to the different structures, tissues and organs characteristic of the plant or animal. The number of cells in the adult organism is enormous with humans having about ten million million (10^{13}) cells, each cell being the end result of a much smaller number of different cell *lineages* several of which may give rise to a tissue. In the adult many of these lineages will have stopped dividing when the tissue has reached its mature size, but others such as the blood-producing cells and those in the gut and skin, go on dividing throughout life with compensating numbers of cells breaking down and being lost. By contrast, nerve cells stop dividing in infancy and others only divide if the tissue is damaged in some way. About 10^{16} cell divisions will occur during a human lifetime.

Cell division

A cell is made up of two main parts, an outer *cytoplasm* and an inner *nucleus*. The cytoplasm is bounded by a membrane and sometimes a cell wall and is the main place where proteins are made. These proteins either combine together to make structures like the membrane or cell wall or they can be *enzymes*, the molecules with high specificity which facilitate chemical reactions within the cell. Both structural proteins and enzymes can be kept in the cell or exported into the environment outside. The nucleus also has a membrane surrounding it but, more importantly, it contains the *chromosomes* which determine the nature of the proteins which an organism makes. The chromosomes carry the instructions in the form of the genetic code of the DNA, not only for each cell but also for the whole organism. In other words, the genetic information in any cell is, normally, identical to that of the fertilized egg cell. A higher organism is, therefore, capable of controlling how the information contained in the chromosomes is used in the many different cell lineages which function together in the whole organism.

A cell which is about to divide is normally twice the size of a *resting* undividing cell. The nuclear membrane will then break down and the chromosomes, normally invisible, can be seen as thread-like structures and are found to be divided into two but held together somewhere along their length at a place called the *centromere*. The paired chromosomes then shorten and arrange their centromeres along a line across the centre of the cell, thereafter half the chromosomes move to one end of the cell while the other half moves to the other. The cell then divides by pinching off its cytoplasm between the two groups of chromosomes which gradually become less visible. Two new nuclear membranes appear and two new cells. The process is called *mitosis*.

The numbers and kinds of chromosomes

The fertilized egg has a chromosome number characteristic of the particular species and called the *diploid number*. It is 46 in Man. The unfertilized human egg has only 23 chromosomes, the *haploid number*, which result from two special cell divisions which occur before egg formation and are called *meiotic* or *reduction* divisions. A similar pair of divisions occur in the male, so the sperm also has 23 chromosomes. On fertilization the two haploid sets combine into a single nucleus. If you remove a dividing cell, preserve it from decay and stain it, it is possible to identify each chromosome under the microscope. It turns out that in Man there are 22 pairs of chromosomes plus one large chromosome, called the X, and a small one called the Y (see Fig. 10.1). All male mammals carry one X and one Y and it is known that the Y carries a gene which determines maleness. Females on the other hand carry two X chromosomes. We are now in a position to understand how sex determination occurs. Every unfertilized egg will, as a result of the meiotic divisions, contain one X chromosome. Every sperm on the other hand has an equal chance of carrying either an X or a Y. Depending on which kind of sperm actually fertilizes the egg, the resulting individual will have either two Xs and be female or an X and a Y and be male.

Excluding the X and Y, all the other chromosomes can be matched together in pairs and are numbered from one to 22 in order of size, one being the largest. This has an important advantage to the individual; if one chromosome is damaged it is almost always possible for the other of the pair to provide the missing functions. A mutation which occurs in one of the pair and not in the other is called a *recessive*, but if it occurs in both it is *dominant*. It is also now possible to add something more to our discussion of ordinary cell division or mitosis. The doubled number of chromosomes seen just before the chromosomes divide is, in effect, 92 and when the chromosomes separate the cell has

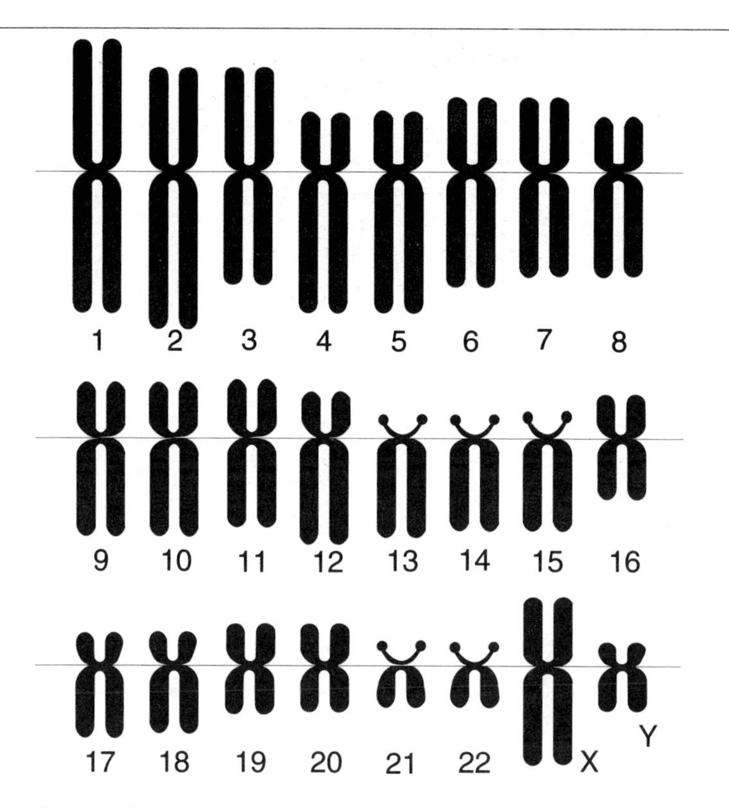

Fig. 10.1 **Schematic drawing of the human chromosome set**. The mitotic chromosomes are arranged in pairs placed on each side of the centromere attachment site, aligned on the horizontal lines. In life each chromosome is about to move apart to form the nucleus of the two daughter cells. The female has two Xs and the male an X and a Y.

to ensure that a complete set of 46 passes to each daughter nucleus, but it does not have to make certain that those derived from the mother and the father are kept separate. In fact they appear to be mixed at random.

The cells which form the gonads and of which some will become eggs or sperm do, in fact, separate off from the other cells quite early after fertilization, forming the *germline*. Their mitotic divisions are like any other so that an egg or sperm will also contain a random assortment of chromosomes from each parent, with the two final meiotic divisions providing the haploid set.

The chemical structure of the gene

We have already mentioned that the chromosomes carry the genes and that genes are made of a chemical polymer called DNA. We need to know something of this structure and how it behaves before understanding how radiation and, indeed, chemicals can alter this structure and damage it. A molecule of DNA consists of a repeating sugar-phosphate chain to which two purines, adenine and guanine, and two pyrimidines, cytosine and thymine, can be

DNA, deoxyribose nucleic acid

In its double-stranded form, DNA is the genetic material of most organisms and organelles, although phage and viral genomes may use single-stranded DNA, single-stranded RNA or double-stranded RNA. The two strands of DNA form a double-helix, the strands running in opposite directions, as determined by the sugar-phosphate 'backbone' of the molecule.

The four bases project towards each other like the rungs of a ladder, with a purine always pairing with a pyrimidine, according to the *base-pairing rules*, in which thymidine pairs with adenine and cytosine with guanine. In its common molecular form the helix is 2.0 nm in diameter with a pitch of 3.4 nm (10 base pairs).

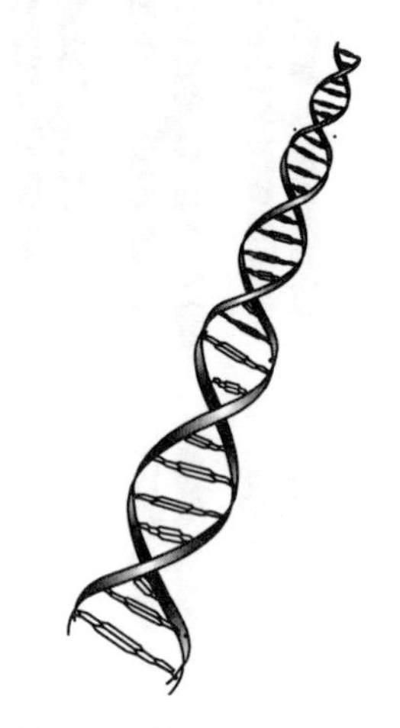

Fig. 10.2 **The double helix**.

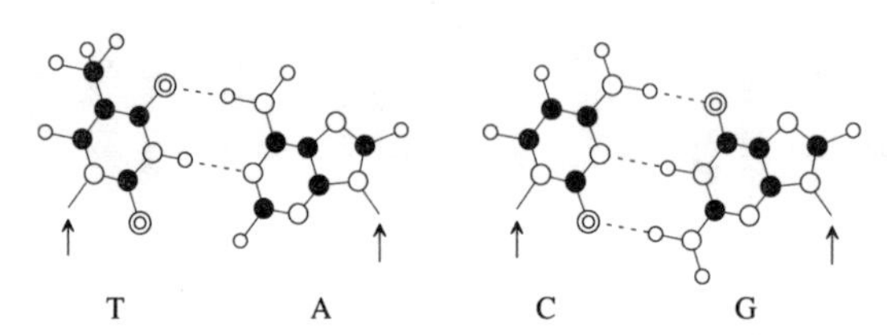

Fig. 10.3 **The DNA base pairs**. The filled circles are carbon atoms, the open are nitrogen (large) and hydrogen (small) and the double circles are oxygen. The arrows indicate the helix attachment sites which are about 1.1 nm apart. T,A,C and G are the base abbreviations. The dashed lines show the hydrogen bonds.

attached to the sugar in any order and to which is hydrogen-bonded a second similar *complementary* chain, such that the *base pairing* rules are satisfied (see panel on p. 90). The rules are that adenine pairs with thymidine and cytosine pairs with guanine.

Man and other mammals contain more than a thousand million of these base pairs in all their chromosomes together, which is called their *genome*. Each chromosome appears to contain just one long thread running from end to end, but interaction with proteins called histones causes the DNA-protein complex to form coiled-coil structures which can condense to form the chromosomes which we see.

When the structure was discovered in 1953 by Crick and Watson, based on the crystallographic evidence of Franklin and Wilkins, it was immediately apparent that the pairing feature provided a mechanism for the exact copying essential if the information is to be passed intact to succeeding generations. Later a

genetic code

The rules which relate the four bases of the DNA or RNA with the 20 amino acids found in proteins. There are 64 possible different 3-base sequences (triplets) using all permutations of the four bases. One triplet uniquely specifies one amino acid (except for AUG when acting as an initiating codon in bacteria), but each amino acid can be coded by up to six different triplet sequences. The code is therefore *degenerate*. *Initiating codons* specify the start of a polypeptide chain and the triplets known as *Ochre, Amber* and *Opal* are *stop codons* which terminate the chain. Initiating codons are AUG and GUG in bacteria with the former specifying the amino acid *n*-formylmethionine at the beginning of the chain and methionine within it. In eukaryotes, AUG is the only initiator and always translates as methionine.

The evidence suggests that the code is universal, applying from the simplest to the most evolutionarily advanced organism, although minor variations have been found particularly in the mitochondrial DNAs.

In the following table the amino acids and the bases are specified by their three and one letter symbols respectively.

Ala	GCU, GCC, GCA, GCG	**Lys**	AAA, AAG
Arg	CGU, CGC, CGA, CGG AGA, AGG	**Met**	AUG
		Phe	UUU, UUC
Asn	AAU, AAC	**Pro**	CCU, CCC, CCA, CCG
Asp	GAU, GAC	**Ser**	UCU, UCC, UCA, UCG AGU, AGC
Cys	UGU, UGC		
Glu	GAA, GAG	**Thr**	ACU, ACC, ACA, ACG
Gln	CAA, CAG	**Trp**	UGG
Gly	GGU, GGC, GGA, GGG	**Tyr**	UAU, UAC
His	GAU, GAC	**Val**	GUU, GUC, GUA, GUG
Ile	AUU, AUC, AUA	***OCHRE***	UAA
Leu	UUA, UUG, CUU, CUC CUA, CUG	***AMBER***	UAG
		OPAL	UGA

number of enzymes were discovered which are able to unravel the DNA and copy it in the manner predicted. Another implicit feature was that the order or sequence of the bases must somehow specify the sequence of amino acids in the final product, the proteins. Again it was demonstrated later that the DNA made an intermediate product, called *messenger RNA* which could pass out of the nucleus to the cytoplasm and there direct how the 20 amino acids can be put together on special structures called *ribosomes*. Soon after the code was discovered linking DNA base sequence and the amino acid (or peptide) sequence in proteins.

The genetic code

As there are 20 amino acids and only four bases it is likely that the bases have to be grouped together in order to specify an amino acid; there are only 16 combinations of two bases, insufficient to specify the 20 amino acids but there are 64 combinations of three bases and in fact three bases are used in nature as shown in the table on p. 91. It will be seen that many amino acids are specified by more than one of the possible triplets. The code is therefore called *degenerate*. Also three triplets are reserved as *stop codons*, i.e. they are used to terminate the polypeptide chain as it is being produced. The code is also linear, each adjacent group of three specifying an adjacent pair of amino acids without any marker to indicate the change from one amino acid to another.

The considerable technical achievement in working out the details of the genetic code was matched during the same period by the discovery that the sequence of peptides provided all the information that was needed to determine the complex three-dimensional structure of a protein. It was shown that gently unravelling a protein by, e.g., breaking hydrogen and other bonds between different parts of the chain would result in a flexible polypeptide chain. Restoring the chemical conditions to their normal state caused the polypeptide chain to fold and form a functional protein. In other words, for any chain of perhaps several hundred peptides in length, the natural folded structure is the most thermodynamically stable.

Although we have considerably shortened and simplified the story so far, we are in a position to appreciate how damage to DNA and the consequent alteration of protein sequences might affect an individual or, if the germ line were damaged, any offspring. There is one important range of problems, however, which we have not considered. These are related to the control of *gene expression*. How does a cell operate differently and have a different structure in the liver and in the kidney? Parallel with this is an understanding of an additional layer of complexity in the DNA; it is not just a string of triplets specifying a vast number of proteins.

Differentiation and control

The mechanisms by which some genes are switched on and others switched off as a cell proceeds to become a functioning tissue has not proved easy to understand and we know little of how these changes are orchestrated in the formation of a complex organ like the liver. What is known, however, is that genes can be switched on and off in response to signals, which can be proteins but are often smaller hormone-like molecules. These molecules bind to specific sites on the DNA often close to the *coding sequence*, i.e. the sequence which uses the genetic code to specify the order of the amino acid sequence. These sites are usually just a few bases long and there may be several of them associated with the synthesis of a particular protein.

In addition to these control sites, which have remained very stable during evolution, there are long stretches of DNA which appear to have no function at all. They not only exist between the genes but also within the coding sequences, so that there is a curious state of affairs in which long stretches of DNA *transcribe* (or specify) a messenger RNA (mRNA), only to have large parts of the mRNA cut out, leaving just the length needed for the protein. The parts left behind are called *exons* and those removed, which can be much longer, are called *introns*. In addition, many animals and plants have significant percentages, even up to 50% of their DNA in the form of very short repeating units a few bases long, often found at special places like the chromosome ends.

The result of this complexity is that damage to the genetic material, whether caused by radiation or by environmental chemicals can have very different effects, depending on the site at which it occurs. Such differences will be over and above those stemming from the function of the protein. Control sequences which, as already mentioned, have altered surprisingly little during evolution, might be expected to be very sensitive but are a small target. Coding sequences are large and damage to them results in abnormal proteins, but whether this is serious will depend on the site and type of mutation, because parts of the protein chain may only be concerned in preserving a fairly generalized shape. A mutation in the active site, where shape must be rigorously preserved, will be much more serious. Again a mutation which inadvertently makes a chain terminating codon or triplet will specify a shortened, usually non-functional, protein. There is also a class of mutations which alter the *reading frame*, i.e. they add or remove one or two bases only with the result that all subsequent triplets are misread (see panel on p. 91). Introns and short repeating sequences on the other hand are both evolutionarily very unstable and unlikely to cause problems to an individual if altered by radiation or mutagenic chemicals.

Damage repair

Because the genetic and developmental integrity of an organism is so important it is not at all surprising to find that complex multiple systems have evolved to make good, as far as possible, any damage to the DNA. In the best studied experimental organism, the bacterium *Escherichia coli*, there are over 200 different molecules engaged in the recognition and repair of such damage.

An example of how they work is the repair of the effects of ultraviolet light on DNA. This causes adjacent thymines in the DNA to become linked together to form a *thymidine dimer*, which left unrepaired will result in an AT changing to a GC pair (see Fig. 10.4). The simplest way of repairing this damage is for the cell to use an enzyme, found in all cells, which in the presence of blue light simply breaks the dimer links and returns the DNA to its undamaged state. But there are also two other methods. In the first of these, several enzymes recognize the damage, cut the DNA strand close by and remove around 30 bases, including the dimer, and then copy the opposite strand to form a new strand, finally closing the gap. The second method is *inducible*, the repair enzymes appearing only after large doses of UV. They also excise and replace DNA, but can introduce further errors in the process and appear to be a method of last resort. The first two repair systems are essentially error-free and are also found in higher organisms.

The presence of repair systems has several consequences. Not only will the damage rate be much reduced but one might expect to find significant differences in the way in which the repair processes are able to function. For example, a rapidly dividing cell may not allow sufficient time for the repair to occur. Even more important, any damage to the genes specifying the repair enzymes should produce catastrophic results. The rare inherited disease of Man, *xeroderma pigmentosum*, is an example in which patients have a greatly increased sensitivity to UV light and also develop multiple skin cancers. This genetic defect is located in the DNA associated with the repair enzymes for UV damage.

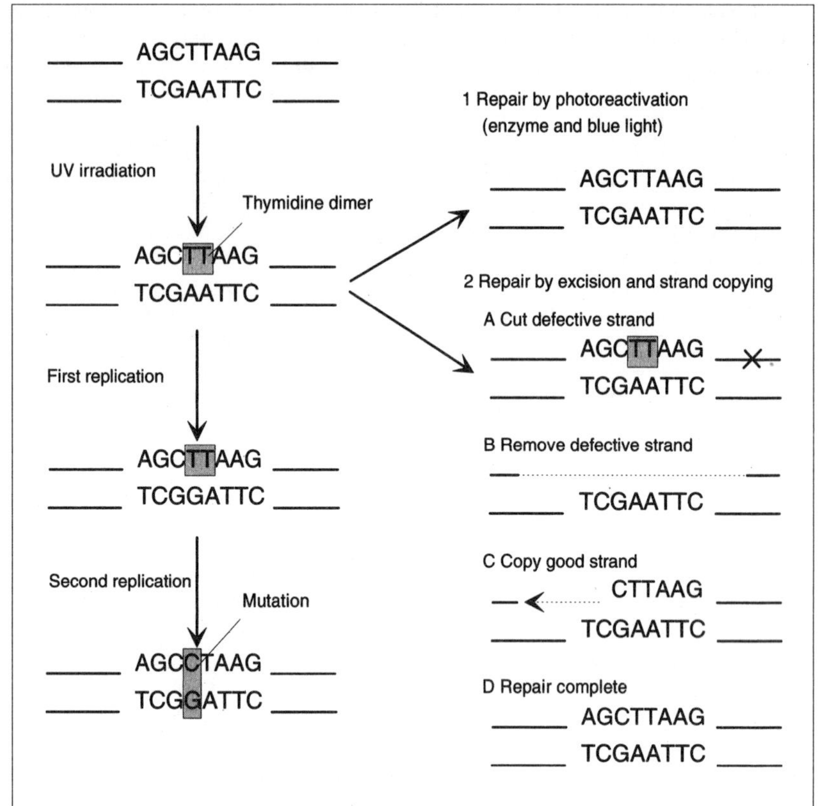

Fig. 10.4 **Ultraviolet radiation induced mutations and their repair**. *Left*, the sequence by which an AT base pair is converted to a GC pair. *Right*, two methods of repair, photoreactivation and excision repair.

Mutations and mutation rate

When genes were first discovered by Gregor Mendel in 1856 their location in the cell was unknown and there was no hint of their chemical nature. It was not until 50 years later that a number of cytologists suggested that the chromosomes behaved in the way which would be expected from the carrier of the genes. Later T.H. Morgan was able to show with the aid of the giant salivary gland chromosomes of the fruit fly, *Drosophila*, that the genes were arranged linearly along the chromosomes. During the early part of this period genes could be considered as 'factors' which behaved in a coherent fashion if the appropriate breeding experiments were performed without consideration of their physical form. Nevertheless the *natural mutation rate* for a given gene could be measured in *Drosophila* and later in maize and mice, and was found to be very low. Very roughly, the chance of a mutation occurring in a natural population was less than one in 100 000. Such a low rate made it difficult to do many genetic experiments. It became much easier when Hermann Muller discovered in 1926 that X-rays markedly increased the mutation rate in *Drosophila*, by a factor of 100- to 1000-fold. This not only provided abundant new experimental material but laid the foundation for radiobiology.

In the laboratory the *heterozygote* animal with only one chromosome of the pair carrying the mutation can be bred by crossing to normal laboratory stock and selecting for the heterozygote which often carries a less severe form of the lesion which in the *homozygote* kills the organism. As knowledge and experimental techniques improved, mutations which affect the structure and function of proteins like haemoglobin and enzymes were isolated in addition to those causing the grosser changes in bodyform and viability on which the earlier geneticists relied. Indeed the mutants produced by X-rays were often accompanied by gross chromosomal alterations or *aberrations* like the loss of part of an arm or the joining of part of an arm of one chromosome to that of another. A process known as chromosomal *translocation*. These changes, which are easy to see in the microscope, have often been used as a measure of the biological effects of radiation.

So far we have been considering *genetic* mutations, i.e. the mutations which arise in the germ line and are transmitted to the sperm or eggs. Such mutations are likely to affect any stage in the development of subsequent individuals, are comparatively easy to study, and are the basis of the science of genetics. They can also have an insidious effect in populations where many individuals can carry and pass to their offspring *recessive mutations*, i.e. the organism is heterozygous for the particular condition, without immediate effect. It is only when the proportion of individuals affected is so high that there is a good chance of both parents carrying the same mutation, that obvious damage will be recognized.

There are also mutations which do not affect the progeny, only occur in the tissues of an animal or plant and have proved more difficult to investigate. These are called *somatic* or body mutations and are important because some of them result in cancer. Indeed, it is generally considered that an increased incidence of cancer is the only hazard likely from low doses of radiation

Carcinogenesis and mutagenesis

In the late 18th century Percival Pott noticed that boy chimney sweepers frequently died of cancer of the scrotum when they reached puberty and that this disease was virtually unknown in the rest of the population. Soot was in fact the first cancer-causing agent ever identified, although much later other chemicals associated with industrial processes like dye manufacture were shown to be carcinogenic. It was not until the late 1920s that E.L. Kennaway was able to produce cancer in mice with aromatic hydrocarbons like benzpyrene and dibenzanthracene. Previous to this a connection between radiation and cancer had been noted and, as we have mentioned, X-rays were found to produce mutations in *Drosophila* in 1926. In the 1930s it became generally accepted that carcinogens were also mutagens, although it was not until 1940 that Charlotte Auerbach discovered chemical mutagenesis by showing that nitrogen mustard and mustard gas produced mutations in *Drosophila*.

Carcinogenesis is a much more difficult process to understand for a number of reasons which we will presently consider but first we have to establish a central finding about cancer. Does cancer arise from an event in a single cell or do

X chromosome inactivation and clones of cancer cells

Every female mammal receives one X chromosome from the mother and one from the father but males receive an X from the mother and a Y from the father. Certain rare humans have two Xs and a Y (XXY) and develop abnormally, implying that the presence of two Xs are harmful. Consistent with this, one of the X chromosomes in normal females becomes condensed and inactivated rather early in development when there are about 20 cells in the embryo. The condensed X can be seen in a microscope and is known as the Barr body and its presence indicates that the cell is female.

On the X chromosome there is a gene for an enzyme called glucose-6-phosphate dehydrogenase (G6PD) which is concerned with glucose metabolism. There is a variant form that confers a selective advantage on people who carry it by increasing their resistance to malaria. It is therefore quite common in certain populations. Moreover its presence can be detected in the cytoplasm of any cell containing it. Women who carry this variant on one of their X chromosmes are found to be *mosaic* for this trait in all their normal tissues, i.e. some of their cells carry the normal enzyme and others, close by, the abnormal as in the lower part of Fig. 10. 5. This strongly suggests that the initial condensation of one of the Xs occurs at random and persists thereafter.

Tumours which occur in such individuals are, on the other hand, always of one type or the other and likely to have been descended from one cell and are therefore *monoclonal.*

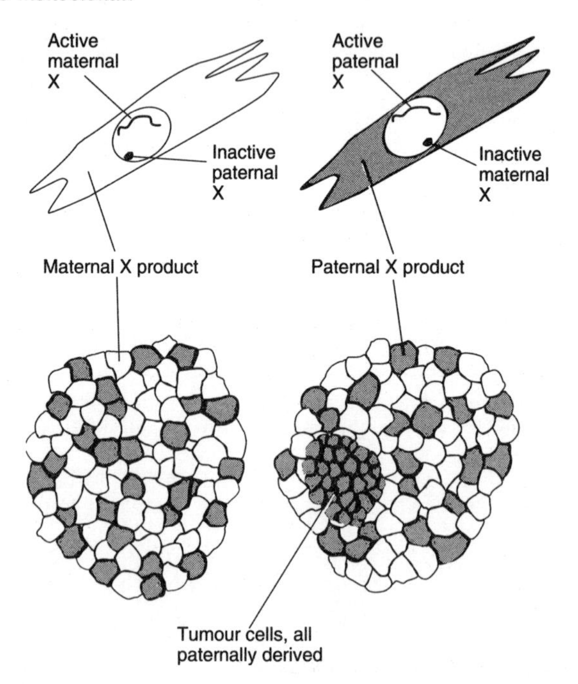

Fig. 10.5 *Upper,* two cells each with a different X chromosome inactivated. *Lower left,* tissue with individual cells mosaic for X chromosome product with, on the *right,* a uniform clone of tumour cells growing in a similar tissue.

several cells have to cooperate for the first step. Nearly all the evidence suggests that cancers start as a lesion in a single cell; in other words they are clones with the same genetic make-up. The evidence for this requires us to return to the X and Y chromosomes of Man and other mammals (see panel on p. 96).

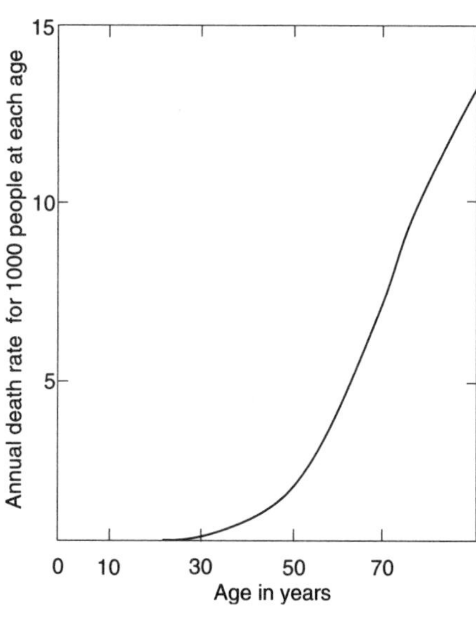

Fig. 10.6 **The annual death rate from cancer per 1000 population.**

Cancers start therefore in single cells and, like genetic mutations, can be induced experimentally in animals by radiation and chemical mutagens. So are there any differences which distinguish carcinogenesis? We have already mentioned that the chance of a given gene mutating is about one in 100 thousand and that a medium sized mammal has about a million times this number of cells, so if the process were exactly the same any individual would accumulate thousands of cancer clones during a lifetime. Cancer is in fact very rare, considering the number of cells at risk. It has a number of other well-known characteristics: it is predominantly a disease of old age with an individual's chances of getting cancer being about 50 times as great at 80 than at 30 years old (see Fig. 10.6); there is usually a lag of many years between the initial cause and the diagnosis of cancer, with workers in some chemical factories getting cancer many years after they have left and the incidence of lung cancer in women only increasing some 20 years after women in general became heavy smokers; further, in experimental animals many chemicals are known which do not induce cancer on their own but only do so after a *later* application of another substance, called a *promoter*. A possible example in humans is the discovery that the cancer associated with asbestos, which has almost eliminated its use worldwide, very rarely occurs in people who do not smoke, a chemical from cigarettes possibly acting as the promotor.

Other important evidence about the initial events in cancer have come more recently from the new ability to look at precise sequences of DNA, changes in which can sometimes be shown to be correlated with certain cancers. In particular, work with certain small DNA tumour viruses has been very important.

Viral carcinogenesis and oncogenes

Viruses are associated with cancer in at least two ways. The virus, HIV I, the Human Immunodeficiency Virus which causes AIDS, eventually destroys the immune system allowing the establishment of infections and cancers which would otherwise be mild in effect or very rare, but are often lethal in AIDS.

Kaposi's sarcoma is an example of a rare cancer which is very common in AIDS sufferers and indicates that probably some early cancer cells are eliminated by a process called *immunosurveillance.* Of more general significance is the role of some small viruses which cause tumours in animals. These viruses often have just a few DNA sequences in their genome but one of them is a sequence called an *oncogene*, which closely matches a DNA sequence in normal cells termed a *proto-oncogene.* There are many such sequences known and they seem to have a regulatory function in normal animals but their function can be altered if they are moved to abnormal sites. A good example is the locally common cancer of children found in parts of Africa and called *Burkitt's lymphoma.* It has been shown that the *myc* oncogene sequence has been made to form its product by being placed next to an antigen receptor gene in the children with Burkitt's. The viruses which carry oncogenes are known to be able to fit themselves into many different places in the genetic material of animals and it is therefore probable that this is how they can cause cancer. It also has to be remembered that radiation and chemicals can cause chromosomes to break and rearrange themselves.

Viruses therefore, like radiation and some chemicals, may be able to initiate the eventual formation of a clone of cancer cells. But the time course suggests that there is a succession of events which have to occur in a cell before it can found such a clone. In fact the shape of the cancer incidence curve in Fig.10.6 could be explained most easily by about four such events. The first may be a somatic mutation, but thereafter other alterations, probably in the genome but perhaps elsewhere in the cell are essential. Viral sequences, radiation and chemicals may all play a part in these later events as well as in initiation. One such sequence of events could be the need to damage several related genes, with the chances of this occurring being very low but slowly increasing as the time from the initial event increases during a lifetime.

Cancers

So far we have treated the problem of cancer as though it was just one disease; it is not. Cancers vary considerably depending on the site at which the original clone of cells occurred although all share the common ability to grow and divide in cellular situations where normally they are under tight control. For example a liver cell will stop dividing once its adult size has been reached unless the liver is damaged and has to be repaired. Otherwise it is only in liver cancer that these cells start to divide again. Once established, cancer cells behave in many different ways, growing slowly or fast or requiring certain hormones or other factors for growth. In particular, many cancers stay at their original site and never split off cells able to found new colonies elsewhere. Cancers which stay in the same place are called *benign* and can be removed if they grow too large. The skin discolourations found in the elderly are of this type. Cancers which are able to shed cells into the blood or lymph vessels, which then go on to find new places to found a colony, are far more dangerous, particularly because the new cells can attract blood vessels to supply their nutrients. Such cancers are *malignant* and the new colonies which they found elsewhere in the body are *metastases.*

Although high doses of radiation can kill an animal or plant directly by e.g. destroying cell membranes and directly killing sensitive cells like those dividing in the blood forming and immune system, it has been recognized for many years that it is the longer term effects of radiation which are harder to investigate and quantify. This is therefore of natural concern to both those who work with radiation and the general public. This chapter has, therefore, concentrated on damage to the germ line, which can affect and indeed persist in subsequent generations and on the way in which carcinogens can increase the chance of an individual contracting cancer. The next chapter considers both these effects in relation to the actual doses and types of radiation which an individual can receive from nuclear and related sources of electromagnetic and particulate radiation.

Chapter 11

Biological Effects

In an earlier chapter we considered the range of electromagnetic waves that are packaged in photons and have an enormous spread of energies over the *electromagnetic spectrum.* Although Man has learnt how to produce all of them for industrial, medical or research purposes, they also occur naturally in nature.

There are several other kinds of radiation that occur without Man's assistance. Alpha particles, helium nuclei, occur in the radioactive decay of, mainly, the heavier elements beyond lead in the periodic table and beta particles, which can carry either a positive or negative charge, are emitted during the radioactive decay of unstable isotopes of any element. There are also cosmic rays composed of protons and heavy nuclei from outer space. These collide with nuclei in the atmosphere, producing other nuclei and smaller charged and uncharged particles called mesons, some of which reach the Earth's surface. Mesons decay to form electrons and neutrinos, and do not appear to be a health hazard, but the heavier nuclei formed by cosmic rays in the upper atmosphere contribute significantly to the background radiation.

Man-made radiation results from many processes: by far the greatest amount comes from medical X-rays and similar diagnostic and therapeutic procedures; much less results from various industrial processes like the production of nuclear fuel and running nuclear power stations. Radioactive substances are also used in research and, importantly, occur in the fallout from nuclear bombs. The distinguishing feature of Man's contribution is the presence of radioactive isotopes of a wide range of elements which are the direct result of nuclear fission. Many of these isotopes, like those of strontium and plutonium are able to replace elements within the body and remain in bone, for example, while emitting ionizing radiation. Their presence is a more insidious danger than direct radiation which can be readily detected and which can be easily reduced by shielding.

Living tissues as the target

Living tissue contains a high proportion of water and in addition the main constituents of the protoplasm are the light elements hydrogen, carbon, nitrogen and oxygen with additional phosphorus and sulphur. This preponderance of light nuclei has a considerable effect on the way radiation interacts with living tissue.

Charged particles

Electrons, protons and fission products will interact with atomic electrons through the Coulomb force and, provided their energy is high enough, will strip off the outer valence electrons, ionizing the atoms and altering their reactivity.

Such charged particles will collide many thousands of times before stopping and it is possible to measure the *stopping power* of a tissue or any other substance in terms of the amount of energy lost per unit thickness of material. Stopping power is not linear along the length of the track and increases as the velocity of the particle falls and as increasingly greater amounts of energy are transferred to any electrons in its path.

Different particles have different masses and different velocities but it is useful to be able to compare the amount of biological damage which they may cause on a common basis. A useful way of doing this is to measure the *linear energy transfer rate* or LET. This is defined as the average rate at which the energy of the particle is dissipated and transferred to the tissue per unit track length and measured in keV per micrometre. It depends only on the velocity and charge of the particle and is independent of its mass.

Neutrons

Neutrons have mass and varying velocity but no electric charge. They also have a short half-life (636 seconds), so they do not persist in the absence of nuclear fission. High-velocity neutrons (*fast neutrons*) lose most of their energy by direct collisions with other nuclei. The amount of energy lost and transferred to the other nucleus depends on the mass of the latter. If the nucleus is heavy very little energy is transferred but the light nuclei in living tissue, particularly hydrogen, can accept nearly all of a neutron's energy with the consequent production of charged particles able, themselves, to lose their energy quickly in multiple ionizations. Gamma rays are also produced by the scattering of neutrons by nuclei. Thermal neutrons can also be captured and produce other charged particles but they are less dangerous because most of their energy is normally dissipated by heating the tissue.

Photons

Photons react with matter in a way very different from charged particles. Instead of making hundreds or thousands of collisions in a short distance as they lose energy, photons make just one collision often losing most of their energy at that point. The distance travelled can be very long; radiowaves can travel thousands of kilometres through the atmosphere and visible light tens or hundreds before the photon will be absorbed. What happens then depends on the energy of the photon. If the energy is less than that required to keep an electron bound to the nucleus then no ionization takes place but energy is transferred to the electrons which shift to a larger orbit and then decay to lower orbits emitting less-energetic photons, which in turn may repeat the process, again travelling relatively long distances. See p. 86 for a further discussion. This boundary, below which ionization is impossible, corresponds to about 5 eV and is the energy of visible light photons.

At energies greater than 5 eV and up to about a 100 keV the predominant reaction is the photoelectric effect in which the energies are large enough to force the orbital electrons into higher-energy orbits and to displace a valence electron from the nucleus with the production of further less energetic photons as the orbital electrons return to their initial state. Above 100 keV the predominant reaction is Compton scattering in which the photon collides with any electron, transferring some of its energy to it, and in which both photon and electron emerge from the atom at high velocity. It is these ejected electrons, which go on to cause multiple further ionizations that cause most of the biological damage. At still higher energies the photon converts all its energy, in the presence of an atomic nucleus into an electron-positron pair.

Because their photons transfer energy to nearby atoms by means of secondary electrons and photons, gamma and X-rays are called *indirectly ionizing* radiation. Similarly neutrons are indirectly ionizing. Alpha particles and protons are *directly* or *densely ionizing* because their energy is transferred over a short distance with a large peak deposited near the end of the track.

So far all the processes described occur in the first 10^{-16} of a second. They are followed by several other processes ending after about 10^{-12} of a second with a trapped or hydrated electron able to diffuse considerable distances and to react with other molecules. This ends the physical stage of irradiation. It is followed by the much longer chemical stage, in which free radicals and other excited molecules are produced and which react over about a millisecond.

Free radicals

Whatever the exact way in which primary radiation reacts with living tissue, the common result is the formation of atoms or molecules which are electrically neutral and have an unpaired electron in their valence (outer) electron shell. These are called *free radicals* and are exceedingly reactive. In living tissue with its preponderance of water, the main free radicals are H· and OH· , the former a strong reducing agent and the latter, which predominates, a strong oxidizing agent, which we can expect to cause abnormal chemical reactions in living tissue.

Chemical radiation protection

Cells can escape radiation death and experimental animals radiation sickness and premature ageing by the administration of molecules containing the highly reactive sulphydryl group, —SH, during the chemical stage of irradiation and the formation of free radicals. The mechanism involves the transfer of the hydrogen of the suphydryl group to the free radical, thereby neutralizing it. Another important protective agent is the enzyme, *superoxide dismutase* (abbrev. *SOD*), which removes the superoxide radical anion ($O_2^- \cdot$). It is thought that this enzyme may perhaps have evolved in order to remove this highly damaging anion. Radiation damage to cells is increased by certain promotor substances (see p. 95), but this increase is abolished by the addition of superoxide dismutase.

The units for measuring radiation

Before discussing the origins of the radiation we all receive and the biological effects that radiation can cause, we need to define the units of measurement applied to radiation so that we can compare different kinds of radiation and relate the results from experiment to the real world outside.

Since 1975 discussion and reporting of radiation have been bedevilled by two sets of units. A new internationally recognized set of Système International d'Unités (SI units) and an older nomenclature which had its problems, but is deeply ingrained in the literature and many people's minds. In each system two

Radiation units

The four special units defined here progressively relate the initial radioactive disintegration to its final effect on a given tissue.

Activity and exposure

The SI unit is the *becquerel* (symbol Bq) named after A. Henri Becquerel who first discovered radioactivity in 1896 in an uranium salt. It is simply one disintegration per second. This is an exceedingly small amount of radioactivity and substances of *megabecquerel* (10^6 Bq) activity will be routinely handled in a biological laboratory and *gigabecquerels* (10^9 Bq) and even *terabecquerels* (10^{12} Bq) would be common in the nuclear industry. For example, one gram of plutonium-238 has an activity of over half a terabecquerel (six hundred thousand million alpha particles emitted per second). The older unit was the *curie* (symbol Ci), based on the number of disintegrations per second from one gram of radium and equal to 3.7×10^{10} becquerels. Marie Curie discovered radium in 1898. Exposure is now defined in terms of coulombs per kilogram of dry air, with the coulomb being the charge transported by one ampere flowing for one second. The older unit is the röntgen, after W.K. Röntgen, the discoverer of X-rays in 1895. It is defined as the radiation which causes 2.58×10^{-4} coulombs of electric charge in one kilogram of air. Exposure therefore takes into account the effectiveness with which a given kind of radiation will ionize a standard material, like air.

Absorbed dose and dose equivalent

Although exposure can be related to substances like water or air with similar properties to living tissue, it does not define the actual dose received by the tissue. The special SI unit for absorbed dose is the *gray* (symbol Gy), named after L.H. Gray, the British radiation biologist, and equal to the absorption of one joule of energy in one kilogram of tissue. The older unit was the *rad*, exactly one hundred times smaller than the gray. A distinction may be made between the total absorbed dose and the dose from the secondary charged ionizing particles liberated by an incident uncharged ionizing particle. This is called the the *kerma* (K) dose and has the same unit as the absorbed dose.

We now come to *dose equivalent*, a quantity used in *radiological protection*. This is the absorbed dose multiplied by a *quality factor*, depending on the kind of radiation, and a *weighting factor*, depending on the sensitivity of the tissue irradiated. The unit is the *sievert* (symbol Sv), named after Rolf Sievert, a Swedish radiobiologist who was once chairman of the International Committee for Radiological Protection. Its units are, like the gray, joules per kilogram. All electrons, positrons and X-rays have a quality factor of one but it can range up to 20 for radiation which transfers 100 or more keV per micrometre of tissue (high LET radiation). Thermal neutrons also have a quality factor of one but for fast neutrons and alpha particles it is now 20. The weighting factor is one for the whole body but less for individual tissues as shown in Table 4 on p. 112. The older unit was the *rem*, an abbreviation for *röntgen equivalent man* and with the absorbed dose measured in rads.

It follows that for whole body irradiation by electrons, positrons or X-rays, the sievert and the gray are equal.

Related units which are frequently used include person (or man) grays, the average dose multiplied by the population at risk.

of the units are applicable to any kind of radiation source. They are a measure of the *activity* of the source, i.e. the number of disintegrations per unit time and the *exposure*, the number of ionizations produced in air by the radioactive source. Another two are related to the *energy deposited* in a material, with the last measuring the effect of the energy deposited in a particular biological tissue. Both sets of units are defined in the panel on p. 103.

Another term frequently used is the *relative biological effectiveness* (abbrev. *RBE*). This is the ratio of the dose of a standard reference radiation required to give a specific biological effect over that of the test radiation needed to give the same effect. The standard radiation is usually cobalt-60 gamma radiation or 250 kV X-rays. The specific biological effect could be the killing of half the cells in a bacterial culture or the production of a given number of mutations.

Radiobiology and radiological protection.

A distinction has to be made between terms like relative biological effectiveness and units like the gray, which are the result of measurements made on tissue or cells and are properly the domain of radiobiology and terms like quality and weighting factor and the unit, sievert. The latter are radiological protection terms and are fixed by committee on the basis of the best radiobiological evidence available; quality factors and weighting factors are designed to have an inbuilt margin of safety.

Radiation from the environment

1. Background radiation

The radiation which we all experience and which would have been the only radiation experienced by our ancestors in the last century, comes from three natural sources: cosmic rays, already mentioned, the radioactive elements present during the formation of the Earth, and solar radiation, mainly in the form of ultraviolet radiation.

In addition to the mesons which continually bombard the Earth, there are a number of heavy radioactive elements which occur in or are produced by cosmic rays when they interact with the atmosphere. They have half-lives, generally between a few days and thousands of years. They include carbon-14, formed during the collision of a proton with a nitrogen-14 nucleus, tritium-3 and two isotopes of phosphorus (see Table 1). Despite their comparatively short half-lives they are being continually replenished and so constitute a fairly stable and uniform background of radioactivity.

Table 1. **Cosmogenic radionuclides.**

Isotope	Half-life	Concentration (Bq / m^3)
Tritium-3	12.3 y	0.17
Carbon-14	5760 y	0.033
Beryllium-7	53 d	0.017
Phosphorus-32	14.3 d	3.3×10^{-4}
Phosphorus-33	25 d	2.6×10^{-4}
Sulphur-35	87 d	2.6×10^{-4}

The *primordial* radioactive elements, on the other hand, have enormously long half-lives, as would be expected from the time of their likely origin being near that of the universe. Half-lives vary from a thousand million years to ten thousand times as long (see Table 2). Their distribution in the Earth's crust is by no means uniform. Areas in Brazil and India with large deposits of sands containing thorium-232 have a hundred or so times the natural background compared to London or the east coast of the USA.

Ultraviolet radiation from the Sun can cause burns in skin but also produces thymidine dimers in the DNA of those cells just below the surface that continually divide to replenish the cells lost from the surface. The DNA damage can result in skin cancers of which there is a high incidence in sunny places with a mainly Caucasian population such as Queensland, Australia.

Table 2 **Primordial radionuclides.**

Isotope	Half-life (years)	Abundance (pts per 10^6)
Potassium-40	1.3×10^9	300
Rubidium-87	4.7×10^{14}	75
Thorium-232	1.4×10^{10}	12
Uranium-238	4.5×10^9	4
Samarium-147	1.2×10^{11}	1

There is therefore a natural background of radiation which varies considerably depending on where you live and to which we are all subject. Altitude will determine the amount of cosmic rays and their products, as well as the ultraviolet energy received. The minerals in the rocks and from which houses are built will determine the radiation background from radionuclides in minerals. The latter is well illustrated in Fig. 11.1 which sums the natural radiation from gamma

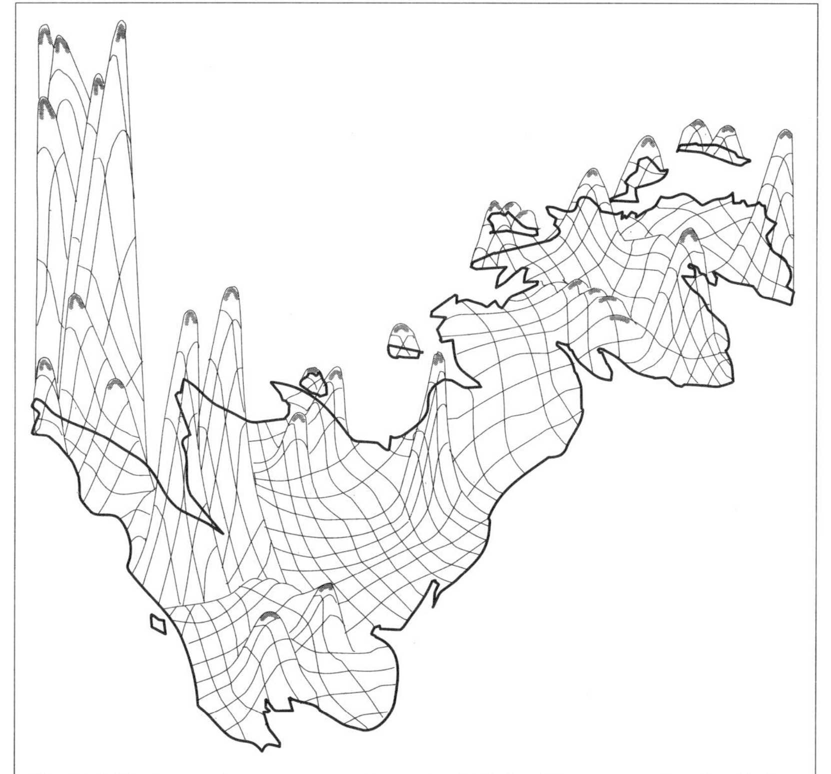

Fig. 11.1 **Radon and gamma-ray doses in Britain.** Measurements of radiation received have been added together and drawn as contours over the map.

rays and from radon gas and plots it over the map of mainland Britain. The average figure for the same population is 1.87 millisievert per year.

2. Man-made radiation

The use of radionuclides and X-rays for diagnostic and therapeutic purposes account for 80% of the man-made radiation received by an individual in the UK, with fallout from nuclear weapons testing contributing about 4% and nuclear discharges less than 1%. Radiation received at work, including those in the nuclear industry, in all types of mining and workers in the medical and dental health professions account for another 4%, with a miscellaneous group the remaining 5%. These numbers are illustrated graphically in Fig. 11.2.

Radionuclides are used for diagnosis in gamma cameras which allow internal organs to be studied without surgery. The main radionuclide now used is technetium-99 which provides a high spatial resolution when used with thin sodium iodide crystals (see p. 76). Radioactivity is of the order of 500 megabecquerels per examination with an effective dose of 1 to 10 mSv.

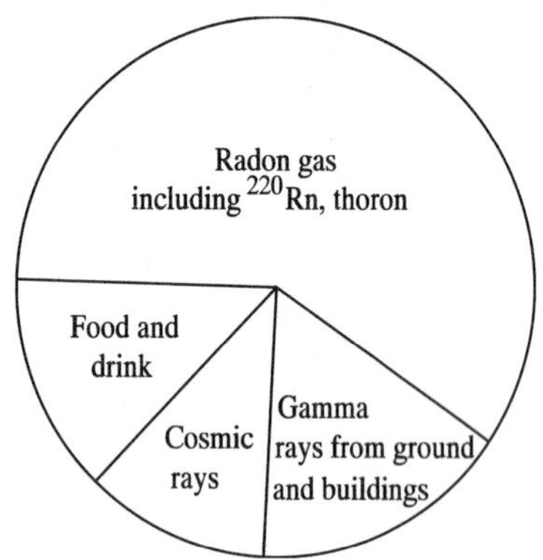

Fig. 11.2 **Sources of natural ionizing radiation.**

The problem with such average figures is that, just as natural background radiation varies considerably, so do occupationally related doses. A good example concerns radiographers. In 1960 it was reported that the children of medical radiographers who worked during pregnancy had a significantly higher incidence of leukaemia. This resulted in a considerable change in working practice and a large improvement in the shielding of medical X-ray equipment, with the result that in UK clinical practice the average annual dose is now 0.7 millisievert (mSv). Contrast this with industrial radiographers who receive on average 1.7 mSv and with radiographers working on pipeline and off-shore oil installations who receive about 27 mSv. Again this should be compared with the dose received by people who live near nuclear discharge areas and eat fish contaminated with caesium-137. They receive an annual dose around 1 mSv.

These doses are higher than those received by the general public but should be compared with the natural background radiation which averages 1.87 mSv per person per year in the UK. Further, the International Commission for Radiological Protection has recommended that the additional radiation dose above background received during a lifetime should not exceed 50 mSv.

We will now discuss the effects of large radiation doses before returning to the more controversial subject of low doses and their effects on cancer incidence.

Acute radiation damage

There is little controversy about the immediate effects of acute exposure to radiation. These are due to the free radicals rupturing the lipid membranes of the cells which then spill their contents into the spaces between the cells and die. If enough cells are killed then whole tissues, like the lining of the gastrointestinal tract and the bloodforming cells in the bone marrow, become disrupted and cease to function. Death of the individual may then occur either because of the loss of vital organs or because of secondary infections from, for example, bacteria in the intestines or less directly, from the breakdown of the immune system and secondary infections like pneumonia. It will all depend on the dose received.

For very high doses of more than ***100 gray***, corresponding to about 10^{16} gamma-ray photons, it seems in the few cases that have occurred and been recorded, that the central nervous system is directly affected with loss of co-ordination and breathing problems, with death occurring in one or two days.

For doses between ***9 and 100 grays***, it is the damage to the gastrointestinal tract that predominates with nausea, vomiting and diarrhoea. Later the individual becomes progressively dehydrated and emaciated with death supervening in one or two weeks.

Lower doses, between ***three and nine grays***, cause damage primarily to the bone marrow and other haematopoietic tissues. This leads to loss of appetite and hair, to haemorrhages, inflammation and emaciation and secondary infections like pneumonia leading to death in many cases. Bone marrow transplants may save the patient and some of the symptoms found and clinical care needed will be like those encountered in the 'rescue' of patients undergoing aggressive chemotherapy and radiotherapy for certain cancers.

Doses less than ***three gray*** are rarely lethal but cause the symptoms of loss of appetite and hair, haemorrhage and diarrhoea and will need the kind of clinical care already discussed for more seriously irradiated individuals.

Longer-term risks

When looking at long-term risks there is general agreement that the incidence of cancer is the main, if not the only, consequence which needs consideration. This is partly because cancer is a well-defined clinical condition in contrast to less specific possible effects such as decreased longevity. There is good evidence in animals and some evidence in humans that exposure to sublethal doses of radiation shortens life, not necessarily by inducing cancer. It may be that widespread somatic mutations cause minor malfunctions which are made worse during the normal processes of aging. These have been difficult to document; it is not easy to determine whether any specific incapacity, other than cancer, is related to radiation exposure at least for the lower exposures found in the workplace or from the environment.

There are two main ways of estimating cancer risk: *additive* risk is the number of *excess* cases found for a given unit of time and of dose for the population at risk; relative or *multiplicative* risk is the ratio of risk in an irradiated population to that in a control group and does not state the number of individuals involved. If there was just one extra case of cancer in 10 000 people each of whom had received an average dose of one gray, the additive risk would be 10^{-4} for the period covered by the investigation.

In addition to the immediate effects described for high radiation doses, the survivors have an increased risk of leukaemia and other cancers. This is best documented in the follow-up studies of the Hiroshima and Nagasaki survivors. In the period of 5 to 10 years after the bombs exploded there was an excess of

13 cases of leukemia per 10 000 survivors which then fell to four after 15 years. On the other hand, there were less than five excess cases of all cancers at other sites in each five year period up to 1965. Thereafter excess cancers at other sites increased to nearly 20 in the next five year period. All the Japanese cases received more than two grays in 1945 and excess cases means the number over and above that found in the unirradiated population. See p. 183 for a fuller account of the 'Lifetime Study' project.

Chronic low-level irradiation

It took 25 years of continuous monitoring of a very large Japanese population who received a dose of around 2 grays in 1945 to establish the fairly small numbers of excess cancer cases reported above. It is therefore not surprising that there is controversy over the contribution made to cancer incidence by natural background radiation and man-made radiation. Of course the population involved is much greater, but so is the variation caused by population movement, the geology of people's homes, their radiological history, nuclear fallout and so on. We also need to estimate the incidence of cancer not the result of radiation and the other factors in the environment which may alter this incidence.

Cancer epidemiology

In Chapter 10, p. 95 some of the factors related to the induction of cancer are discussed and the points were made that it is probable that cancer may be initiated by a mutation in one of the body's cells, but that a number of other ill-understood events must also occur before a cell or its descendents becomes cancerous. Further, cancer often does not appear until years after the initiating event, as was found for the bomb survivors, and it is predominantly a disease of old age. We did not consider the epidemiology of the disease, in particular, the widespread variations for cancers at different sites in different populations. This is of some importance because if it can be shown there is a proportion of cancer unaccounted for by diet, smoking or other habits, then this might provide an upper level for the effect of background radiation and for the presumably smaller excess contributed by Man.

Table 3 on p. 109, compiled by Sir Richard Doll, shows the incidence of the common cancers in the populations with the highest and lowest rates. The incidence rates vary up to 300-fold and it is difficult not to conclude that much of cancer is environmentally related and indeed Doll himself believes that at least 90% of all cancers are of environmental origin. Although it is generally the more developed societies which have the highest incidence, there is no obvious correlation with variations in background or indeed man-made radiation, although the data does not address this point. Most cancer epidemiologists appear to consider that factors in the diet and other habits and practices, like smoking and circumcision, account for the differences.

Additional evidence about the importance of the environment comes from the changes in cancer incidence which follow emigration. For example, the incidence of stomach cancer among Japanese in Japan is over six times higher than that in Californian whites, but it fell to 4.6 times for those who emigrated to California earlier this century and to 3.0 in their sons. Similarly, Jews born in

Table 3 **Variations in the incidence of common cancers.**

Type of cancer	Region of highest incidence	Highest incidence %	Region of lowest incidence	Lowest incidence %
		Men		
Skin	Queensland	over 20	Bombay	less than 0.1
Oesophagus	Northeast Iran	20	Nigeria	0.6
Lung	Great Britain	11	Nigeria	0.3
Stomach	Japan	11	Uganda	0.5
Liver	Mozambique	8	Norway	0.1
Prostate	US (Blacks)	7	Japan	0.35
Colon	Connecticut	3	Nigeria	0.3
Mouth	India	over 2	Denmark	less than 0.08
Rectum	Denmark	2	Nigeria	0.1
Bladder	Connecticut	2	Japan	0.5
		Women		
Cervix	Colombia	10	Jewish Israelis	0.3
Breast	Connecticut	7	Uganda	0.5

Europe retained their European incidence rate but their children born in Israel had a much lower rate, more typical of the indigenous population and of the Jews who came from Africa and other parts of Asia.

These figures cannot rule out the possibility that much of the variation is due to differences in background or man-made radiation but they argue against such an interpretation. This is because, firstly, there are instances where the immediate cause of the variation is known, smoking being the obvious example and, secondly, radiation is known to be a cause, but not the only cause, of leukaemia which is not a disease with a high variation in natural incidence between large populations.

Dosage threshold

The major problem with very low radiation doses is whether their biological effects remain proportional to the absorbed dose or whether there is a threshold below which radiation has no effect. The argument for the threshold revolves around the ability of a cell to repair damage. As we discussed earlier on p. 93, there are varied and widespread repair mechanisms of both radiation and chemical carcinogens and it is well known that large and potentially lethal doses of X-rays can be safely given to mice provided the dose is divided and spread over several days. Similarly, therapeutic doses of X-rays have fewer side effects if they are divided or *fractionated.* The implication is that some of the effects can be repaired. On the other hand an animal once irradiated is more susceptible to subsequent irradiation.

Definitive experiments are almost impossible because in order to obtain a result which is significant statistically you need two populations of mice, one given a uniform low dose and another none at all. This would require the continued observation of several million laboratory mice over the mouse's lifetime and would be exceedingly expensive and unrewarding at least to their carers. The regulatory authorities therefore assume that these effects are essentially *stochastic*, that is that they follow statistically at random with a given probability after irradiation. There is therefore no threshold and biological limits for radiation have been set accordingly. In addition there are some non-stochastic effects with a threshold such as the production of opacity in the eye and erythema of the skin.

Genetic effects

Cancer is initiated by damage to the genetic material of the somatic cells but effects lasting for generations, can result from similar, but probably not identical damage to the DNA of the *germ line*, i.e. the cell lineage which splits off from the somatic cells rather early in development of an individual and results in the formation of the gonads. The range of damage is likely to be larger, not only involving cancer but malformations and even more subtle inheritable defects. The gonads are known to be more sensitive to radiation than the whole body but they are of course a relatively small target. There is also another factor which is important. Almost a fifth of human conceptuses abort naturally in the first three months of pregnancy and it is known that many of these have chromosomal abnormalities. Early development is very sensitive to any damage to the genetic material and acts as a kind of filter for gross abnormalities.

But, as we have seen, most mutations will affect only one chromosome of the pair and be *recessive*. This will cause the accumulation of recessive mutations in a population without obvious effect until two parents, both with the same recessive mutation, have progeny who may then be abnormal. Furthermore, if the population is large, such children may not be born until many generations later. Low levels of radiation are likely therefore to increase the proportion of damaged genes in a population. Any increase in this *genetic load* would be a very serious long-term effect of radiation.

One way of studying this is to look for the abnormal proteins made by recessive genes. It should be possible to identify some at least of both the normal functional proteins and the damaged ones in e.g. serum. This approach has been taken in the Lifetime Sudy project at Hiroshima and Nagasaki (see p. 183). The problem is to know if you are looking at a representative sample.

This study has shown that children born to mothers exposed to high doses of radiation (around 2 Sv) at Hiroshima and Nagasaki, but not irradiated *in utero*, do not have a higher incidence of childhood leukaemias compared to unirradiated controls. Neither does the incidence of a number of other markers, including the occurrence of abnormal proteins, increase. For these and similar reasons irradiation at near background levels has not been thought to be a significant factor for the children of workers who may have been exposed to the normal permitted levels in, particularly, the nuclear industry.

Leukaemia clusters

This conclusion has had to be reconsidered in the light of recent findings concerning *leukaemia clusters* near nuclear installations. Clusters of childhood leukaemia cases have been noted over the years, including a part of Italy and in the southern Argentine remote from nuclear installations, but a television programme in 1983 called attention to such a cluster near the nuclear reprocessing facilities at Sellafield (formerly Windscale) in the UK. Several investigations failed to find any evidence connecting these cases to the nuclear facilities until in 1990, M.J. Gardner and his colleagues reported that there was a correlation between the cases of leukaemia and the *fathers'* irradiation history following a very careful case-control study. Each child who developed leukaemia was matched with eight controls whose names appeared just before and after in the birth register and residing in the area (West Cumbria) and eight other controls who were born in the same parish.

Of the 46 cases investigated nine had fathers working in the Sellafield plant. Other cases had fathers working in chemicals, farming, iron and steel (5 cases each) and farming (two cases). A number of factors which might effect leukaemia incidence were recorded for each case and control, like antenatal X-rays and virus infections, eating local shell fish and playing on local beaches, but no significant correlation could be found. There was only a high relative risk for children of those fathers who had worked in the Sellafield plant and who had received more than 100 mSv total exposure prior to conception. These four children had had a relative risk six to eight times their controls. For the five other children whose fathers received a lower dose the relative risk was around one.

This result is very important but still somewhat controversial for two reasons. One is straightforward, the dosage received by the fathers is about twenty times less then that received by the parents in the Hiroshima and Nagasaki studies. The second is more subtle. The onset of the two main kinds of childhood leukaemia is very closely associated with a number of specific chromosomal and molecular changes. These changes involve genes concerned with cell growth, differentiation and proliferation and they occur only in the tumour cells; other tissues are normal. There is also evidence that if they had occurred in the foetus, normal development would have been impossible but no other abnormalities have been reported in these children.

These reservations do not of themselves invalidate the findings in the Gardner report. There may be a very sensitive site in the germ line of the father which, if damaged, facilitates the later chromosomal and molecular changes in the child's tumour cells but it does suggest that a wider study involving childhood leukaemia and paternal irradiation is needed, and that other possible causal agents such as chemical carcinogens, which may have been passed to the mother, need to be considered. Investigations on other groups of workers from other nuclear sites have been undertaken and some have been reported. In particular the results from the cluster reported earlier around the Dounreay nuclear site in northern Scotland do not support the Windscale study. Of the eight leukaemia patients who lived within 25 km of Dounreay only two had fathers employed by the nuclear plant and neither of these had received doses in excess of 100 mSv, the threshold above which the Gardner report had found a correlation.

Dosage standards

The previous discussion shows the extreme difficulty of establishing what should be the limits for radiation exposure above the natural background. The standards that are used worldwide have been set by the *International Commission for Radiological Protection* (ICRP), a respected international organization with its headquarters in Vienna. The standards distinguish between radiation workers and the general public, who are not continuously monitored.

Radiation workers

The ICRP standard for radiation workers is now 20 mSv per year provided that:

> No practice involving radiation exposure should be adopted unless its adoption can be demonstrated to bring about a positive net benefit.
> All radiation doses should be kept as low as is reasonably achievable, economic and social factors being taken into account, the ALARA concept.
> All radiation doses should be limited to those set by the ICRP.

It is also assumed that all radiation workers are routinely monitored and that records are kept. The 20 mSv dose limit is for whole body radiation with the dose limits for individual organs higher, corresponding to the weighting factor already mentioned on p. 103. These are set out in table 4.

Table 4 **The ICRP recommended annual dose limits.** The remainder below refers to 10 other organs and tissues, each having a weighting factor of 0.005.

Tissue or organ	Dose limit (mSv)	Weighting factor
Gonads	80	0.25
Red bone marrow	167	0.12
Colon	167	0.12
Lung	167	0.12
Stomach	167	0.12
Bladder	400	0.05
Breast	400	0.05
Liver	400	0.05
Oesophagus	400	0.05
Thyroid	400	0.05
Skin	2000	0.01
Bone surfaces	2000	0.01
Remainder	400	0.05
Whole body	20	1.00

There are a number of derived limits for radiation workers, the most important of which concerns the maximum amount of radioactive air that can be inhaled. This must allow not only for the number of disintegrations of the radionuclide but also its chemical toxicity. These range from 800 000 Bq per cubic metre for tritium to 0.08 Bq per cubic metre for plutonium-239, with natural uranium and radium considered to be about 10 times less toxic than plutonium.

General public

The dose rates recommended by the ICRP for the general public are 20 times more stringent, namely an annual whole-body dose rate of 1 mSv, with the proviso that in some years 5 mSv may be allowed provided the lifetime dose does not exceed 1 mSv per year. This figure is over and above the average background radiation exposure which averages 1.87 mSv for the UK population as a whole from both the primordial radionuclides like radium and uranium, and from the cosmogenic radionuclides which are continually replenished by the action of cosmic rays on the atmosphere.

Conclusion

This and the preceding 10 chapters attempt to provide enough information, albeit in a condensed form, for the reader to find an answer to many questions about the nuts and bolts of the subject. And even, if so inclined, to form a judgement about the controversies which surround nuclear energy and radioactivity. I have not intended to present a view of my own, except my strongly held belief that knowledge of the engineering, the physics and the biology are all required if that judgement is to be in any way balanced.

a- Prefix signifying *on, in*. Also shortened form of *ab-*, **ad**-, **an**-, **ap**-.

a Symbol for (1) **acceleration**; (2) relative activity; (3) **linear absorption coefficient**; (4) **amplitude**.

α See under **alpha**. Symbol for (1) **absorption coefficient**; (2) attenuation coefficient; (3) **acceleration, angular acceleration**; (4) fine structure constant; (5) helium nucleus.

A Symbol for **ampere**

A Symbol for (1) area; (2) **absolute temperature**; (3) relative atomic mass (atomic weight); (4) magnetic vector potential; (5) Helmholtz function.

Å Symbol for Ångström,

A-bomb See **atomic bomb**.

absolute filter A filter which removes most particulate matter from gases.

absolute instrument An instrument which measures a quantity directly in absolute units, without the necessity for previous calibration.

absolute pressure Pressure measured with respect to zero pressure, in units of force per unit of area.

absolute temperature A temperature measured with respect to **absolute zero**, i.e. the zero of the **Kelvin thermodynamic scale of temperature**, a scale which cannot take negative values. See **kelvin**.

absolute units Units derived directly from the fundamental units of a system and not based on arbitrary numerical definitions. The differences between absolute and international units were small; both are now superseded by the definitions of the SI.

absolute zero The least possible temperature for all substances. At this temperature the molecules of any substance possess no heat energy. A figure of –273.15°C is generally accepted as the value of absolute zero.

absorbed dose The energy absorbed by the patient from the decay of a radionuclide given for diagnostic or therapeutic purposes. The unit is a gray (*Gy*). 1 Gy = 1 J kg^{-1}.

absorber Any material which converts energy of radiation or particles into another form, generally heat. Energy transmitted is not absorbed. Scattered energy is often classed with absorbed energy. See **total absorption coefficient, true absorption coefficient**.

absorber rod See **control rod**.

absorbing material Any medium used for absorbing energy from radiation of any type.

absorptance A measure of the ability of a body to absorb radiation; the ratio of the radiant flux absorbed by the body to that incident on the body. Formerly *absorptivity, absorptive power*.

absorption band A dark gap in the continuous spectrum of white light transmitted by a substance which exhibits selective absorption.

absorption coefficient (1) At a discontinuity (*surface absorption coefficient*), (*a*) the fraction of the energy which is absorbed or (*b*) the reduction of amplitude, for a beam of radiation or other wave system incident on a discontinuity in the medium through which it is propagated, or in the path along which it is transmitted. (2) In a medium (*linear absorption coefficient*), the natural logarithm of the ratio of incident and emergent energy or amplitude for a beam of radiation passing through unit thickness of a medium. (The *mass absorption coefficient* is defined in the same way but for a thickness of the medium corresponding to unit mass per unit area.) NB *True absorption coefficients* exclude scattering losses, *total absorption coefficients* include them. See **atomic absorption coefficient**.

absorption discontinuity See **absorption edge**.

absorption edge The wavelength at which there is an abrupt discontinuity in the intensity of an absorption spectrum for electromagnetic waves, giving the appearance of a sharp edge in its photograph. This transition is due to one particular energy-dissipating process becoming possible or impossible at the limiting wavelength. In X-ray spectra of the chemical elements, the K-absorption edge for each element occurs at a wavelength slightly less than that for the K-emission spectrum. Also *absorption discontinuity*.

absorption lines Dark lines in a continuous spectrum caused by absorption by a gaseous element. The positions (i.e. the wavelengths) of the dark absorption lines are identical with those of the bright lines given by the same element in emisssion.

absorption lines

absorption spectrum The system of absorption bands or lines seen when a selectively absorbing substance is placed between a source of white light and a spectroscope. See **Kirchhoff's law**.

absorptivity, absorptive power See **absorptance**.

abundance, abundance ratio For a naturally occurring element, the proportion or percentage of one isotope to the total. Percentages are given for most of the elements important in nuclear engineering. Also *abundance ratio*.

acceleration The rate of change of velocity, expressed in metres (or feet) per second squared. It is a vector quantity and has both magnitude and direction.

accelerator See **particle accelerators** p. 206.

acoustics The science of mechanical waves including production and propagation properties.

actinic radiation Ultraviolet waves, which have enhanced biological effect by inducing chemical change; basis of the science of photochemistry.

actinides The elements including and after actinium in the periodic table, with atomic numbers from 89–104. The name indicates that they have similar properties to actinium.

actino- Prefix from Gk. *aktis*, ray.

actinobiology The study of the effects of radiation upon living organisms.

action Time integral of kinetic energy (E) of a conservative dynamic system undergoing a change, given by

$$2\int_{t_1}^{t_2} E\,dt\,.$$

activation Induction of radioactivity in otherwise non-radioactive atoms, e.g. in a cyclotron or reactor.

activation cross-section Effective cross-sectional area of a target nucleus undergoing bombardment by neutrons etc for radioactivation analysis. Measured in *barns*. Cf. **nuclear cross-section**.

active lattice Regular pattern of arrangement of fissionable and non-fissionable materials in the core of a lattice reactor. » p. 26.

active materials (1) General term for essential materials required for the functioning of a device, e.g. iron or copper in a relay or machine; electrode materials in a primary or secondary cell; emitting surface material in a valve or photocell; phosphorescent and fluorescent material forming a phosphor in a CRT, or that on the signal plate of a TV camera. (2) Term applied to all types of radioactive isotopes.

activity The number of disintegrations occurring in a known amount of a radionuclide and therefore a measure of radioactivity. The unit is a **becquerel** (*Bq*).

ad- Prefix from L. *ad*, to, at.

adenine *6-aminopurine*, one of the five bases in nucleic acids in which it pairs with thymine in DNA and uracil in RNA. See **genetic code**.

adhesion Intermolecular forces which hold matter together, particularly closely contiguous surfaces of neighbouring media, e.g. liquid in contact with a solid. US *bond strength*.

adiabatic Without loss or gain of heat.

adiabatic change A change in the volume and pressure of the contents of an enclosure without exchange of heat between the enclosure and its surroundings. Also *adiabatic expansion*.

adiabatic curve The curve obtained by plotting P against V in the adiabatic equation.

adiabatic equation PV^{γ} = constant, an equation expressing the law of variation of pressure (P) with the volume (V) of a gas during an adiabatic change, γ being the ratio of the specific heat of the gas at constant pressure to that at constant volume. The value of γ is approximately 1.4 for air at STP.

adiabatic expansion See **adiabatic change**.

adiabatic process Process which occurs without interchange of heat with surroundings.

adiactinic Said of a substance which does not transmit photochemically-active radiation, e.g. safelights for darkroom lamps.

adsorbate Substance adsorbed at a phase boundary.

advanced gas-cooled reactor Carbon-dioxide-cooled, graphite-moderated reactor using slightly enriched uranium oxide fuel clad in stainless steel, in use in the UK. Abbrev. *AGR*. » p. 32.

advantage ratio The ratio between the radiation received at any point in a nuclear reactor and that measured at a reference position.

aerial radiometric surveying See panel on p. 117.

aether See **ether**.

afterheat Heat which comes from fission products in a reactor after it has been shut down.

age equation See **age theory**.

ageing Change in the properties of a substance with time, e.g. the deterioration of fuel cladding or pressure vessel in a nuclear reactor due to irradiation. Artificial ageing would be the simulation of such processes by increasing the rate of irradiation to obtain information more rapidly.

age theory In nuclear reactor theory, the slowing-down of neutrons by elastic collisions. The *age equation* relates the spatial distribution of neutrons to their energy. The equation is given by

$$\nabla^2 q - \frac{\partial q}{\partial t} = 0,$$

where q is the slowing-down density and τ is the **Fermi age**. It was first formulated by Fermi who assumed that the slowing-down process was continuous and so is least applicable to media containing light elements.

AGR Abbrev. for **Advanced Gas-cooled Reactor**.

airborne radiometric surveying See **aerial radiometric surveying** p. 117.

air dose Radiation dose in röntgens delivered at a point in free air.

airlift The use of a jet of air or neutral gas to move solid or liquid material during processing, to avoid pumps particularly in 'maintainence-free' radioactive environments.

air monitor Radiation (e.g. γ-ray) measuring instrument used for monitoring contamination or dose rate in air. » p. 74.

air wall Wall of ionization chamber designed to give the same ionization intensity inside the chamber as in an open space. This means that the wall is made of elements with atomic numbers similar to those of oxygen and nitrogen. » p. 74.

Al Symbol for **aluminium**.

ALARA Abbrev. for *As Low As Reasonably Achievable*. Said e.g. of radiation levels, decontamination etc.

aliquot A small sample of a material assayed to determine the properties of the whole. Term often applied to radioactive material.

allo- Prefix from Gk. *allos*, other.

allobar A mixture of isotopes of an element differing in proportion from that naturally occurring.

allochromy Fluorescent reradiation of light of different wavelength from that incident on a surface.

allowed band Range of energy levels permitted to electrons in a molecule or crystal. These may or may not be occupied.

allowed transition Electronic transition between energy levels which is not prohibited by any quantum selection rule.

alpha chamber Ionization chamber for measurements of α-radiation intensity. Also *alpha counter tube*.

alpha counter Tube for counting α-particles, with pulse selector to reject those arising from β- and γ- rays.

alpha counter tube See **alpha chamber**.

alpha decay Radioactive disintegration resulting in emission of α-particle. Also *alpha disintegration*.

alpha-decay energy The sum of the kinetic energies of the α-particle emitted and the recoil of the product atom in a radioactive decay. Also *disintegration energy*.

alpha disintegration See **alpha decay**.

alpha emitter Natural or artificial radioactive isotope which disintegrates through emission of α-rays.

alpha particle Nucleus of helium atom of mass number four, consisting of two neutrons and two protons and so doubly positively charged. Emitted from natural or radioactive isotopes. Often written *α-particle*. » p. 10.

alpha radiation **Alpha particles** emitted from radioactive isotopes.

alpha rays Stream of **alpha particles**.

alpha-ray spectrometer Instrument for measuring energy distribution of α-particles emitted by a radioactive source.

alternating gradient synchrotron A *synchrotron* modified by having magnetic-field gradients around the orbit alternating towards and away from the centre of the orbit. This produces a focusing effect which reduces beam divergence caused by the 3mutual repulsion of the particles in the beam. Proton energies of up to 500 GeV and electron energies of about 10 GeV can be achieved. See **particle accelerators** p. 206.

aluminium A metallic element. Symbol Al, at.no. 13, r.a.m. 26.98, valency 3, one stable isotope, ^{27}Al, atomic radius (6-fold co-ordination) 0.57 Å. It has seven sotopes:

A	Abundance %	half-life	decay mode
24		2.07 s	ε
25		7.18 s	ε
26		0.72 My	ε
27	100		
28		2.24 m	β^-
29		6.6 m	β^-
30		3.7 s	β^-

The familiar silver-white metal of everyday use is the third commonest element in the Earth's crust after oxygen and silicon and the commonest metallic element (8.8% by mass); it does not occur as native Al, but always in silicate or other minerals. Bauxite is the principal ore of aluminium, which is

aerial radiometric surveying

Bomb testing, nuclear accidents like that at Chernobyl in the Ukraine and the crashing of the small reactor from a Soviet reconnaissance satellite in the Canadian North-West Territories (see **space reactors** p. 239), all cause radioactive material to be spread over terrain which is often difficult and inaccessible. One method of discovering the extent of the contamination has been to send out teams with appropriate radiation detectors. This was both laborious and too slow to give accurate information to people who might be in immediate danger. It was also inaccurate because local features may well mask gamma emitters in a narrow valley and equipment which can be hand carried may not be able to distinguish the energies of the various gamma emitters present, and thus information about the source of the fallout will be lost.

However, gamma radiation from many nuclear disintegrations can penetrate several hundred metres through air. It is therefore possible to fly an aircraft low over the ground and equip it to detect and distinguish gamma rays of different energies. The method is much used in geophysical surveying where it is necessary to resolve gamma rays coming from disintegrations in naturally occurring radium, thorium and potassium. Very large volumes (up to nearly 50 litres) of sodium iodide crystal, cut into strips, can be used as scintillators (see p. 76). These are 'looked at' by several photomultipliers whose signals are fed to multichannel analysers to distinguish the energies of the gamma rays received from a wide area; typically 90% of the gamma rays can be recorded from an area with linear dimensions about five times the aircraft's height above the ground.

Similar methods, optimized to detect the gamma rays from fission products, like caesium-137, can provide a fast and accurate survey following any future nuclear incident. Fig. 1 shows the result of such a survey.

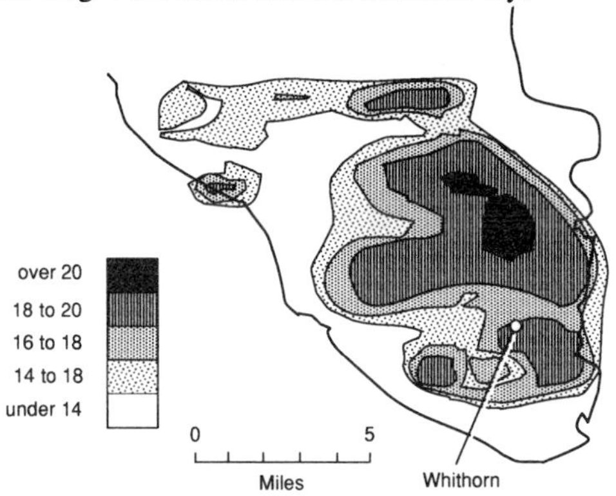

Fig. 1 **Simplified survey results from part of Galloway in Scotland**. The values are kiloBq per square metre of caesium-137.

prepared on a vast scale for the metal's many uses. In seawater it is only present to the extent of 0.01 ppm. Its corrosion resistance is good with a high thermal conductivity. It has been used as a cladding in reactors fuelled with highly enriched uranium and run at low temperatures (< 100°C). It has the following properties.

σ_{cap} mb	mp °C	Therm. conduct. $W\,m^{-1}K^{-1}$	density $g\,cm^{-3}$
230	660	237	2.7

US *aluminum.*

amagat The unit of density of a gas at 0°C and a pressure of one atmosphere; usually 1 amagat = 1 mole/22.4 dm^3.

amorphous metal A material with good conductivity, electrical and thermal, and with other metallic properties but with atomic arrangements that are not periodically ordered as in crystalline metal solids, e.g. metallic glass.

amorphous semiconductor Semiconductors prepared in the amorphous state. They tend to have much lower electrical conductivities than those of their crystalline counterparts.

amp Deprecated term for **ampere**.

amperage Current in amperes, more esp. the rated current of an electrical apparatus, e.g. fuse or motor.

ampere SI unit of electric current. Defined as that current which, if maintained in two parallel conductors of infinite length, of negligible cross-section, and placed 1 metre apart in vacuum, would produce between the conductors a force equal to 2×10^{-7} newtons per metre of length. One of the SI fundamental units. See **SI units** p. 237.

ampere hour Unit of charge, equal to 3600 coulombs, or 1 ampere flowing for 1 hour.

ampere-hour capacity Capacity of an accumulator battery measured in ampere-hours, usually specified at a certain definite rate of discharge. Also applicable to primary cells.

Ampère's law Law stating that the magnetic field (H) in the neighbourhood of a conductor, length l, carrying current I is;

$$\int H dl = I.$$

Ampère's rule A rule for the direction of the magnetic field associated with a current. The direction of the field is that of an advancing right-hand screw when turning with the current. Alternatively, if the conductor is grasped with the right hand, the thumb pointing in the direction of the current, the fingers will curl around the conductor in the direction of the field.

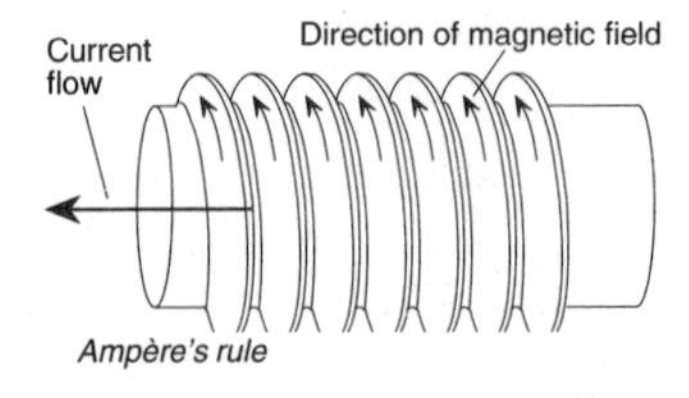

Ampère's rule

Ampère's theory of magnetization A theory based on the assumption that the magnetic property of a magnet is due to currents circulating in the molecules of the magnet.

ampere-turn SI unit of magneto-motive force, which drives flux through magnetic circuits, arising from 1 ampere flowing round one turn of a conductor. Abbrev. *At*.

ampere-turns per metre SI unit of magnetizing force or magnetic field intensity.

amphi- Prefix from Gk. meaning both, on both sides (or ends) or around.

amplitude The maximum value of a periodically varying quantity during a cycle; e.g. the maximum displacement from its position of a vibrating particle, the maximum value of an alternating current (see **peak value**), or the maximum displacement of a sine wave.

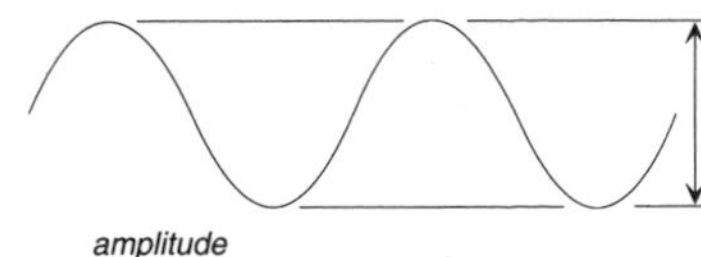
amplitude

amu Abbrev. for **atomic mass unit**.

an- Prefix from Gk. *an*, not. See **ap-**.

ana Prefix from Gk. *ana*, up, anew.

anchor ring See **torus**.

Ångström Unit of wavelength for electromagnetic radiation covering visible light and X-rays. Equal to 10^{-10} m. The unit is also used for interatomic spacings. Symbol Å. Largely superseded by nanometre (10^{-9} m). Named after the Swedish physicist, A.J. Ångström (1814-74). Also *angstrom*.

angular accelaration The rate of change of angular velocity; usually expressed in radians per second squared.

angular momentum The moment of the linear momentum of a particle about an axis. Any rotating body has an angular momentum about its centre of mass, its *spin angular momentum*. The angular momentum of the centre of mass of a body relative to an external axis is its *orbital angular momentum*. In atomic physics, the orbital angular momentum of an electron is *quantized* and can only have values which are exact multiples of **Dirac's constant**. In particle physics, the angular momentum of particles which appear to have spin energy is quantized to values that are multiples of half the Dirac constant.

anion Negative ion, i.e. atom or molecule which has gained one or more electrons in an electrolyte, and is therefore attracted to an anode, the positive electrode. Anions include all non-metallic ions, acid radicals and the hydroxyl ion. In a primary cell, the deposition of anions on an electrode makes it the negative pole. Anions also exist in gaseous discharge.

aniso- Prefix from Gk. *an*, not; *isos*, equal.

anisotropic A crystalline material with physical properties which depend upon their direction relative to crystal axes. These properties normally include elasticity, conductivity, permittivity and permeability.

annealing Process of maintaining a material at a known elevated temperature to reduce dislocations, vacancies and other metastable conditions, e.g. steel or glass. In ferrous alloys the metal is held at a temperature above the upper critical temperature for a variable time and then cooled at a predetermined rate, depending on the alloy and the particular properties of hardness, machinability etc that are needed. The term is usually qualified, e.g. *quench annealing, isothermal annealing, graphitizing*. In nuclear engineering it refers, additionally, to the process of removing the dislocations and swelling which occurs in e.g. graphite, under neutron bombardment, particularly at low temperatures. See **Wigner effect**. » p. 47.

annihilation Spontaneous conversion of a particle and its antiparticle into radiation, e.g. positron and electron yielding two γ-ray photons each of energy 0.511 MeV.

annihilation radiation The radiation produced by the annihilation of an elementary particle with its corresponding antiparticle.

anomalous dispersion The type of dispersion given by a medium having a strong absorption band, the value of the refractive index being abnormally high on the longer wave side of the band, and abnormally low on the other side. In the spectrum produced by a prism made of such a substance the colours are, therefore, not in their normal order.

anomalous scattering See **scattering**.

anomaly Any departure from the strict characteristics of the type.

ante- Prefix from L. *ante*, before.

antero- Prefix from *anterior*, former.

anthropogenic The equivalent of *man-made*, the opposite of *natural* used particularly of radiation and nuclear particles.

anti- Prefix from Gk. *anti*, against.

antibaryon Antiparticle of a baryon, i.e. a hadron with a baryon number of −1. The term **baryon** is often used generically to include both.

antibonding orbital Orbital electron of 2 atoms, which increases in energy when the atoms are brought together, and so acts against the closer bonding of a molecule.

anticathode The anode target of an X-ray tube on which the cathode rays are focused, and from which the X-rays are emitted.

antilepton An antiparticle of a **lepton**. Positron, positive muon, antineutrinos and the tau-plus particle are antileptons.

antimony Metallic element. Symbol Sb, at.no. 51, r.a.m. 121.75, rel.d. 6.6, mp 630°C. Used in alloys for cable covers, batteries etc; also as a donor impurity in germanium. Has several radioactive isotopes which emit very penetrating γ-radiation. These are used in laboratory neutron sources. Abundance in Earth's crust 0.2 ppm.

antimuon Antiparticle of a **muon**.

antineutrino Antiparticle of the **neutrino**. As there are four types of neutrino there are also four types of antineutrino.

antineutron Antiparticle with spin and magnetic moment oppositely orientated to those of a neutron.

antinode At certain positions in a standing wave system of acoustic or electric waves or vibrations, the location of maxima of some wave characteristic, e.g. amplitude, displacement, velocity, current, pressure, voltage. At the *nodes* these would have minimum values.

antiparticle A particle that has the same mass as another particle but has opposite values for its other properties such as charge, baryon number, strangeness etc. The antiparticle to a fundamental particle is also fundamental, e.g. the electron and positron are particle and antiparticle. Interaction between such a pair means simultaneous **annihilation**, with the production of energy in the form of radiation.

antiproton Short-lived particle, half-life 0.05 μsec, identical with proton, but with negative charge; annihilating with normal proton, it yields mesons. Also *negative proton*.

antiquark The antiparticle of a **quark**.

antisymmetric Said of a pattern or waveform in which symmetry is complete except for one particular feature, e.g. sign of electric charge, direction of current or of components in a waveform. A system containing several electrons must be described quantum mechanically by an *antisymmetric eigenfunction*.

ap- Another form of **an-**.

aperiodic Said of any potentially vibrating system, electrical, mechanical or acoustic, which, because of sufficient damping, does not vibrate when impulsed. Used particularly of the pointers of indicating instruments, which having no natural period of oscillation, do not oscillate before coming to rest in the final position, and so give their ultimate reading as fast as possible.

apex The top or pointed end of anything. Adj. *apical*.

apo- Prefix from Gk. *apo*, away.

Appleton layer See **F-layer**.

applied mathematics Originally the application of mathematics to physical problems, differing from physics and engineering in being concerned more with mathematical rigour and less with practical utility. More recently, also includes numerical analysis, statistics and probability, and applications of mathematics to biology, economics, insurance etc.

applied potential tomography A system of medical imaging based on the measurement of the electrical impedance at a frequency of about 50 kHz between many electrodes placed around the body.

applied power The power *applied* to an electrical transducer is not equal to the actual power received, because of the reflection arising from non-equality of impedance matching. The *applied power* is the power which would be received if the load matched the source in impedance.

Ar Symbol for **argon**.

arc therapy X-ray therapy in which the rotation of the beam is limited to avoid sensitive tissues, e.g. the lungs or gonads.

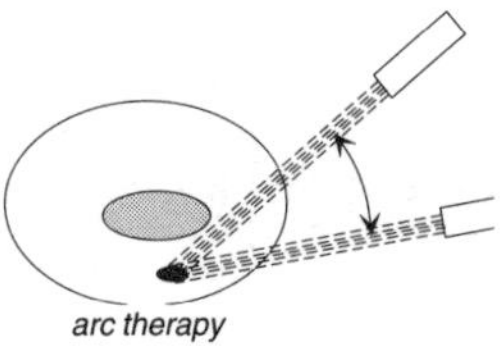

arc therapy

area monitoring The survey and measurement of types of ionizing radiation and dose levels in an area in which radiation hazards are present or suspected. See **aerial radioactive surveying** p. 117.

argon An element. Symbol Ar, at.no. 18, r.a.m. 39.948. A rare gas, it constitutes about 1% by volume of the atmosphere. It has 11 isotopes:

A	Abundance %	half-life	decay mode
34		0.84 s	ε
35		1.78 s	ε
36	0.337		
37		35.0 d	ε
38	0.063		
39		269 y	β^-
40	99.60		
41		1.83 h	β^-
42		33 y	β^-
43		5.4 m	β^-
44		11.9 m	β^-

The isotope, argon-40, is formed by the radioactive decay of the isotope potassium-40. The ratio potassium-40 : argon-40 in a rock or mineral is used for the determination of the age in years (abbrev. *K – Ar method).*

argon laser A laser using singly-ionized argon. It gives strong emission at 488.0, 514.5 and 496.5 nm.

arithmetic The science of numbers, including such processes as addition, subtraction, multiplication, division and the extraction of roots.

artefact, artifact A simple man-made stone, wood or metal implement.

arteriography The radiological examination of arteries following direct injection of a **contrast medium**, e.g. coronary arteriography, renal arteriography, carotid arteriography.

artifact An error in an image which has no counterpart in reality.

artificial daylight Artificial light having approximately the same spectral distribution curve as daylight, i.e. having a colour temperature of about 4000 K.

artificial disintegration The transmutation of non-radioactive substances brought about by the bombardment of the nuclei of their atoms by high-velocity particles, such as α-particles, protons or neutrons.

artificial radioactivity Radiation from isotopes after high-energy bombardment in an accelerator by α-particles, protons and other light nuclei, or by neutrons in a nuclear reactor. Discovered by I. Curie-Joliot and F. Joliot in 1933. See **activation**.

aspect ratio In a Tokamak type of fusion machine, the ratio of the major to minor radii of a torus. » p. 40.

Aston whole-number rule Empirical observation that relative atomic masses of isotopes are approximately whole numbers. See **mass spectrometer**.

asymmetry The condition of any object which does not have an axis or plane of symmetry.

atm Abbrev. for **standard atmosphere**. See **atmospheric pressure**.

atmolysis The method of separation of the components of a mixture of two gases, which depends on their different rates of diffusion through a porous partition. The basis for the enrichment of uranium-235 by the diffusion process. » p. 57.

atmospheric pressure The pressure exerted by the atmosphere at the surface of the Earth due to the weight of the air. Its standard value is

$$1.013\,25 \times 10^5\ \mathrm{N m^{-2}},$$

1.013 25 bar or 14.7 lbf $\mathrm{in^{-2}}$, the values of the **standard atmosphere**. Variations in the

atomic structure

The chemical behaviour of the various elements arises from the differences in the electron configuration of the atoms in their normal electrically neutral state. Each atom consists of a heavy nucleus with a positive charge produced by a number of **protons** equal to its atomic number. There are an equal number of electrons outside the nucleus to balance this charge. The nucleus also contains electrically neutral **neutrons**. Protons and neutrons are collectively referred to as *nucleons*.

The Sommerfeld model, modified by the wave mechanical concept of orbitals, describes the electron configuration of the atom. Electrons are fermions which must conform to the *Pauli exclusion principle* which governs the way in which electrons can fill the available *orbitals* because no two electrons in the same atom can be in the same quantum state, i.e. have the same set of four quantum numbers. The principal quantum number (n) indicates the shell to which the orbital belongs and varies from 1 (K-shell) closest to the nucleus to 7 (Q-shell), the most remote.

For a given principal quantum number n, there are n allowed values of l, the orbital angular momentum quantum number; for each value of l, there are $(2l+1)$ allowed values of m_l, the magnetic angular momentum quantum number; for each value of m_l, there are two values of m_s, the magnetic spin number. This makes a total of $2n^2$ orbitals ($2n^2$ electrons per shell) for a given value of n, and, as the Pauli principle allows only one electron for each set of four quantum numbers n, l, m_l, m_s, this limited number of allowed orbitals makes up the electron shell for a given n.

In general, the closer an electron is to the nucleus the greater the coulomb attraction and so the greater the binding energy retaining the electron in the atom. Inner filled shells are therefore relatively inert and the chemical properties of the atom are determined by the electron arrangement in the outermost shell.

Nuclear binding forces tend to give greatest stability when the neutron number and the proton number are approximately equal. Due to electrostatic repulsion between protons, the heavier nuclei are most stable when more than half their nucleons are neutrons; elements with more than 83 protons are unstable and undergo radioactive disintegration. Those with more than 92 protons are not found naturally on Earth, but can be synthesized in particle accelerators and nuclear reactors. These are the transuranic elements which have short half-lives. Most elements exist with several stable **isotopes** and the chemical atomic weight gives the average of a normal mixture of these isotopes. » p. 10.

atmospheric pressure are measured by means of the barometer.

atomic absorption coefficient For an element, the fractional decrease in intensity of radiation per number of atoms per unit area. Symbol μ_α. Related to the linear absorption coefficient μ by

$$\mu = \frac{1}{V}\sum_{i} n_i (\mu_\alpha)_i ,$$

where the material contains n_i atoms of element i in a volume V.

atomic absorption spectroscopy A method in which light from a standard source is passed through a flame into which a sample of the substance under investigation has been introduced. The outer electrons of the sample are excited and emit energy at characteristic wavelengths which, in turn, absorb those from the standard source. The resulting spectrum can identify the elements present and indicate their relative proportions.

atomic bomb A bomb in which the explosive power, measured in terms of equivalent TNT, is provided by nuclear fissionable material such as U-235 or Pu-239. The bombs dropped on Hiroshima and Nagasaki (1945) were of this type. Also *A-bomb, atom* *bomb*, *fission bomb*. Terms using 'atom' are deprecated here because the energy released is of nuclear origin. See **hydrogen bomb**. » p. 65.

atomic clock A clock whose frequency of operation is controlled by the frequency of an atomic or molecular process. The inversion of the ammonia molecule with a frequency of 23 870 Hz provides the basic oscillations of the *ammonia clock*. The difference in energy between two states of a caesium atom in a magnetic field giving a frequency of 9 192 631 770 Hz is the basis of the *caesium clock* which has an accuracy of better than 1 in 10^{13}.

atomic disintegration Natural decay of radioactive atoms, as a result of radiation, into chemically different atomic products.

atomic displacement cross-section The probability of a neutron displacing an atom from its place in a crystalline solid. Measured in *barns* as for other **cross-sections**. Important in determining the lifetime of graphite moderator and structural parts of reactors.

atomic frequency A natural vibration frequency in an atom used in the atomic clock.

atomic mass unit Exactly one twelfth the mass of a neutral atom of the most abundant isotope of carbon, ^{12}C. $u = 1.660 \times 10^{-27}$ kg. Before 1960 u was defined in terms of the mass of the ^{16}O isotope and u was 1.6599×10^{-27} kg. Abbrevs. *u*, *amu*. Also *mass unit*. See **relative atomic mass**.

atomic number The order of an element in the periodic (Mendeleev) chemical classification, and identified with the number of unit positive charges in the nucleus (independent of the associated neutrons). Equal to the number of external electrons in the neutral state of the atom; it determines its chemistry. Symbol Z.

atomic scattering Scattering of radiation, usually electrons or X-rays, by the individual atoms in the medium through which it passes. The scattering is by the electronic structure of the atom in contrast to *nuclear scattering* which is by the nucleus.

atomic scattering factor The ratio of the amplitude of coherent scattered X-radiation from an atom to that of a single electron placed at the atomic centre. The atomic scattering factor depends on the electron-density distribution in the atom and is a function of the scattering angle.

atomic spectrum Electronic transitions between the discrete energy states of an atom involve either the emission or absorption of *photons*. Such emission or absorption spectra are called *line spectra*. The spectrum is characteristic of the atom involved. See **absorption lines**.

atomic structure See panel on p. 121.

atomic transmutation The change of one type of atom to another as a result of a nuclear reaction. The transmutation can be produced by high-energy radiation or particles and is most easily produced by neutron irradiation. The change in atomic number means the chemical nature of the atom has been changed, e.g. gold can be transmuted into mercury; the converse of the ancient alchemist's goal. Also *transformation*. » p. 10.

attenuation General term for reduction in magnitude, amplitude or intensity of a physical quantity, arising from absorption, scattering or geometrical dispersion. The latter, arising from diminution by the inverse-square law, is not generally considered as attenuation proper.

attenuation of X-rays The term covering both absorption and scattering of X-rays as they pass through an object.

atomic weight See **relative atomic mass**. Abbrev. *at.wt*.

Auger effect An atom ionized by the ejection of an inner electron can lose energy either by the emission of an X-ray photon as an outer electron makes a transition to the vacancy in the inner shell *or* by the ejection of an outer electron, the *Auger effect*. The energies of the Auger electrons emitted are characteristic of the atomic energy levels.

Auger yield For a given excited state of an atom of a given element, the probability of de-excitation by Auger process instead of by X-ray emission.

aut-, auto- Prefixes from Gk. *autos*, self.

automation Industrial closed-loop control system in which manual operation of controls is replaced by servo operation.

avalanche Self-augmentation of ionization. See **Townsend avalanche**. » p. 75.

Avogadro number The number of atoms in 12 g of the pure isotope ^{12}C; i.e. the reciprocal of the *atomic mass unit* in grams. It is also by definition the number of molecules (or atoms, ions, electrons) in a **mole** of any substance and has the value $6.022\,52 \times 10^{23}$ mol^{-1}. Also *Avogadro constant*. Symbol N_A or L.

axes Pl. of *axis*.

axial ratio Ratio of major to minor axis of polarization ellipse for wave propagated in waveguide, polarized light etc.

azimuthal power instability Abnormal neutron behaviour which results in uneven nuclear conditions in the reactor.

B

β For β-particles, see under **beta**. Symbol for (1) phase constant; (2) ratio of velocity to velocity of light.

B Symbol for (1) **boron**; (2) susceptance in an a.c. circuit (unit = siemens; measured by the negative of the reactive component of the admittance); (3) magnetic flux density in a magnetic circuit (unit = tesla= Wb m^{-2} = V s m^{-2}).

Babinet's principle The radiation field beyond a screen which has apertures, added to that produced by a complementary screen (in which metal replaces the holes, and spaces the metal) is identical to the field which would be produced by the unobstructed beam of radiation, i.e. the two diffraction patterns will also be complementary.

backfitting Making changes to nuclear (and other) plants already designed or built, e.g. to cater for changes in safety criteria.

background A general problem in physical measurements, which limits the ability to detect or accurately measure any given phenomenon. Background consists of extraneous signals arising from any cause which might be confused with the required measurements, e.g. in electrical measurements of nuclear phenomena and of radioactivity, it would include counts emanating from amplifier noise, cosmic rays, insulator leakage etc.

background radiation Radiation coming from sources other than that being observed.

back scatter The deflection of radiation or particles by scattering through angles greater than 90° with reference to the original direction of travel. Cf. **forward scatter**.

bacteria A large group of unicellular or multicellular organisms, lacking chlorophyll, multiplying rapidly by simple fission, some species developing a highly resistant resting ('spore') phase. They are the principle members of the *prokaryotes*, whose genetic material is not packaged into chromosomes contained in a nucleus. Instead they have a single circle of DNA free in the cytoplasm, which reproduces by a 'rolling circle' mechanism. They lack mitochondria, oxidative phosphorylation occurring across the plasma membrane, and they are bounded by a cell wall situated outside and separated from the plasma membrane by a periplasmic space. The taxonomy of bacteria is difficult, not only because of the obvious problems of size and rapid mutation but also because a phylogenetic classification, so powerful in Botany and Zoology, is largely unprofitable in these micro-organisms. Instead morphological, chemical and pathogenic characteristics are used. In particular, bacteria can be divided into those which stain with Gram's method and have a simpler kind of cell wall and those which do not. Colony colour and shape on agar containing a defined nutrient or growth in the presence of antibiotics and other chemicals are often used to screen and identify cultures. In shape they can be spherical, rod-like, spiral or filamentous. Some are motile by means of flagella and they occur in every natural habitat often in large numbers, as much as 10^8 per gram. Since Pasteur's time they have been actively studied and classified in relation to disease but they are also important in the production of chemicals, enzymes and antibiotics. This commercial production of bacterial products has become increasingly important as *genetic manipulation* has allowed the production of substances like human insulin and growth hormones in bacteria.

bacteriophage A **virus** which infects bacteria. Also *phage*.

Balmer series A group of lines in the hydrogen spectrum discovered by H.H. Balmer in 1885. It was not until 20 years later that it was shown that they correspond to excitation levels of the hydrogen electron. Their positions are given by the formula:

$$\nu = R_{\mathrm{H}}\left(\frac{1}{2^2} - \frac{1}{n^2}\right),$$

where n has various integral values, ν is the wave number and R_{H} is the hydrogen Rydberg number (= 1.096 775 8×10^7 m^{-1}). A number of similar series called Lyman, Paschen etc are found in other parts of the electromagnetic spectrum. » p. 84.

band spectrum Molecular optical spectrum consisting of numerous very closely spaced lines which are spread through a limited band of frequencies.

bar The unit of pressure or stress, 1 bar = 10^5 N m^{-2} or pascals = 750.07 mm of mercury at 0°C and lat. 45°. The *millibar* (1 mbar = 100 N m^{-2} or 10^3 dyn cm^{-2}) is used for barometric purposes. (NB Standard atmospheric pressure = 1.013 25 bar.) The *hectobar* (1 hbar = 10^7 N m^{-2}, approx. 0.6475 tonf in^{-2}) is used for some engineering purposes.

barite See **barytes**.

barium A heavy element in the second group of the periodic system, an alkaline earth metal. Symbol Ba, at.no. 56, r.a.m. 137.34, mp 725°C. In most of its compounds it occurs as Ba^{2+} and is present to the extent of

390 ppm in the Earth's crust. Its mass makes it an effective absorber of high-energy particles and it is used as barytes in loaded concrete for this purpose. It has 15 isotopes:

A	Abundance %	half-life	decay mode
127		12.7 m	ε
128		2.43 d	ε
129		2.2 h	ε
130	0.11		
131		12.0 d	ε
132	0.10		
133		10.7 y	ε
134	2.42		
135	6.59		
136	7.85		
137	11.2		
138	71.7		
139		82.9 m	β^-
140		12.7 d	β^-
141		18.3 m	β^-

It is a brittle and expensive metal, difficult to machine and giving off toxic dust. It has nevertheless been used for nuclear purposes and has the following properties:

σ_{cap} mb	mp °C	Therm. conduct. $W\,m^{-1}K^{-1}$	density $g\,cm^{-3}$
9.2	1280	201	1.85

barn Unit of effective cross-sectional area of nucleus equal to 10^{-28} m^2. So called, because it was pointed out that although one barn is a very small unit of area, to an elementary particle the size of an atom which could capture it was 'as big as a barn door'. See **cross-section.** » p. 12.

baryon A *hadron* with a baryon number of +1. Baryons are involved in strong interactions. Baryons include neutrons, protons and hyperons.

baryon number An intrinsic property of an elementary particle. The baryon number of a baryon is +1, of an antibaryon −1. The baryon number of mesons, leptons and gauge bosons is zero. Baryon number is conserved in all types of interaction between particles. **Quarks** have a baryon number of +1/3 and antiquarks of −1/3.

barytes, barite Barium sulphate; it typically has tabular orthorhombic crystals and is a common mineral in association with lead ores. Used to load drilling muds and increase the density of concrete in radiation shields.

barytes concrete See **loaded concrete**.

base load The steady load, more economically produced by nuclear power with peaks of output carried by other methods.

base pairing rule The bases in the opposite strands of DNA are stable when adenine pairs with thymine and guanine with cytosine. » p. 90.

base unit The International System of Units (SI) is a coherent system based on seven base units. All derived units are obtained from the base units by multiplication without introducing numerical factors, and approved prefixes are used in the construction of submultiples and multiples. There is only one base or derived unit for each physical quantity. The base units are *metre*, *kilogram*, *second*, *ampere*, *kelvin*, *candela* and *mole*. See **SI Units** p. 237.

basi- Prefix from Gk. *basis*, base.

batch process Any process or manufacture in which only a limited amount of material or number of articles are treated at any one time, as opposed to a continuous process.

Be Symbol for **beryllium**.

beam A collimated, or approximately unidirectional, flow of electromagnetic radiation (radio, light, X-rays), or of particles (atoms, electrons, molecules). The angular beam width is defined by the half-intensity points.

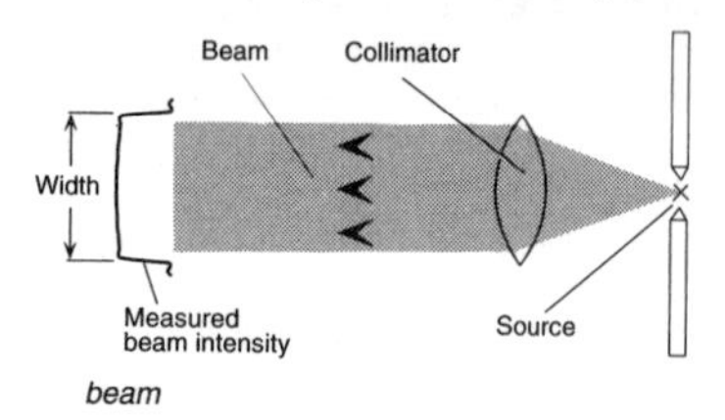

beam

beam hole Hole in shield of reactor, or that around a cyclotron, for extracting a beam of neutrons or γ-rays or to insert equipment or samples for irradiation.

beauty See **bottomness**.

becquerel SI unit of radioactivity; one becquerel is the activity of a quantity of radioactive material in which one nucleus decays per second. Symbol Bq. Replaces the **curie**.

$$1\ \text{Bq} = 2.7\times10^{-11}\ \text{Ci}.$$

It is a very small unit and commonly used with the standard SI prefixes, a gigabecquerel (GBq or 10^9 Bq) being often needed. It and other related radiation units are defined on p. 103.

beryllium A steely uncorrodible white metallic element (discovered by Wöhler, 1828). Gk. and L. *beryl*, the old mineral name. Symbol Be, at. no. 4, r.a.m. 9.0122, mp 1281°C, bp 2450°C, rel.d. 1.93. It is a rare element both cosmically and in the Earth, where its abundance is only 2 ppm. It has five isotopes:

A	Abundance %	half-life	decay mode
7		53.3 d	ε
8		0.07 s	α
9	100		
10		1.6 My	β^-
11		13.8 s	β^-

It is a brittle and expensive metal, difficult to machine and giving off toxic dust. It has nevertheless been used for nuclear purposes and has the following properties:

σ_{cap} mb	mp °C	Therm. conduct. $W\,m^{-1}K^{-1}$	density $g\,cm^{-3}$
9.2	1280	201	1.85

best available technology A US term for the process giving the maximum abatement of pollution without regard to cost or proven necessity. Abbrev. *BAT*. Also *best available control technology (BACT)*.

best practical environmental option Concept recognizing that treatment of pollutants in one medium of the environment (air, land, water) may simply transfer them into another. Thus removing sulphur dioxide from flue gases may cause the calcium sulphate produced to have polluting effects at its disposal sites. Also the *cross-media approach*, it is a co-ordinated approach to pollution pathways, media and disposal routes. Abbrev. *BPEO*.

best practical means Term with statutory force since 1863, and the basis for control of atmospheric pollution in the UK. Defined as the best practicable means with regard to local conditions, financial implications and current technical knowledge, and includes the provision, maintenance and correct use of plant. Abbrev. *BPM*.

best technical means available European Commission term requiring consideration of the economic availability of the means of pollution abatement. It approaches the UK term **best practical means**. Abbrev. *BTMA*.

beta decay Radioactive disintegration with the emission of an electron or positron accompanied by an uncharged antineutrino or neutrino. The mass number of the nucleus remains unchanged but the atomic number is increased by one or decreased by one depending on whether an electron or positron is emitted. See **electron capture**.

beta detector A radiation detector specially designed to measure β-radiation.

beta disintegration See **beta decay**.

beta disintegration energy For electron emission, β^-, it is the sum of the energies of the particles, the neutrino and the recoil atom. For positron emission, β^+, there is in addition the energy of the rest masses of two electrons. » p. 10.

beta particle An electron or positron emitted in beta decay from a radioactive isotope.

beta rays Streams of β-particles.

beta-ray spectrometer A spectrometer which determines the spectral distribution of the energies of β-particles from radioactive substances or secondary electrons.

beta thickness gauge Instrument measuring thickness, based on absorption and backscattering (reflection) by material or sample being measured of β-particles from a radioactive source.

betatopic Said of atoms differing in atomic number by one unit. One atom can be considered as ejecting an electron (β-particle) to produce the other one.

betatron Machine used to accelerate electrons to energies of up to 300 MeV in pulsed output. The electrons move in an orbit or constant radius between the poles of an electromagnet, and a rapidly alternating magnetic field provides the means of acceleration. See **particle accelerator** p. 206.

beta value In fusion, the ratio of the outward pressure exerted by the plasma to the inward pressure which the magnetic field is capable of exerting. Also *plasma beta*.

BeV See **GeV**.

Bevatron A synchrotron at Berkeley, California, USA, which gives a beam of 6.4 GeV protons.

billion In the US and now more generally, a thousand million, or 10^9. Previously elsewhere, a million million or 10^{12}.

billion-electron-volt See **GeV**.

binding energy of a nucleus All nuclei have *rest* masses less than the total rest mass of their constituent protons and neutrons. The mass difference m is the *mass decrement* or *mass defect*. This arises because all nucleons bound to the nuclei must have negative energy (potential well). So if a system of free nucleons are combined to form a nucleus the total energy of the system must decrease by an amount B, the binding energy of the nucleus. The decrease B is accompanied by a decrease in the mass, the mass decrement, $M = B/C^2$ (**mass-energy equation**). The binding energy per nucleon is B/A where A is the atomic mass number. See **packing fraction**, **fusion**, **fission**. » p. 13.

biochemistry The chemistry of living

things; physiological chemistry.

biological half-life Time interval required for half of a quantity of radioactive material absorbed by a living organism to be eliminated naturally.

biological hole A cavity within a nuclear reactor in which biological specimens are placed for irradiation experiments.

biological shield The heavy concrete barrier placed round a nuclear reactor or other plant to protect workers from radiation.

biological warfare The use of bacteriological (biological) agents and toxins as weapons. Cf. **war gas**.

bionics The various phenomena and functions which characterize biological systems with particular reference to electronic systems.

biosphere That part of the Earth (upwards at least to a height of 10 000 m, and downwards to the depths of the ocean, and a few hundred metres below the land surface) and the atmosphere surrounding it, which is able to support life.

black body A body which completely absorbs any heat or light radiation falling upon it. A *black body* maintained at a steady temperature is a full radiator at that temperature, since any black body remains in equilibrium with the radiation reaching and leaving it.

black-body radiation Radiation that would be radiated from an ideal black body. The energy distribution is dependent only on the temperature and is described by **Planck's radiation law**. See **Stefan-Boltzmann law**, **Wien's laws for radiation from a black body**.

black-body temperature The temperature at which a **black body** would emit the same radiation as is emitted by a given radiator at a given temperature. The *black-body temperature* of carbon-arc crater is about 3500°C, whereas its true temperature is about 4000°C.

black box A generalized colloquial term for a self-contained unit of electronic circuitry; not necessarily black. It should produce a defined output for a defined input, without the operator needing to know its contents.

blanket (1) Region of **fertile** material surrounding the core in a breeder reactor in which neutrons coming from the core breed more fissile fuel, e.g. uranium-233 from thorium. (2) The lithium surrounding a fusion reactor core within which fusion neutrons are slowed down, heat is transferred to a primary coolant and tritium is bred from lithium. » p. 40.

blast wave See **shock wave**.

BNFL Abbrev. for *British Nuclear Fuels* plc. Organization involved in uranium enrichment, fabrication of fuel elements, reprocessing of irradiated nuclear fuel and production of plutonium. It also operates experimental reactors.

Bo Symbol for **boron**.

body-section radiography *Cross-section radiography*. See **tomography, emission**.

Bohr atom Concept of the atom, with electrons moving in a limited number of circular orbits about the nucleus. These are *stationary states*. Emission or absorption of electromagnetic radiation results only in a **transition** from one orbit (state) to another.

Bohr magneton Unit of magnetic moment, for electron, defined by

$$\mu_B = eh/4\pi m_e c,$$

where e = charge, h = Plancks's constant, m_e = rest mass and c = velocity of light, so that

$$\mu_B = 9.27 \times 10^{-27}\ \mathrm{J\,T^{-1}}.$$

The *nuclear Bohr magneton* is defined by

$$\mu_N = \frac{eh}{4\pi Mc} = \frac{\mu_B}{1836} = 5.05 \times 10^{-27}\ \mathrm{J\,T^{-1}},$$

M being the rest mass of the proton.

Bohr radius According to the Bohr model of the hydrogen atom, the electron when in its lowest energy state, moves round the nucleus in a circular orbit of radius

$$a_0 = \frac{4\pi\varepsilon_0\hbar^2}{m_e e^2} = 0.5292 \times 10^{-10}\ \mathrm{m}$$

where $\hbar$ is Planck's constant divided by 2π, m_e is the mass of the electron, e is the electronic charge and ε_0 the permittivity of free space. The Bohr radius is a fundamental distance in atomic phenomena.

Bohr-Sommerfeld atom Atom obeying modifications of Bohr's laws suggested by Sommerfeld and allowing for possibility of elliptic electron orbits.

Bohr theory A combination of the Rutherford model of the atom with the quantum theory. The Bohr theory is based on four postulates: (1) An electron in an atom moves in a circular orbit about the nucleus under the influence of the coulombic attraction between the electron and the nucleus. (2) It is only possible for an electron to move in an orbit for which its orbital angular momentum is an integral multiple of $h/2\pi$, where h is Planck's constant. (3) An electron moving in such an orbit does not radiate electromagnetic energy and so its total energy, E, remains constant. (4) Electromagnetic radiation is emitted if an electron makes a transition from an orbit of energy E_i to one of *lower* energy E_f, and the frequency of the emitted radiation is

$$\nu = (E_i - E_f)/h.$$

See **atomic structure** p. 121.

boiling The very rapid conversion of a liquid into vapour by the violent evolution of bubbles. It occurs when the temperature reaches such a value that the saturated vapour pressure of the liquid equals the pressure of the atmosphere. Also *ebullition*.

boiling point The temperature at which a liquid boils when exposed to the atmosphere. Since, at the boiling point, the saturated vapour pressure of a liquid equals the pressure of the atmosphere, the boiling point varies with pressure; it is usual, therefore, to state its value at the standard pressure of 101.325 kN m^{-2}. Abbrev. *bp*.

boiling-water reactor Light-water reactor in which the water surrounding the fuel is allowed to boil into steam which drives the turbines directly. Abbrev. *BWR*. » p. 28.

Boltzmann equation A fundamental diffusion equation based on particle conservation. The rate of losses, including leakage out of the region of interest and the rate of disappearance by reactions of all kinds, is equal to the rate of production from sources within the region and the rate of scattering into the region.

Boltzmann principle Theory concerning the statistical distribution of large numbers of small particles when subjected to thermal agitation and acted upon by electric, magnetic or gravitational fields. In statistical equilibrium, the number of particles n per unit volume in any region is given by

$$n = n_0 \exp(-E/kT),$$

where k = Boltzmann's constant, T = absolute temperature, E = potential energy of particle in given region, n_0 = number per unit volume when $E = 0$.

Boltzmann's constant Given by

$$k = R/N = 1.3805 \times 10^{-23}\ \mathrm{J\ K^{-1}},$$

where R = ideal gas constant, N = Avogadro constant.

bombardment Process of directing a beam of neutrons or high-energy charged particles on to a target material in order to produce nuclear reactions.

bond Link between atoms, considered to be electrical and arising from electrons as distributed around the nucleus of atoms so bonded.

bond strength See **adhesion**.

bone tolerance dose The dose of ionizing radiation which can safely be given in treatment without bone damage.

Boral sheet A composite made of boron carbide crystals dispersed in aluminium and also faced with aluminium. Used as a neutron absorber.

Born-Oppenheimer approximation An approximation used in considering the electronic behaviour of molecules. The problems of the electronic and nuclear motion are treated separately.

boron An amorphous yellowish-brown element discovered by Davy, 1808, also Gay-Lussac and Thénard. Symbol B, at.no. 5, r.a.m. 10.811, rel.d. 2.5, mp 2300°C. Can be formed into a conducting metal. Most important in reactors, because of large cross-section (absorption) for neutrons; thus, boron steel is used for control rods. The isotope ^{10}B on absorbing neutrons breaks into two charged particles ^{7}Li and ^{4}He which are easily detected, and is therefore useful for detecting and measuring neutrons. Its abundance in the Earth's crust is 9 ppm and in seawater 4.8 ppm. It has six isotopes:

A	Abundance %	half-life	decay mode
8		0.77 s	ε
9		0.85 s	α
10	19.8		
11	80.2		
12		20.4 ms	β^-
13		17.4 ms	β^-

boron chamber, boron counter Counter tube containing boron fluoride, or boron-covered electrodes, for the detection and counting of low-velocity neutrons, which eject α-particles from the isotope boron-10.

Bose-Einstein statistics The statistical mechanics law obeyed by a system of particles whose wave function is unchanged when two particles are interchanged.

boson A particle which obeys Bose-Einstein statistics but not the Pauli exclusion principle. Bosons have a total spin angular momentum of n, where n is an integer and is the Dirac constant. Photons, α-particles and all nuclei having an even mass number are bosons.

bottomness A property that characterizes *quarks* and therefore hadrons. The bottomness of leptons and gauge bosons is zero. Bottomness is conserved in strong and electromagnetic interactions between particles but not in weak interactions. Also *beauty*.

Bouguer law of absorption The intensity p of a parallel beam of monochromatic radiation entering an absorbing medium is decreased at a constant rate by each infinitesimally thin layer *db*,

$$\frac{-\,dp}{p} = k\,db,$$

where *k* is a constant that depends on the nature of the medium and on the wavelength.

bound state Quantum mechanical state of a system in which the energy is *discrete* and the wave function is localized, e.g. that of an electron in an atom, where transitions between the bound states give rise to atomic spectrum lines.

Bq Symbol for **becquerel**, the SI unit of radioactivity.

Bragg angle The angle the incident and diffracted X-rays make with a crystal plane when the Bragg equation is satisfied for maximum diffracted intensity.

Bragg curve Graph giving average number of ions per unit distance along beam of initially monoenergetic α-particles (or other ionizing particles) passing through a gas.

Bragg equation If X-rays of wavelength λ are incident on a crystal, diffracted beams of maximum intensity occur in only those directions in which constructive interference takes place between the X-rays scattered by successive layers of atomic planes. If *d* is the interplanar spacing, the Bragg equation,

$$n\lambda = 2d\sin\theta,$$

gives the condition for these diffracted beams; θ is the angle between the incident and diffracted beams and the planes, and *n* is an integer. Also applied to electron, neutron and proton diffraction.

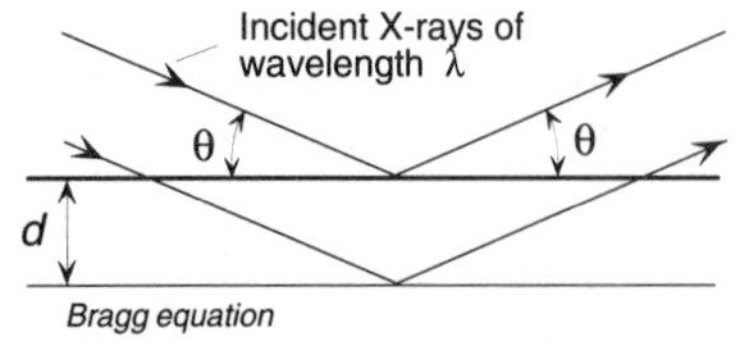

Bragg equation

Bragg rule An empirical relationship according to which the mass stopping power of an element for α-particles (also applicable to other charged particles) is proportional to (relative atomic mass)$^{-0.5}$.

branching The existence of two or more modes or branches by which a radionuclide can undergo radioactive decay, e.g. copper-64 can undergo β^-, β^+ and electron-capture decay.

breakeven In fusion, when the power produced exceeds the power input for heating and confinement.

breeder Fusion reactor in which further fuel (tritium) is bred from lithium *blanket* surrounding the fusion chamber. » p. 40.

breeder reactor A fission reactor which produces more fissile material than is consumed in its operation. **Fast reactors** can be so designed. » p. 36.

breeding ratio The number of fissionable atoms produced in fertile material per fissionable atom destroyed in a nuclear reactor. Symbol b_r. ($b_r - 1$) is known as the *breeding gain*.

Breit-Wigner formula A theoretical expression for the dependence of the cross-section σ of a particular nuclear reaction, involving species A and B, on the energy *E* of the bombarding particle and the width Γ of the resonant energy E_R. σ is proportional to

$$\sigma(A, B) = (2l+1)\frac{\lambda^2}{4\pi}\cdot\frac{\Gamma_A\Gamma_B}{(E - E_R)^2 + (\frac{1}{2}\Gamma)^2}.$$

The formula has been used with considerable success for many nuclear reactions particularly those involving neutron bombardment.

bremsstrahlung Electromagnetic radiation emitted when a charged particle changes its velocity. Thus when electrons collide with a target and suffer large decelerations, the X-radiation emitted constitutes the continuous *X-ray spectrum*.

brevi- Prefix from L. *brevis*, short.

Brewster windows Windows attached in certain designs of gas laser to reduce the reflection losses which would arise from the use of external mirrors. Their operation depends on the setting of the windows at the Brewster angle to the incident light.

British Standards Institution A national organization for the preparation and issue of standard specifications.

British Standards specification A specification of efficiency, grade, size etc, drawn up by the British Standards Institution, referenced so that the material required can be briefly described in a bill or schedule of quantities. The definitions are legally acceptable.

British Thermal Unit The amount of heat required to raise the temperature of 1 lb of water by 1 Fahrenheit degree (usually taken as 60°–61°F). Abbrev. *BTU*. Equivalent to 252 calories, 778.2 ft lbf, 1055 J. 10^5 Btu = 1 therm.

broad beam Said of a gamma- or X-ray beam when scattered radiation makes a significant contribution to the radiation intensity or dose rate at a point in the medium traversed by the beam.

Broglie wavelength See **de Broglie wavelength**

Brownian movement Small movements of light suspended bodies such as galvanometer coils, or a colloid in a solution, due to statistical fluctuations in the bombardment by surrounding molecules of the dispersion medium.

BSI Abbrev. for **British Standards Institution**.

BTU Abbrev. for *Board of Trade Unit* = 1 kW h.

bubble chamber A device for making visible the paths of charged particles moving through a liquid. The liquid, often liquid hydrogen, heated above its normal boiling point, becomes superheated if the applied pressure is suddenly released. Bubbles of vapour are formed on the ions produced by the passage of charged particles through the chamber. The tracks of the particles are revealed by the bubbles, which if suitably illuminated, can be photographed. See **cloud chamber**.

buckling A term in reactor diffusion theory giving a measure of curvature of the neutron density distribution. The *geometric buckling* depends only on the shape and dimensions of the assembly while the *material buckling* provides a measure of the multiplying properties of an assembly as a function of the materials and their disposition.

buffer tank A closed tank that cushions the explosive expulsion of liquid from a system connected to it by controlling the gas pressure in the tank.

build-up Increased radiation intensity in an absorber over what would be expected on a simple exponential absorption model. It results from scattering in the surface layers and increases with increasing width of the radiation beam.

bulk test Test for materials having a high attenuation for use as a radiation shield.

bundle See **fuel assembly**.

bundle divertor See **divertor**.

burial site Place for the deposition, usually in suitable containers, of radioisotopes after use, contaminated material or radioactive products of the operation of nuclear reactors. Also *graveyard.* » p. 62.

burnable poison Neutron absorber introduced into a reactor system to reduce initial reactivity but becoming progressively less effective as burnup proceeds. This helps to counteract the fall in reactivity as the fuel is used up. Boron-10, which is transmuted into helium by neutron capture, has been used in the form of borosilicate glass placed in empty control-rod guides.

burnup (1) In nuclear fuel, amount of fissile material burned up as a percentage of total fissile material originally present. (2) Of fuel element performance, the amount of heat released from a given amount of fuel, expressed as megawatt- (or gigawatt-) days per tonne.

burst (1) A defect, often very small, in fuel cladding or sheathing which allows fission products to escape. (2) Unusually large pulse arising in an ionization chamber caused by a cosmic-ray shower.

burst-can (-cartridge) detector An instrument for the early detection of ruptures of the sheaths of fuel elements inside a reactor. Also *leak detector*.

burst cartridge, burst slug Fuel element with a small leak, emitting fission products.

Butex TN for diethylene glycol dibutyl ether, formerly used for separating U and Pu from fission products. Cf. **Purex**.

BWR Abbrev. for **Boiling Water Reactor**.

C

c Symbol used for the velocity of electromagnetic radiation *in vacuo*. Its value, according to the most accurate recent measurements, is

$$2.997\,924\,56 \times 10^8 \text{ m s}^{-1}.$$

C Symbol for (1) **carbon**; (2) coulomb; (3) used following a temperature (e.g. 45°C) to indicate the **Celsius scale** or **Centigrade scale**.

cadmium A white metallic element. Symbol Cd, at.no. 48, r.a.m. 112.40, mp 320.9°C, bp 767°C, rel.d. 8.648. It is a rare element, the Earth's crust containing only 0.16 ppm. It has 15 isotopes:

A	Abundance %	half-life	decay mode
104		58 m	ε
105		56 m	ε
106	1.25		
107		6.5 h	ε
108	0.89		
109		463 d	ε
110	12.5		
111	12.8		
112	24.1		
113	12.2		
114	28.7		
115		53.4 h	β^-
116	7.5		
117		2.4 h	β^-
118		50.3 m	β^-

These isotopes vary considerably in their nuclear capture cross-sections; cadmium-113 has a σ_a of 19 800 barns but on capturing a neutron transforms into cadmium-114 with a much lower cross-section, so that its absorbing properties reduce during the lifetime of the reactor. Natural cadmium has nevertheless been used for control rods in nuclear reactors despite its low mechanical strength by cladding it in stainless steel.

cadmium red line Spectrum line formerly chosen as a reproducible standard of length. Wavelength = 643.8496 nm.

cadmium-sulphide detector Radiation detector equivalent to a solid-state ionization chamber, but with amplifier effect (due to hole trapping).

caesium, cesium A metallic element. Symbol Cs, at.no. 55, r.a.m. 132.905, mp 28.6°C, bp 713°C, rel.d. 1.88. It is a rare element with an abundance of 2.6 ppm in the Earth's crust. It has nine isotopes:

A	Abundance %	half-life	decay mode
130		29.2 m	ε
131		9.69 d	ε
132		6.47 d	ε
133	100		
134		2.06 y	β^-
135		3 My	β^-
136		13.1 d	β^-
137		30.2 y	β^-
138		32.2 m	β^-

The radioactive isotope caesium-137 is used in radiotherapy and as a medical tracer, and the resonance frequency of the natural isotope caesium-133 has been used as a standard for the measurement of time.

caesium unit Source of radioactive caesium (*half-life* 30 years) mounted in a protective capsule.

calandria Closed vessel penetrated by pipes so that liquids in each do not mix. In evaporating plant the tubes carry the heating fluid and in certain types of nuclear reactor, e.g. CANDU reactors, the sealed vessel is called a calandria.

calibration The process of determining the absolute values corresponding to the graduations on an arbitrary or inaccurate scale on an instrument.

calorescence The absorption of radiation of a certain wavelength by a body, and its re-emission as radiation of shorter wavelength. The effect is familiar in the emission of visible rays by a body which has been heated to redness by focusing infrared heat rays on to it.

calorie The unit of quantity of heat in the CGS system. The 15°C calorie is the quantity of heat required to raise the temperature of 1 g of pure water by 1°C at 15°C; this equals 4.1855 J. By agreement, the International Table calorie (cal_{IT}) equals 4.186 J exactly, the thermochemical calorie equals 4.184 J exactly. There are other designations, e.g. gram calorie, mean calorie, and large or kilocalorie (= 1000 cal, used particularly in nutritional work). The calorie has now been largely replaced by the SI unit of the joule (J).

calorimeter The vessel containing the liquid used in calorimetry. The name is also applied to the complete apparatus used in measuring thermal quantities. » p. 43.

calorimetry The measurement of thermal constants, such as specific heat, latent heat, or calorific value. Such measurements usually necessitate the determination of a quantity of heat, by observing the rise of temperature it produces in a known quantity of water or other liquid.

campaign The period often of several months between starting and closing a batch of operations in a nuclear fuel reprocessing plant.

can Cover for reactor fuel rods, usually metallic (aluminium, magnox, stainless steel, zircaloy). Also *cartridge*, *jacket*. See **cladding**.

canal Water-filled trench into which the highly active elements from a reactor core can be discharged. The water acts as a shield against radiation but allows objects to be easily inspected.

candela Fundamental SI unit of luminous intensity. If, in a given direction, a source emits monochromatic radiation of frequency 540×10^{12} Hz, and the radiant intensity in that direction is 1/683 watt per steradian, then the luminous intensity of the source is 1 candela. Symbol **cd**.

candle Unit of luminous intensity. See **candela**.

CANDU Type of thermal nuclear power reactor developed by and widely used in Canada. It uses natural (unenriched) uranium oxide fuel canned in zircaloy and heavy water as moderator and coolant. » p. 29.

canyon US for long narrow space often partly underground with heavy shielding for essential processing of wastes from reactors.

capillarity A phenomenon associated with surface tension, which occurs in fine bore tubes or channels. Examples are the elevation (or depression) of liquids in capillary tubes and the action of blotting paper and wicks. The elevation of liquid in a capillary tube above the general level is given by the formula

$$h = \frac{2T\cos\theta}{\rho g r},$$

where T is the surface tension, θ is the angle of contact of the liquid with the capillary, ρ is the liquid density, g is the acceleration due to gravity, r is the capillary radius.

capture Any process in which an atomic or nuclear system acquires an additional particle. In a nuclear radiative capture process there is an emission of electromagnetic radiation only, e.g. the emission of γ-rays subsequent to the capture of a neutron by a nucleus.

carbon An amorphous or crystalline (graphite and diamond) element. Symbol C, at.no. 6, r.a.m. 12.011, mp above 3500°C, bp 4200°C. Its allotropic modifications are diamond and graphite. In the Earth's crust its abundance is 180 ppm and in seawater 28 ppm. As CO_2 it also makes up 335 ppm by volume of the atmosphere. It has seven isotopes:

A	Abundance %	half-life	decay mode
9		0.13 s	ε
10		19.2 s	ε
11		20.4	ε
12	98.89		
13	1.11		
14		5730 y	β^-
15		2.45 s	β^-

Carbon-14, ^{14}C, is an important source of natural terrestrial radioactivity and is used as a dating method for organic material. See **radiometric dating** p. 222.

carbon dating *Radiocarbon dating*. Atmospheric carbon dioxide contains a constant proportion of radioactive carbon-14, formed by cosmic radiation. Living organisms absorb this isotope in the same proportion. After death it decays with a half-life 5.57×10^3 years. The proportion of carbon-12 to the residual carbon-14 indicates the period elapsed since death. See **radiometric dating** p. 222.

carbon-dioxide laser A laser in which the active gaseous medium is a mixture of carbon dioxide and other gases. It is excited by glow-discharge and operates at a wavelength of 10.6 µm. Carbon-dioxide lasers are capable of pulsed output with peak power up to 100 MW or continuous output up to 60 kW.

carcinogenesis The inducement in a biological cell of the change which will eventually cause it to become a cancer. » p. 95.

Carnot's theorem No heat engine can be more efficient than a reversible engine working between the same temperatures. It follows that the efficiency of a reversible engine is independent of the working substance and depends only on the temperatures between which it is working.

carrier Non-active material mixed with, and chemically identical to, a radioactive compound. Carrier is sometimes added to carrier-free material.

cartridge See **can**.

cascade The arrangement of stages in an enrichment or reprocessing plant in which the products of one stage are fed either *forwards* to the next closely similar or identical stage or *backwards* to a previous stage, eventually resulting in two more or less pure products at each end of the cascade. The classic examples are gaseous or centrifugal enrichment plants. See p. 56. An *ideal* cascade is the arrangement of stages in series and in parallel which gives the highest yield for a given number of units (e.g. centrifuges) and a given separation factor. It has the shape shown in the figure with more units in parallel in the middle of the cascade where the feedstock enters and the number tapering off towards the extremities. A *squared-off* cascade has the number of stages arranged as rectangular blocks approximately like the ideal arrangement.

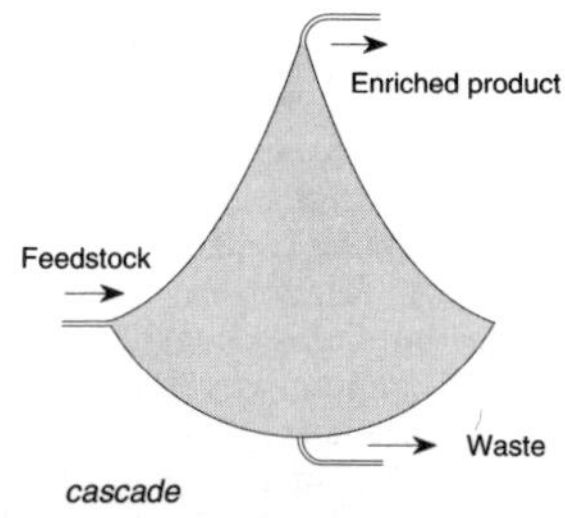

cascade

cascade particle Particle formed by a cosmic ray in a **cascade shower**.

cascade shower Manifestations of cosmic rays in which high-energy mesons, protons and electrons create high-energy photons, which produce further electrons and positrons, thus increasing the number of particles until the energy is dissipated.

cask, casket See **flask**.

cata- See **kata-**.

catcher foil Aluminium sheet used for measuring power levels in nuclear reactor by absorption of fission fragments.

cation Ion in an electrolyte which carries a positive charge and which migrates towards the cathode under the influence of a potential gradient in electrolysis. It is the deposition of the cation in a primary cell which determines the *positive terminal*.

CAT scanner See **computer-aided tomography**.

cave Well-shielded enclosure in which highly radioactive materials can be kept and manipulated safely. Also *hot cell*.

cavity radiation The radiation emerging through a small hole coming from a constant temperature enclosure. Such radiation is identical to **black-body radiation** at the same temperature, no matter what the nature of the inner surface of the enclosure.

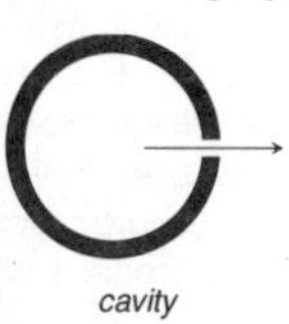
cavity

cc Abbrev. for *cubic centimetre*, the unit of volume in the CGS metric system. Also cm^3.

Cd Symbol for **cadmium**.

cd Symbol for **candela**.

Ce Symbol for **cerium**.

cell (1) Unit of homogeneous reactivity in reactor core. (2) Small storage or work place for 'hot' radioactive preparations.

Celsius scale The SI name for **Centigrade scale**. The original Celsius scale of 1742 was marked zero at the boiling point of water and 100 at the freezing point, the scale being inverted by Strömer in 1750. Temperatures on the International Practical Scale of Temperature are expressed in degrees Celsius. See **Kelvin thermodynamic scale of temperature**.

cent Unit of **reactivity** equal to one-hundredth of a **dollar**.

centi- Prefix from L. *centum*, meaning one-hundredth.

Centigrade scale The most widely used method of graduating a thermometer. The fundamental interval of temperature between the freezing and boiling points of pure water at normal pressure is divided into 100 equal parts, each of which is a *Centigrade degree*, and the freezing point is made the zero of the scale. To convert a temperature on this scale to the Fahrenheit scale, multiply by 1.8 and add 32; for the Kelvin equivalent add 273.15. See **Celsius scale**.

centimetre-gramme-second unit See **CGS unit**.

centipoise One-hundredth of a **poise**, the CGS unit of viscosity. Symbol cP; $1\ \text{cP} = 10^{-3}\ \text{N m}^{-2}\ \text{s}$.

central potential A spherically symmetric potential in which the potential depends only on the *distance* from some centre; the orbital angular momentum is constant for a single particle moving in such a potential. In quantum mechanics, the Schrödinger equation can be solved for such a system; the hydrogen atom is an example in which an electron moves in the central coulomb potential provided by the nuclear charge.

centre of gravity The particles making up a body all experience gravitional forces. The resultant of these forces act through a single point, the *centre of gravity*. In a uniform gravitional field, such as that near the Earth's surface, the centre of gravity coincides with the **centre of mass**.

centre of mass The point in an assembly of mass particles where the entire mass of the assembly may be regarded as being concentrated and where the resultant of the external forces may be regarded as acting for considerations not concerned with the rotation of the assembly. The point is defined vectorially by

$$\bar{\mathbf{r}} = \frac{\sum m\,\mathbf{r}}{\sum m},$$

where m is the mass of a particle at a position $\mathbf{r}$ and the summation extends over the whole assembly.

centrifuge enrichment The enrichment of the uranium-235 isotope by the high-speed centrifugation of uranium hexafluoride gas. It uses the mass difference between U-235 and U-238 to effect the separation. The separation factor per stage is much greater than for gaseous diffusion enrichment. » p. 56.

ceramic A non-organic and non-metallic substance, often oxides or carbides.

ceramic fuel Nuclear fuel with high resistance for temperature, e.g. uranium dioxide, uranium carbide.

Cerenkov counter Radiation counter which operates through the detection of Cerenkov radiation.

Cerenkov radiation Radiation emitted when a charged particle travels through a medium at a speed greater than the velocity of light in the medium. This occurs when the refractive index of the medium is high, i.e. much greater than unity, as for water.

cerium A rare earth metal. Symbol Ce, at.no. 58, r.a.m. 140.12, 6.7, mp 795°C, rel.d. at 20°. It is the commonest of the lanthanide elements, having an abundance in the crust of 66 ppm. It has 13 isotopes:

A	Abundance %	half-life	decay mode
133		5.4 h	ε
134		76 h	ε
135		17.6 h	ε
136	0.19		
137		9.0 h	ε
138	0.25		
139		137.2 d	ε
140	88.5		
141		32.5 d	β^-
142	11.1		
143		33.0 h	β^-
144		284 d	β^-
145		2.98 m	β^-

cermet Ceramic articles bonded with metal. Composite materials combining the hardness and high temperature characteristics of ceramics with the mechanical properties of metal, e.g. cemented carbides and certain reactor fuels.

cesium See **caesium**.

CGS unit, Centimetre-Gram-Second unit, A unit based on the centimetre, the gram and the second as the *fundamental units* of length, mass and time. For most purposes superseded by **SI units** p. 237.

change of state A change from solid to liquid, solid to gas, liquid to gas, or vice versa.

channel Passage through reactor core for coolant, fuel rod or control rod.

channelling effect The reduced absorption of a radiation-absorbing material with voids relative to similar homogeneous material. Expressed numerically by the ratio of the attenuation coefficients. Particularly important in a moderator which can exist in two phases, e.g. water and steam. Also used for escape of radiation through flaws in shielding of reactor etc. Also *streaming effect*. » p. 29.

characteristic radiation Radiation from an atom associated with electronic transitions between energy levels; the frequency of the radiation emitted is characteristic of the particular atom.

characteristic spectrum Ordered arrangement in terms of frequency (or wavelength) of radiation (optical or X-ray) related to the atomic structure of the material giving rise to them.

characteristic X-radiation X-radiation consisting of discrete wavelengths which are characteristic of the emitting element. If arising from the absorption of X- or γ-radiation, may be called *fluorescence X-radiation*. See **characteristic radiation**.

charge Fuel material in nuclear reactor.

charge, discharge machine A device for inserting or removing fuel in a nuclear reactor without allowing escape of radiation and, in some reactors, without shutting the reactor down. Also *fuelling machine*, *refuelling machine*. » p. 34.

charge exchange Exchange of charge between neutral atom and ion in a plasma. After its charge is neutralized, high-energy ions can normally escape from plasma – hence this process reduces plasma temperature.

charge face In a reactor, that face of the biological shield through which fuel is loaded. » p. 34.

charge independent Said of nuclear forces between particles, the magnitude and sign of which do not depend on whether the particles are charged. See **nuclear force**, **short-range forces**.

charge-mass ratio Ratio of electric charge to mass of particle; of great importance in physics of all particles and ions.

charm A property characterizing *quarks* and

therefore hadrons. The charm of leptons and gauge bosons is zero. Charm is conserved in strong and electromagnetic interactions between particles but not in weak interactions.

chemical binding effect A variation in the cross-section of a nucleus for neutron bombardment depending on how the element is combined with others in a chemical compound.

chemical engineering Design, construction and operation of plant and works in which matter undergoes change of state and composition.

chemical shift A shift in position of a spectrum peak due to a small change in chemical environment. Observed in the *Mössbauer effect* and in *nuclear magnetic resonance*.

chemical shim A means of reducing the initial high reactivity at the start-up of a reactor by introducing an element with a high neutron capture cross-section which gradually changes under neutron bombardment to an isotope of lower cross-section. Boron and gadolinium have been used. See **burnable poison**.

Cherenkov radiation See **Cerenkov radiation**.

chopper Device consisting of a rotating mechanical shutter made of a sandwich of aluminium and cadmium sheets which provides bursts of neutrons for a **time-of-flight spectrometer**. See **neutron spectrometer, neutron velocity selector.**

chopping Cutting spent fuel elements into lengths suitable for passing into a dissolver cell for extracting uranium and plutonium. Also *shearing*. » p. 61.

chrom-, chromo-, chromat-, chromato- Prefixes from Gk. *chroma, chromatos*, colour.

chromosome In eukaryotes the deeply staining rod-like structures seen in the nucleus at cell division, made up of a continuous thread of DNA which with its associated proteins (mainly *histones*) forms higher order structures called *nucleosomes* and has special regions, the *centromere* at the point of attachment to the spindle fibres and the *telomere* at the ends. Normally constant in number for any species, there are 22 pairs and 2 sex chromosomes in the human. In micro-organisms the DNA is not associated with histones and does not form visible condensed structures. » p. 88.

CHU, Chu Abbrev. for *Centigrade Heat Unit*. The same as the *pound-calorie*.

Ci Obsolete unit of radioactivity. See **curie**.

cine radiography The rapid sequence of X-ray films taken by a camera attached to an image intensifier.

circular magnetization The magnetization of cylindrical magnetic material in such a way that the lines of force are circumferential.

cladding Thin protective layer, usually metallic, of reactor fuel units to contain fission products and to prevent contact between fuel and coolant. See **can**.

classical Said of theories based on concepts established before relativity and quantum mechanics, i.e. largely in conformity with Newton's mechanics and Maxwell's electromagnetic theory. US *non-quantized*.

classical scattering See **Thomson scattering**.

Clausius' inequality For any thermodynamic system undergoing a cyclic process

$$\oint \frac{dQ}{T} \leq 0,$$

where dQ is an infinitesimal quantity of heat absorbed or liberated by the system at the temperature T kelvin. The equality is appropriate to a reversible process.

clean critical assembly A reactor before start-up, when the fuel, moderator and lattice elements have not been irradiated and their compositions are known.

closed circuit A circuit in which there is zero impedance to the flow of any current, the voltage dropping to zero.

cloud chamber A device for making visible the paths of charged particles moving through a gas. The gas, saturated with a vapour, is suddenly cooled by expansion and the vapour condenses preferentially on the ions produced by the passage of charged particles through the chamber. The tracks of the particles are revealed by the drops of liquid which, if suitably illuminated, can be photographed. See **bubble chamber**.

Clusius column Device used for isotope separation by method of thermal diffusion, consisting of a long vertical cylinder with a hot wire up the axis.

cm^3 Symbol for cubic centimetre.

Co Symbol for **cobalt**.

coastdown The slowing down of a turbine or other moving part after its energy supply is cut off.

coated fuel particle Fuel in small particles and coated in dense carbon and silicon carbide to minimize the release of fission products. » p. 31.

cobalt A hard, grey metallic element. Symbol Co, at.no. 27, r.a.m. 58.93, mp 1495°C, rel.d. 8.90, abundance in the Earth's crust

29 ppm. It has 10 isotopes:

A	Abundance %	half-life	decay mode
54		0.19 s	ε
55		17.5 h	ε
56		78.8 d	ε
57		271 d	ε
58		70.8 d	ε
59	100		
60		5.27 y	β^-
61		1.65 h	β^-
62		1.5 m	β^-
63		27.5 s	β^-

cobalt bomb (1) Theoretical nuclear weapon loaded with cobalt-59. The long-life radioactive cobalt-60, formed during fission, would make the surrounding area uninhabitable. (2) Radioactive source comprising cobalt-60 in lead shield with shutter.

cobalt unit Source of radioactive cobalt (half-life 5.3 years) mounted in a protective capsule; operated like the **caesium unit**.

coefficient of absorption See **absorption coefficient**.

coefficient of apparent expansion The value of the coefficient of expansion of a liquid, which is obtained by means of a dilatometer if the expansion of the dilatometer is neglected. It is equal to the difference between the true coefficient of expansion of the liquid and the coefficient of cubical expansion of the dilatometer.

coefficient of elasticity See **elasticity**.

coefficient of expansion The fractional expansion (i.e. the expansion of the unit length, area, or volume) per degree rise in temperature. Calling the coefficients of linear, superficial and cubical expansion of a substance α, β, and γ respectively, β is approximately twice, and γ three times, α.

coefficient of friction See **friction**.

coefficient of reflection See **reflection factor**.

coefficient of restitution The ratio of the relative velocity of two elastic spheres after direct impact to that before impact. If a sphere is dropped from a height on to a fixed horizontal elastic plane, the coefficient of restitution is equal to the square root of the ratio of the height of rebound to the height from which the sphere was dropped. See **impact**.

coefficient of viscosity The value of the tangential force per unit area which is necessary to maintain unit relative velocity between two parallel planes unit distance apart in a fluid; symbol η. That is, if F is the tangential force on the area A and (dv/dz) is the velocity gradient perpendicular to the direction of flow, then

$$F = \eta A(dv/dz) .$$

For normal ranges of temperature, η for a liquid decreases with increase in temperature and is independent of the pressure. Unit of measurement is the **poise**, 10^{-1} N m^{-2} s in SI units and 1 dyne cm^{-2} s in CGS units.

coffin See **flask**.

coherent units A system of coherent units is one where no constants appear when units are derived from base units.

coincidence detection The simultaneous detection of two annihilation **photons** emitted during positron decay.

cold critical The state of a low-power fission reactor in which a chain reaction is sustained without the production of significant heat.

cold-working The operation of shaping metals at or near atmospheric temperature so as to produce *strain-hardening*, i.e. an increased resistance to deformation.

collective dose equivalent The quantity obtained by multiplying the average effective dose equivalent by the number of persons exposed to a given source of radiation. Expressed as a sievert (*Sv*).

collective model of the nucleus Combines certain features of the *shell model* and *liquid-drop model*. It assumes that the nucleons move independently in a real potential but the potential is not the spherically symmetric potential of the shell model. It is instead a potential capable of undergoing deformation and this represents the collective motion of the nucleons as in the liquid-drop model.

colliding-beam experiment A technique in high-energy physics whereby two beams of particles are made to collide head-on. A greater proportion of the energy of the incident particles is available for the creation of new particles in the collision than in a fixed target experiment of similar total energy.

collimation (1) The process of aligning the various parts of an optical system. (The word is falsely derived from the L. *collineare, -atum*, to bring together in a straight line). (2) The limiting of a beam of radiation to the required dimensions. See **collimator** (1).

collimator (1) A device for obtaining a parallel or near parallel beam of radiation or particles. An optical collimator consists of a source, usually a fine slit, at the principal focus of a converging lens or mirror. Penetrating radiation such as X-rays or γ-rays is collimated by a series of holes or slits in a highly absorbing material such as lead. (2)

The lens of a gamma camera imaging system which absorbs photons travelling in inappropriate directions and originating from parts of the body other than the region under examination.

collision An interaction between particles in which momentum is conserved. If also the kinetic energy of the particles is conserved, the collision is said to be *elastic*, if not then the collision is *inelastic*. With particles in nuclear physics, there is no contact unless there is *capture*. Collision then means a nearness of approach such that there is mutual interaction due to the forces associated with the particles.

collisional excitation The transfer of energy when an atom is raised to an excited state by collision with another particle.

colour Each *flavour* of quarks comes in three 'colours': red, blue and green; and for antiquarks, three anticolours. All hadrons are 'colourless' *either* one red plus one blue plus one green quark *or* one quark and its oppositely coloured antiquark.

colour filter Film of material selectively absorbing certain wavelengths, and hence changing spectral distribution of transmitted radiation.

colour temperature That temperature of a black body which radiates with the same dominant wavelengths as those apparent from a source being described. See **Planck's law**.

column Laboratory or industrial cylindrical vessels of glass or metal, in which solvent extraction or other procedures are carried out. » p. 61.

commissioning The process prior to a contractor handing over equipment to a purchaser in which the system is tested to see if it conforms to specification.

committed dose equivalent The calculated dose equivalent of a given radiation dose integrated over a lifetime, assumed to be 50 years.

common-mode failure Failure of two or more supposedly independent parts of a system (e.g. a reactor) from a common external cause or from interaction between the parts.

complementarity In quantum mechanics the wave and particle models are complementary. There is a correspondence between particles of momentum p and energy E and the associated wavetrain of frequency $\nu = E/h$ and wavelength $\lambda = h/p$, where h is Planck's constant. A measurement proving the wave character of radiation on matter cannot prove the particle character in the same measurement, and conversely.

complete radiator See **black body**.

complex wave A non-sinusoidal waveform which can be resolved into a fundamental with superimposed harmonics. See **Fourier principle**.

compound nucleus In certain nuclear reactions such as that between uranium-235 and a thermal neutron, the bombarding particle forms a highly excited unstable *compound nucleus* with the target nucleus (uranium-236). This compound nucleus decays to complete the reaction.

Compton absorption That part of the absorption of a beam of X-rays or γ-rays associated with Compton scattering processes. In general, it is greatest for medium-energy quanta and in absorbers of low atomic weight. At lower energies **photoelectric absorption** is more important, and at high energies **pair production** predominates.

Compton effect Elastic scattering of photons by electrons, i.e. scattering in which both momentum and energy are conserved. If λ_s and λ_i are respectively the wavelengths associated with scattered and incident photons, the Compton shift is given by

$$\lambda_s - \lambda_i = \lambda_0 (1 - \cos\theta),$$

where θ is the angle between the directions of the incident and scattered photons and λ_0 is the *Compton wavelength* (λ_0 = 0.002 43 nm) of the electron. The effect is only significant for incident X-ray and γ-ray photons.

Compton recoil electron An electron which has been set in motion following an interaction with a photon (**Compton effect**).

Compton scatter A change in the direction of travel of a photon due to the interaction between the photon and the tissue. This is the major cause of loss of resolution in radionuclide imaging.

Compton wavelength Wavelength associated with the mass of any particle, such that

$$\lambda = \frac{h}{mc},$$

where λ = wavelength, h = Planck's constant, m = rest mass and c = velocity of light.

computed tomography See **computer-aided tomography**.

computer (1) A device or set of devices that can store data and a program that operates on the data. A general-purpose computer can be programmed to solve any reasonable problem expressed in logical and arithmetical terms. The first general-purpose automatic computing machine called the *analytical engine* was only partly built. The first fully operational general-purpose computer, electromechanical and using binary digits, was probably the Z3, built in Germany by Konrad Zuse in 1941. Electronic stored program computers are *digital*. (2) Widely used to mean an *electronic digital computer*. An electronic device or set of devices that can store *data* and *programs*, and executes the programs by performing simple *bit* operations at high speed. (3) Also used to refer to a *computer system* which incorporates at least one *digital computer*. (4) An abstract model for a computing machine. (5) A *calculator*.

computer-aided tomography A method of reconstructing cross-sectional images of the body by using rotating X-ray sources and detectors which move around the body and record the X-ray transmissions throughout the 360° rotation. One detector of a bank is shown below. A computer reconstructs the image in the **slice**. Also *CAT scanner*, *computed tomography*.

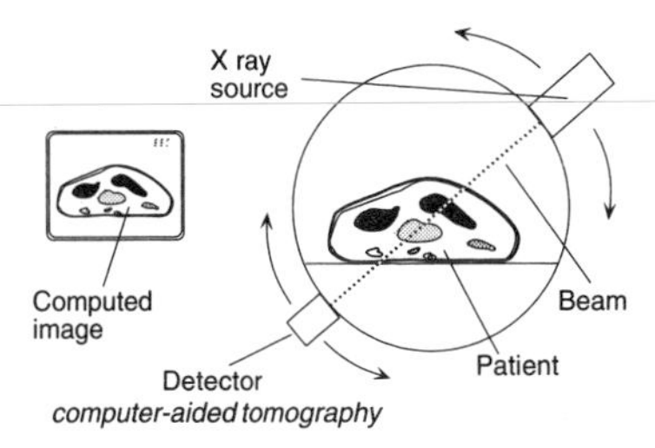

computer-aided tomography

CONCAWE Abbrev. for *CONservation, Clean Air and Water-Western Europe*. An association of oil refining companies operating in Europe that provides a service to international bodies, e.g. the EC, and publishes the results of environmental studies.

conduction of heat The transfer of heat from one portion of a medium to another, without visible motion of the medium, the heat energy being passed from molecule to molecule. See **thermal conductivity**.

conductivity The ratio of the *current density* in a conductor to the electric field causing the current to flow. It is the conductance between opposite faces of a cube of the material of 1 metre edge. Reciprocal of **resistivity**. Unit $\text{ohm}^{-1}\ \text{m}^{-1}$.

conductor A material used for the transference of heat energy by conduction. All materials conduct heat to some extent, but it can be chanelled (like electricity) using materials of different conductivity.

configuration control Control of reactivity of a reactor by alterations to the configuration of the fuel, reflector and moderator assembly.

confinement See **containment**.

conservation laws In the interaction between particles certain quantities remain the same before and after the event. (1) Dynamical quantities such as mass-energy and momentum are conserved. (2) Intrinsic properties of charge and *baryon number* are conserved in nuclear interactions. In addition, in *strong* and *electromagnetic* interactions between elementary particles, the intrinsic properties of strangeness, charm, topness and bottomness are also conserved but not in *weak* interactions.

conservation of energy The total energy of an isolated system is constant. Energy may be converted from one form to another, but is not created or destroyed. If the system has only conservative forces, then the total mechanical energy (kinetic and potential) is constant. See **mass-energy equation**.

conservation of momentum The sum of the momenta in a closed system (i.e. one in which no influences act upon it from outside) is constant and is not affected by processes occurring within the system.

conservative system A system such that in any cycle of operations where the configuration of the system remains unchanged overall, the work done is zero.

constitution Structural distribution of atoms and/or ions composing a regularly coordinated substance. Includes percentage of each constituent and its regularity of occurrence through the material.

contact radiation therapy Radiation from a very short distance, e.g. 20 mm, with voltages around 50 kV.

containment In fusion, the use of shaped magnetic fields or of **inertial confinement** to contain a plasma. Also *confinement*. See **magnetic confinement**. » p. 24.

containment time The time for which a given temperature and pressure can be maintained in a fusion experiment. » p. 24.

contamination meter Particular design of Geiger-Müller circuit for indicating for civil defence purposes the degree of radioactive contamination in an area, esp. for estimating the time for its safe occupation.

continuity The existence of an uninterrupted path for current in a circuit.

continuous spectrum A spectrum which shows continuous non-discrete changes of intensity with wavelengths or particle energy.

contour fringes Interference fringes formed by the reflection of light from the top and bottom surfaces of a thin film or wedge as in *Newton's rings*. The fringes correspond to optical thickness. Also *Fizeau fringes*.

control Maintenance of power level of a reactor at desired setting by adjustments to the reactivity by control rods or other means.

control absorber See **control rod**.

control rod Rod moved in and out of reactor core to vary reactivity. May be neutron-absorbing rod, e.g. boron or cadmium, or less often, a fuel rod. See **regulating rod**, **shim rod**. » p. 19.

control rod worth The change in reactivity of a critical reactor caused by the complete insertion or withdrawal of the control rod.

convection current Current in which the charges are carried by moving masses appreciably heavier than electrons.

convection of heat The transfer of heat in a fluid by the circulation flow due to temperature differences. The regions of higher temperature, being less dense, rise, while the regions of lower temperature move down to take their place. The convection currents so formed help to keep the temperature more uniform than if the fluid was stagnant.

converging-beam therapy A form of **cross-fire technique** in which a number of beams of radiation are used simultaneously to treat a particular region through different entry portals. The figure shows four separate X-ray beams directed at the region. In this and the following entry the techniques are designed to reduce dosage to organs other than the target.

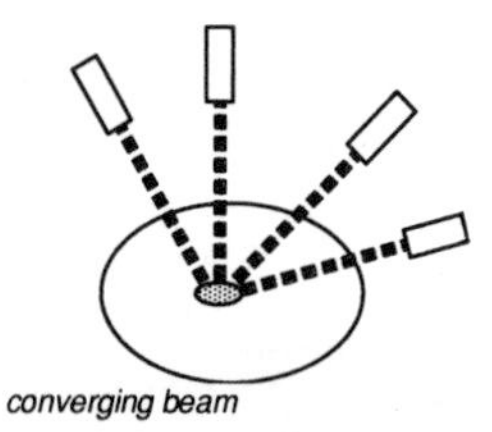

converging beam

converging-field therapy Form of *moving-field therapy* in which the source of radiation moves in a spiral path. The figure shows one X-ray source and beam at the two ends of the spiral path. A mechanical linkage ensures that, whatever the position of the source, the beam is directed at the target.

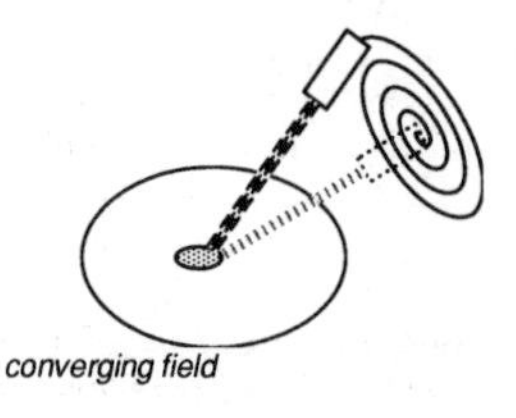

converging field

conversion factor Factor by which a quantity, expressed in one set of units, must be multiplied to convert it to another.

conversion ratio Number of fissionable atoms of e.g. plutonium-239 produced per fissionable atom of uranium-235 destroyed in a reactor. Corresponding conversion gain is defined as R–1. Symbol R.

converter reactor A reactor in which fertile material in the reactor core is converted into a fissile material different from the fuel material. See **breeder reactor**. » p. 36.

coolant, reactor The gas, liquid or liquid metal circulated through a reactor core to carry the heat generated in it by fission and radioactive decay to boilers or heat exchangers. In water-cooled reactors, it is often the moderator. » p. 24.

cooling The decay of activity of irradiated nuclear fuel or highly radioactive waste before it is processed or disposed of.

cooling pond A water filled space in which the initial high radioactivity and thermal output of spent elements can be allowed to dissipate. The water allows both safe inspection and cooling by convection.

coprecipitation The precipitation of a radioisotope with a similar substance, which precipitates with the same reagent, and which is added in order to assist the process.

core (1) That part of a nuclear reactor which contains the fissile material, either dispersed or in fuel pins or cans. (2) In an atom, the nucleus and all complete shells of electrons. In the atoms of the alkali metals, the nucleus, together with all but the outermost of the planetary electrons, may be considered to be a core, around which the valency electron revolves in a manner analogous to the revolution of the single electron in the hydrogen atom around the nucleus. In this manner, the simple Bohr theory may be made to give an approximate representation of the alkali spectra. See **atomic structure** p. 121.

corpuscular radiation A stream of atomic or subatomic particles, which may be charged positively, e.g. α-particles; negatively, e.g. β-particles; or not at all, e.g. neutrons.

corpuscular theory of light The view, held by Newton, that the emission of light consisted of the emission of material particles at very high velocity. Although this theory was discredited by observations of interference and diffraction phenomena, which could only be explained on the wave theory, there has been, to some extent, a return to the corpuscular idea in the conception of the photon.

correspondence principle The predictions of quantum and classical mechanics must correspond in the limit of very large quantum numbers.

cosmogenic Said of an isotope capable of being produced by the interaction of cosmic radiation with the atmosphere or the surface of the Earth. » p. 105.

cosmogonic Relating to the origin of the universe and therefore to radionuclides surviving from that period. Because of confusion with the preceding entry, **primordial** is used here for such radionuclides.

cosmotron Large proton synchrotron using frequency modulation of an electric field; it accelerates protons to energies greater than 3 GeV. See **particle accelerators** p. 206.

coulomb SI unit of electric charge, defined as that charge which is transported when a current of one ampere flows for one second. Symbol C.

coulomb energy Fraction of binding energy arising from simple electrostatic forces between electrons and ions.

coulomb force Electrostatic attraction or repulsion between two charged particles.

coulomb potential Potential calculated from Coulomb's inverse square law and from known values of electric charge. The term is used particularly in nuclear physics to indicate that component of the potential energy of a particle which varies with position as a consequence of an inverse square law of force of *Yukawa potential*.

coulomb scattering Scattering of particles by action of coulomb force.

Coulomb's law Fundamental law which states that the electric force of attraction or repulsion between two point charges is proportional to the product of the charges and inversely proportional to the square of the distance between them. The force also depends on the permittivity of the medium in which the charges are placed. In SI units, if Q_1 and Q_2 are the point charges a distance d apart, the force is

$$F = \frac{1}{4\pi\varepsilon}\frac{Q_1Q_2}{d^2},$$

where ε is the permittivity of the medium. The force is attractive for charges of opposite sign and repulsive for charges of the same sign.

Coulomb's law for magnetism The law which states that the force between two isolated point magnetic poles (theoretical abstractions) would be proportional to the product of their strengths and inversely proportional to the square of their distance apart times the permeability of the medium between them, i.e. F is proportional to

$$\frac{M_1M_2}{\mu d^2},$$

where M_1 and M_2 are the strengths of the two poles, d is their distance apart and μ is the relative permeability of the medium.

count Summation of photons or ionized particles, detected by a counting tube, which passes pulses to counting circuits.

counter, counting tube Device for detecting ionizing radiation by electric discharge resulting from **Townsend avalanche** and operating in proportional or Geiger region. » p. 75.

countercurrent flow In processes involving the transfer of material between two streams, as in the extraction by an organic solvent of uranium and plutonium from the nitric acid solution in a reprocessing plant, the arrangement of flow so that at all stages the more saturated organic phase is in contact with the more saturated aqueous phase, thus ensuring a more rapid distribution between the two phases and greater economy. Achieved by adding the lighter fresh organic phase at the bottom of the column and the saturated aqueous phase at the top. » p. 61. The same effect is obtained for heat flow processes in e.g. steam generators, by introducing the hottest primary water at the opposite end to the feedwater inlet from the turbines.

counter efficiency The ratio of counts recorded by a counter to the number of particles or photons reaching the detector. Counts may be lost due to (a) absorption in the window, (b) passage through the detector without initiating ionization or (c) passage through the detector during dead time that follows the previous count while the instrument recovers.

counter life The total number of counts a nuclear counter can be expected to make without serious deterioration of efficiency.

counter range See **start-up procedure**.

counting chain A system for the detection and recording of ionizing radiation. Consisting essentially of a detector, linear amplifier, pulse height analyser and a device to display or record the counts.

counting tube See **counter**.

count ratemeter A ratemeter which gives a continuous indication of the rate of count of ionizing radiation, e.g. for radiation surveying.

counts The disintegrations that a radionuclide detector records.

coupled control The state of a reactor where the output inherently follows the load,

critical mass

The critical mass is the minimum size of fissionable isotope which will sustain a chain reaction. It will depend on the other substances present, particularly non-fissionable isotopes, and on the geometry of the mass in question. Present-day natural uranium has a very low rate of fission producing about 20 neutrons per second per kilogram. All these neutrons are captured or lost, so that a chain reaction can never occur unless the uranium is enriched for the fissile isotope, uranium-235. The figure shows a plot of the *bare sphere* critical mass against the percentage enrichment of uranium-235.

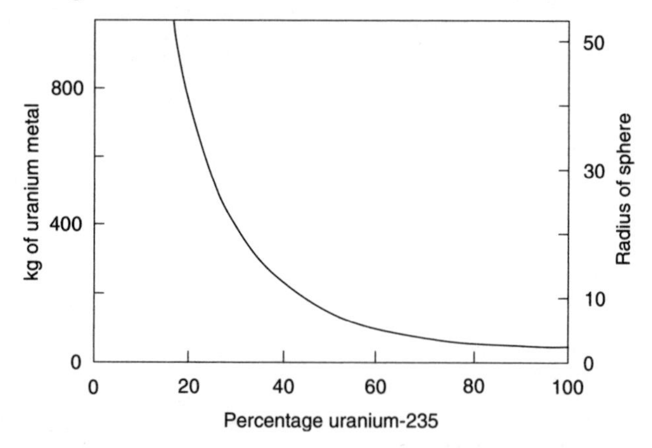

The figure indicates that about 50 kg of 90% enriched uranium in a sphere of about 5 cm radius is enough to sustain the chain reaction. Various factors alter these numbers: any other shape will increase them as will impurities or the inclusion of uranium oxides or carbides. The critical mass will decrease if neutron reflectors, like blocks of tungsten carbide, are placed round the sphere to prevent neutron escape.

The effect of becoming critical by, for example, carefully adding small pieces of tungsten to form a reflecting wall round a subcritical sphere, is dramatic; a fairly inert material is transformed into a substance emitting over a thousand million neutrons per cubic centimetre and giving off an enormous burst of radiation, becoming red hot and extremely dangerous. Serious accidents of this type occurred in 1945 and 1946. Such a critical mass cannot explode because it will distort and melt before that can happen and immediately become subcritical again.

One important consequence is the practical one of maintaining *criticality control* during the handling and processing of fissile material. Large-scale chemical processing equipment must be designed so that fissile compounds cannot accumulate in stagnant 'traps' and the total volume of containers may have to be limited so that unscheduled solvent evaporation cannot result in the build-up of a critical mass of the solute. Similarly, enriched fuel elements must be held in containers which cannot inadvertently allow the fuel to aggregate.

as in pressurized-water reactors. » p. 24.

cP Symbol for **centipoise**.

cps, CPS, c/s Abbrevs. for *cycles per second*, superseded by hertz (Hz).

creep Continuous deformation of metals under steady load. Exhibited by iron, nickel, copper and their alloys at elevated temperature and by zinc, tin, lead and their alloys at room temperature.

crest value See **peak value**.

crit Abbrev. for **critical mass** p. 140.

crith Unit of mass, that of 1 litre of hydrogen at stp, i.e. 89.88 milligrams.

criticality State in a nuclear reactor when multiplication factor for neutron flux reaches unity and external neutron supply is no longer required to maintain power level, i.e. the chain reaction is self-sustaining. See **start-up procedure**. » p. 65.

criticality control The design procedures required to ensure that a **criticality incident** never occurs. See **critical mass** p. 140.

criticality incident The accidental accumulation of fissile material in a plant or during handling which leads to criticality and the sudden emission of dangerous amounts of neutrons, gamma rays and heat.

critical load The acceptable annual dose of sulphur from the emissions of a conventional power station that a particular area can tolerate. Emissions can then be regulated so that *acid deposition* does not exceed this dose.

critical mass See panel on p. 140. Abbrev. *crit*.

critical size The minimum size for a nuclear reactor core of given configuration.

critical temperature The temperature above which a given gas cannot be liquefied.

Crookes dark space A dark region separating the cathode from the luminous 'negative glow' in an electrical discharge in a gas at low pressure. The thickness of the Crookes dark space increases as the pressure is reduced. For air, it is about 1 cm thick at 10 N m^{-2} pressure.

Crookes radiometer A small mica 'paddlewheel' which rotates when placed in daylight in an evacuated glass vessel. Alternate faces of the mica vanes are blackened and the slight rise of temperature of the blackened surfaces caused by the radiation which they absorb warms the air in contact with them and increases the velocity of rebound of the molecules, the sum of whose impulse constitutes the driving pressure.

Crookes tube Original gas-discharge tube, illustrating striated positive column, Faraday dark space, negative glow, Crookes dark space, and cathode glow.

cross bombardment A method of identification of radioactive nuclides through their production by differing reactions.

cross-fire technique The irradiation of a deep-seated region in the body from several directions so as to reduce damage to surrounding tissues for a given dose to that region.

cross-section In atomic or nuclear physics, the probability that a particular interaction will take place between particles. The value of the cross-section for any process will depend on the particles under bombardment and upon the nature and energy of the bombarding particles. Suppose I particles per second are incident on a target area, A, containing N particles and I_r of the incident particles produce a given reaction, then if $I_r < I$,

$$I_r = \frac{I N \sigma}{A},$$

where σ is the cross-section for the reaction; σ can be imagined as a disk of area σ surrounding each target particle. Measured in **barns**. » p. 12.

cryogenic Term applied to low-temperature substances and apparatus.

crypto- Prefix from Gk. *kryptos*, hidden.

crystal Solid substance showing some marked form of geometrical pattern, to which certain physical properties, angle and distance between planes, refractive index etc can be attributed.

crystallography Study of internal arrangements (ionic and molecular) and external morphology of crystal species, and their classification into types.

crystal spectrometer An instrument that uses crystal lattice diffraction to analyse the wavelengths (and energies) of scattered radiation. The radiation can be neutrons, electrons, X-rays or γ-rays and the scattering can be elastic or inelastic.

Cs Symbol for **caesium**.

c/s Abbrev. for *cycles per second*. See **cps**, **hertz**.

CTR Abbrev. for *Controlled Thermonuclear Reactor*, or *Reaction*. See **fusion reactor**.

cumulative distribution In an assembly of particles, the fraction having less than a certain value of a common property, e.g. size or energy. Cf. **fractional distribution**.

cumulative dose Integrated radiation dose resulting from repeated exposure.

curie Unit of radioactivity: 1 curie is defined as 3.700×10^{10} decays per second, roughly equal to the activity of 1 g of radium-226. Symbol Ci. Now replaced by the **becquerel** (*Bq*): 1 Bq = 2.7×10^{-11} Ci.

curie balance A torsion balance for measuring the magnetic properties of non-ferromagnetic materials by the force exerted on the specimen in a non-uniform magnetic field.

current (1) A flow of e.g. water, air etc. (2) Rate of flow of charge in a substance, solid, liquid or gas. Conventionally, it is opposite to the flow of (negative) electrons, this having been fixed before the nature of the electric current had been determined. Practical unit of current is the **ampere**.

current generator Ideally, a current source of infinite impedance such that the current will be unaltered by any further impedance in its circuit. In practice, a generator whose impedance is much higher than that of its load.

curtain Neutron-absorbing shield, usually made of cadmium.

curve fitting The process of finding the best algebraic function to describe a set of experimental measurements. Usually accomplished by using a least-squares process by which the parameters of the function are adjusted to minimize the sum of the squares of the deviations of the observations from the theoretical curve.

cut Proportion of input material to any stage of an isotope separation plant which forms useful product. Also *splitting ratio*. » p. 57.

cycle A series of occurrences in which conditions at the end of the series are the same as they were at the beginning. Usually, but not invariably a cycle of events is recurrent.

cyclo- Prefix from Gk. *kyklos*, circle.

cyclotron Machine in which positively charged particles are accelerated in a spiral path within *dees* in a vacuum between the poles of a magnet, energy being provided by a high-frequency voltage across the dees. When the radius of the path reaches that of the dees, the particles are electrically deflected out of the cyclotron for use in nuclear experiments. See **particle accelerators** p. 206.

cyclotron frequency Any particle of charge q moving perpendicular to a magnetic field of flux density B, moves in a circular path. The number of revolutions around this path per second, the *cyclotron frequency*, is given by

$$f = \frac{B\,q}{2\pi\,m},$$

where m is the mass of the particle. The frequency is independent of the velocity, a result that is used in the *cyclotron* and the *magnetron*.

cyclotron resonance The resonant coupling of electromagnetic power into a system of charged particles undergoing orbital movement in a uniform magnetic field. Used for the quantitative determination of the band parameters in semiconductors.

cyclotron resonance heating Mode of heating of a plasma by resonant absorption of energy based on the waves induced in the plasma at the cyclotron frequency of electrons (abbrev. *ECRH*) or ions (abbrev. *ICRH*).

cylindrical wave A wave where equiphase surfaces form co-axial cylinders.

cytoplasm That part of the cell outside the nucleus but inside the *cell wall* if it exists.

cytosine *6-aminopurine-2-one*. One of the five major bases found in nucleic acids. It pairs with guanine in both DNA and RNA. See **DNA**, **genetic code**. » p. 90.

D

D Symbol for (1) angle of deviation; (2) electric flux density (displacement); (3) diffusion coefficient.

Dalitz pair Electron-positron pair produced by the decay of a free neutral pion (instead of one of the two gamma quanta normally produced).

damped oscillation Oscillation which dies away from an initial maximum asymptotically to zero amplitude, usually with an exponential envelope, e.g. the note from a struck tuning fork. See **damping.**

damping Extent of reduction of amplitude of oscillation in an oscillatory system, due to energy dissipation, e.g. friction and viscosity in a mechanical system, and resistance in an electrical system. With no supply of energy, the oscillation dies away at a rate depending on the **degree of damping**. The effect of damping is to increase slightly the period of vibrations. It also diminishes the sharpness of resonance for frequencies in the neighbourhood of the natural frequency of the vibrator. See **logarithmic decrement**. The figure shows the rapid damping of a sine wave.

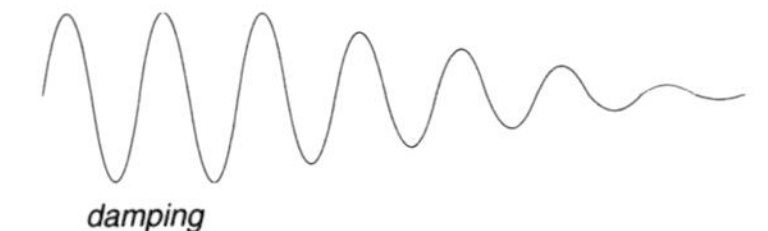

damping

dark current Residual current in a photocell, video camera tube etc, when there is no incident illumination. The current depends on temperature.

dating, radioisotope See **radiometric dating** p. 222.

daughter product A nuclide that originates from the radioactive disintegration of a *parent* nuclide. Also *decay product*.

dead time Time after ionization during which a detector cannot record another particle. Reduced by a *quench* as in Geiger-Müller counters. When the dead time of a detector is variable a fixed electron dead time may be incorporated in subsequent circuits. Also *insensitive time*.

dead-time correction Correction applied to the observed rate in a nuclear counter to allow for the probable number of events occurring during the dead time.

de Broglie wavelength The wavelength associated with a particle by virtue of its motion, i.e. $\lambda = h/p$ where λ is the wavelength, h is Planck's constant and p the particle's relativistic momentum. Only for electrons and other elementary particles can the de Broglie wavelength be large enough to produce observable diffraction effects. See **electron diffraction**.

Debye length Maximum distance at which coulomb fields of charged particles in a plasma may be expected to interact.

Debye-Waller factor The intensities of coherently scattered X-rays or neutrons from a crystal are reduced by the thermal vibrations of the atoms. If it is assumed that the thermal vibrations are isotropic then the intensities are reduced by the Debye-Waller factor,

$$\exp(-2B \sin\theta/\lambda^2)$$

where θ is the angle of scatter, λ the wavelength and B is the temperature factor and generally is in the range 2–3 K.

decade Any ratio of 10:1. Specifically the interval between frequencies of this ratio.

decay The process of spontaneous transformation of a radionuclide.

decay chain The series of radionuclides in which one nucleus disintegrates to form another until a stable, non-radioactive isotope is reached.

decay constant See **disintegration constant.**

decay heat The heat produced by the radioactive decay of fission products in a reactor core. This continues to be produced even after the reactor is shut down. Also *shutdown heating, shutdown power*. » p. 46.

decay law If for a physical phenomenon, the rate of decrease of a quantity is proportional to the quantity at that time, then the decay law is an exponential, i.e.

$$N(t) = N_0 e^{-\lambda t},$$

where $N(t)$ is the quantity at time t, λ is the *decay constant* and N_0 is the value at time $t = 0$. This law holds for the decay of radioactive nuclei and the figure shows a typical decay curve of a nucleus with a half-life of 1.29 minutes.

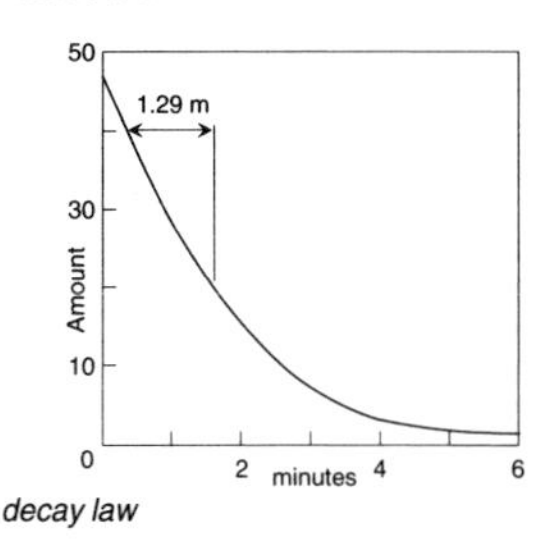

decay law

decay product See **daughter product**.

decay time Time in which the amplitude of an exponentially decaying quantity reduces to e^{-1} (36.8%) of its original value.

deceleration Negative acceleration. The rate of diminution of velocity with time. Measured in metres (or feet) per second squared.

deci- Prefix with physical unit, meaning one-tenth.

decimetre One-tenth of a **metre**.

decommissioning The permanent withdrawal from service of a nuclear facility and the subsequent operations to bring it to a safe and stable condition. » p. 63.

decontamination Removal of radioactivity from area, building, equipment or person to reduce exposure to radiation. More generally, removal or neutralization of bacteriological, chemical or other contamination.

decontamination factor Ratio of initial to final level of contamination for a given process.

decrement Ratio of successive amplitudes in a damped harmonic motion.

deep therapy X-ray therapy of underlying tissues by **hard radiation** (usually produced at more than 180 kVp) passing through superficial layers.

degenerate Said of two or more quantum states which have the same energy. The *degeneracy* is the number of states having a given energy.

degradation Loss of energy of motion caused solely by collision. Deliberate slowing of neutrons in a reactor is **moderation**. In an isolated system, the **entropy** increases.

degree The unit of temperature difference. It is usually defined as a certain fraction of the fundamental interval, which for most thermometers is the difference in temperature between the freezing and boiling points of water. See **Celsius scale**, **centigrade scale**, **Fahrenheit scale**, **international practical temperature scale**, **Kelvin thermodynamic scale of temperature**.

degree of damping The extent of the damping in an oscillatory system, expressed as a fraction or percentage of that which makes the system critically damped.

delayed critical An assembly of fissile material that is critical only after the release of delayed neutrons. Cf. **prompt critical**.

delayed neutron groups Fission products placed into groups with a characteristic decay constant and fractional yield, so that calculations relating to delayed neutrons are made easier. Six groups are commonly used whose parameters vary depending on the fissile nucleus. The table shows, as an example, the six groups associated with uranium-235 fission.

Delayed group	Fractional yield	Decay constant, s^{-1}
1	0.00025	0.0127
2	0.00154	0.0320
3	0.00134	0.128
4	0.00259	0.304
5	0.00089	1.35
6	0.00016	3.63
Total	0.00677,	

It shows that 0.7% of the total neutrons produced are delayed. See **inhour equation**.

delayed neutrons Neutrons arising from fission but not released instantaneously. Fission neutrons are always **prompt neutrons**; those apparently delayed (up to seconds) arise from the breakdown of fission products, not primary fission. Such delays ease the control of reactors. » p. 20.

Delbruck scattering Elastic coherent scattering of gamma rays in the coulomb field of a nucleus. The effect is small and so far has not been conclusively detected.

delta particle Very short-lived hyperon which decays almost instantaneously through the strong interaction.

delta ray Any particle ejected by recoil action from passage of ionizing particles, e.g. in a Wilson cloud chamber.

delta-ray spectrometer See **spectrometer** and **delta ray**.

demountable Said of X-ray tubes or thermionic valves when they can be taken apart for cleaning and filament replacement, and are continuously pumped during operation.

denaturant An isotope added to fissile material to render it unsuitable for military use, e.g. uranium-238 can be added to uranium-233, but denaturing with fertile uranium-238 necessarily produces plutonium in any reactor and the spent fuel must either be reprocessed or stored.

density The mass of unit volume of a substance, expressed in such units as kg m^{-3}, g cm^{-3} or pounds per cubic foot. See **relative density**.

density of gases According to the **gas laws**, the density of a gas is directly proportional to the pressure and the relative molecular mass and inversely proportional to the absolute temperature. At standard temperature and pressure the densities of gases range from 0.0899 g dm^{-3} for hydrogen to 9.96 g dm^{-3} for radon.

density of states The number of electronic states per unit volume having energies in the range from E to $E + dE$; an important concept of the band theory of solids.

depleted uranium Natural uranium from which most of the uranium-235 has been removed. It can be used as a heavy metal in ballistic missiles and as a *fertile* material in a fast-breeder reactor.

depletion Reduction in the proportion of a specific isotope in a given mixture. Cf. **enrichment**.

depression of freezing point A solution freezes at a lower temperature than the pure solvent, the amount of the depression of the freezing point being proportional to the concentration of the solution, provided this is not too great. The depression produced by a 1% solution is called the *specific depression* and is inversely proportional to the molecular weight of the solute. Hence the depression is proportional to the number of moles dissolved in unit weight of the solvent and is independent of the particular solute used.

derived units Units derived from the fundamental units of a system by consideration of the dimensions of the quantity to be measured. See **dimensions, SI units** p. 237.

derived working unit The upper limit to the concentration of a radioactive substance which can be present continuously without contravening the current dose limitations.

desoxyribose nucleic acid, DNA In its double-stranded form the genetic material of most organisms and organelles, although phage and viral genomes may use single-stranded DNA, single-stranded RNA or double-stranded RNA. The two strands of DNA form a double-helix, the strands running in opposite directions, as determined by the sugar-phosphate 'backbone' of the molecule. The four bases project towards each other like the rungs of a ladder, with a purine always pairing with a pyrimidine, according to the *base-pairing rules*, in which thymidine pairs with adenine and cytosine with guanine. In its B molecular form the helix is 2.0 nm in diameter with a pitch of 3.4 nm (10 base pairs). » p. 90.

detector Device in which presence of radiation induces physical change which is observable. See **Geiger Müller counter, germanium radiation detector, nuclear emulsion, proportional counter, scintillation counter** etc. » Chapter 8.

deuterium Isotope of element hydrogen having one neutron and one proton in nucleus. Symbol D, when required to be distinguished from natural hydrogen, which is both ^1H and ^2H because it contains 0.015% of deuterium. Heavy hydrogen is thus twice as heavy as ^1H, but similarly ionized in water. See **deuteron**.

deutero- Prefix from Gk. *deuteros*, (1) denoting second in order, derived from; (2) in chemistry, containing heavy hydrogen (**deuterium**).

deuteron Charged particle, D^+, the nucleus of **deuterium**, a stable but lightly-bound combination of one proton and one neutron. It is mainly used as a bombarding particle accelerated in cyclotrons.

diathermanous Relatively transparent to radiant heat.

dielectric Substance, solid, liquid or gas, which can sustain a steady electric field, and hence an insulator. It can be used for cables, terminals, capacitors etc.

difference of potential See **potential difference**.

differential absorption ratio The ratio of concentration of a radioisotope in different tissues or organs at a given time after the active material has been ingested or injected.

differential cross-section The ratio of the number of scattered particles per unit solid angle in a given direction to the number of incoming particles per unit area.

differential ionization chamber A two compartment system in which the resultant ionization current recorded is the difference between the currents in the two chambers. One version *(compensated ion chamber)* may be used to distinguish between neutrons and gamma radiation. » p. 74.

diffraction The phenomenon, observed when waves are obstructed by obstacles or apertures, of the disturbance spreading beyond the limits of the geometrical shadow of the object. The effect is marked when the size of the object is of the same order as the wavelength of the waves and accounts for the alternately light and dark bands, *diffraction fringes*, seen at the edge of the shadow when a point source of light is used. It is one factor that determines the propagation of radio waves over the curved surface of the Earth and it also accounts for the audibility of sound around corners.

diffraction angle The angle between the direction of an incident beam of light, sound or electrons, and the direction of any resulting diffracted beam.

diffraction grating One of the most useful optical devices for producing spectra. In one of its forms, the diffraction grating consists of a flat glass plate with equidistant parallel straight lines ruled in its surface by a diamond. There may be as many as 1000 per millimetre. If a narrow source of light is viewed through a grating it is seen to be accompanied on each side by one or more spectra. These are produced by diffraction effects from the lines acting as a very large number of equally spaced parallel slits.

diffraction pattern Pattern formed by equal intensity contours as a result of diffraction effects, e.g. in optics or radio transmission.

diffractometer An instrument used in the analysis of the atomic structure of matter by the diffraction of X-rays, neutrons or electrons by crystalline materials. A monochromatic beam of radiation is incident on a crystal mounted on a goniometer. The diffracted beams are detected and their intensities measured by a counting device. The orientation of the crystal and the position of the detector are usually computer-controlled.

diffusion area Term used in reactor diffusion theory. One-sixth of the mean square displacement (i.e. direct distance travelled irrespective of route) between point at which neutron becomes thermal and where it is captured.

diffusion barrier Porous partition for gaseous separation according to molecular weight and hydrodynamic velocities, esp. for the separation of isotopes. Often a fired but unglazed plate. » p. 56.

diffusion constant The ratio of diffusion current density to the gradient of charge carrier concentration in a semiconductor.

diffusion length Square root of **diffusion area**.

diffusion plant Plant used for isotope separation by **gaseous (molecular) diffusion** or **thermal diffusion**. » p. 56.

diffusion theory Simplified neutron migration theory based on **Fick's law of diffusion**. Less accurate than the more detailed **transport theory**.

diffusivity A measure of the rate at which heat is diffused through a material. It is equal to the thermal conductivity divided by the product of the specific heat at constant pressure and the density. Unit $m^2\ s^{-1}$.

digital subtraction angiography A radiological technique where an initial X-ray image is digitized and subtracted from another taken after the injection of *contrast medium*. As only the contrast in the blood vessels is added, high quality images of these blood vessels can be obtained after a small intravenous injection. Avoids catheterization of an artery for *angiography*.

dimensions The dimensions of **derived units** used to express the measurement of a physical quantity are the *powers* to which the fundamental units are involved in the quantity; e.g. velocity has dimensions +1 in length and −1 in time or $[LT^{-1}]$, force has dimensions +1 in mass, +1 in length and −2 in time, $[MLT^{-2}]$.

DIN *Deutsche Institut für Normung*, the German national standards organization; in particular, their system of photographic **speed** rating with logarithmic increments.

dineutron Assumed transient existence of a set of two neutrons in order to explain certain nuclear reactions.

diploid Possessing two sets of chromosomes, one set coming from each parent. Most organisms are diploid. Cf. **haploid**. » p. 88.

dipole Equal and opposite charges separated by a close distance constitute an electric dipole. A bar magnet or a coil carrying a steady current produce a magnetic dipole.

dipole molecule A molecule which has a permanent moment due to the permanent separation of the effective centres of the positive and negative charges.

Dirac's constant *Planck's constant* (*h*) divided by 2π. Usually termed *h-bar*, and written as

$$\hbar.$$

It is the unit in which *electron spin* is constant. See **Planck's law**.

Dirac's theory Using the same postulates as the **Schrödinger equation**, plus the requirement that quantum mechanics conform with the theory of relativity, an electron must have an inherent angular momentum and magnetic moment (1928).

direct cycle The type of nuclear reactor in which the coolant is allowed to boil and pass directly to the turbines, as in the **boiling water reactor**. » p. 28.

direct interaction A mechanism for describing how a nuclear reaction takes place. It assumes that the interaction between bombarding nucleus and target nucleus involves only a few nucleons near the surface of the nuclei. Cf. **compound nucleus**.

direct radiation See **primary radiation**.

disadvantage factor Ratio of average neutron flux in reactor lattice to that within actual fuel element.

discharge (1) Unloading of fuel from a reactor. (2) The abstraction of energy from a cell by allowing current to flow through a load. (3) Reduction of the potential difference at the terminals (plates) of a capacitor to zero. (4) Flow of electric charge through gas or air due to ionization, e.g. lightning, or at reduced pressure, as in fluorescent tubes.

discomposition effect See **Wigner effect**.

dishing Placing depressions at the ends of cylindrical fuel pellets to allow for expansion after irradiation.

disintegration A process in which a nucleus ejects one or more particles, applied, esp. but not only, to spontaneous radioactive decay.

disintegration constant The probability of radioactive decay of a given unstable nucleus per unit time. Statistically, it is the constant λ, expressing the exponential decay $\exp(-\lambda t)$ of activity of a quantity of this isotope with time. It is also the reciprocal of the mean life of an unstable nucleus. Also *decay constant, transformation constant.* See **decay law**.

disintegration energy See **alpha-decay energy**.

dispersion The dependence of wave velocity on the frequency of wave motion; a property of the medium in which the wave is propagated. In the visible region of the electromagnetic spectrum, dispersion manifests itself as the variation of refractive index of a substance with wavelength (or colour) of the light. Dispersion enables a prism to form a spectrum as in the figure. See **anomalous dispersion**.

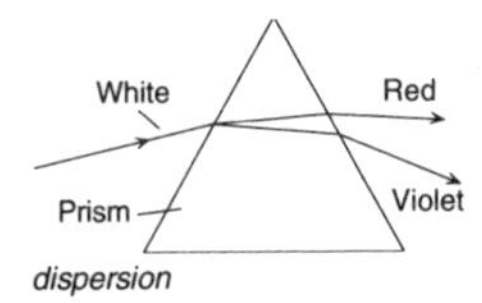

dispersion

dispersion curve A plot of frequency against wavelength for a wave in a dispersive medium.

dispersive medium A medium in which the phase velocity of a wave is a function of frequency.

dispersive power The ratio (ν) of the difference in the refractive indices of a medium for the red and violet to the mean refractive index (n) diminished by unity. This may be written

$$\nu = \frac{n_V - n_R}{n - 1},$$

where n_V and n_R are the refractive indices of violet and red light.

dispersivity quotient The variation of refractive index n with wavelength λ, $dn/d\lambda$.

displacement law Soddy and Fajans formulation that radiation of an α-particle *reduces* the atomic number by 2 and the mass number by 4, and that radiation of a β-particle *increases* the atomic number by 1, but does not change the mass number. It was later found that emission of positron *decreases* the atomic number by 1, but does not change the mass number. Gamma emission and isomeric transition change neither mass nor atomic number. Displacement laws are summarized as follows:

Type of disintegration	Atomic number change	Mass number change
alpha emission	−2	−4
beta electron emission	+1	0
beta positron emission	−1	0
beta electron capture	−1	0
isomeric transition	0	0
gamma emission	0	0

A change in atomic number means displacement in the periodic classification of the chemical elements; a change in mass number determines the radioactive series. » p. 10.

disposal The removal of radioactive waste to a secure place without the intention of recovering it later.

distribution coefficients (1) In the countercurrent columns of a reprocessing plant, the ratio of the total amount of a substance in the organic phase to that in the aqueous phase. Also *spread factor.* (2) Chromaticity coordinates for spectral (monochromatic) radiations of equal power, i.e. for the component radiations forming an equal energy spectrum.

distribution factor A *modifying factor* used in calculating biological radiation doses, which allows for the non-uniform distribution of an internally-absorbed radioisotope.

divergence Initiation of a chain reaction in a reactor, in which slightly more neutrons are released than are absorbed and lost. The rate and extent of the divergence are normally controlled by neutron-absorbing rods of e.g. boron or cadmium. » p. 20.

divergent Term applied to reactor or critical experiment when multiplication constant exceeds unity.

divertor Trap used in thermonuclear device to divert magnetic impurity atoms from entering plasma, and fusion products from striking walls of chamber. Also *bundle divertor.*

DNA Abbrev. for **Desoxyribose Nucleic Acid**.

dollar US unit of reactivity for a reactor defined in terms of the effective neutron multiplication factor if prompt neutrons only are assumed to contribute. The *cent* is a hundredth of a dollar.

Doppler broadening Frequency spread of radiation in single spectral lines, because of the Maxwellian distribution of velocities in the molecular radiators. This also broadens the resonance absorption curve for atoms or molecules excited by incident radiation. It is particularly important in the design of nuclear reactors; the peaks of uranium-238 capture cross-sections, plotted against incident neutron energy, broaden with increasing temperature. The uranium-238 nuclei therefore capture more neutrons, which contributes significantly to reactor stability. » p. 20.

Doppler effect The apparent change of frequency (or wavelength) because of the relative motion of the source of radiation and the observer. For example, the change in frequency of sound heard when a train or aircraft is moving towards or away from an observer. For sound, the observed frequency is related to the true frequency by

$$f_o = \frac{V - V_o(+W)}{V - V_s(+W)} \cdot f_s ,$$

where V_s, V_o are velocities of source and observer, V is the phase velocity of the wave and W is the velocity of the wind. For electromagnetic waves, the *Lorentz transformation* is used to give

$$f_o = \sqrt{\frac{1 - v/c}{1 + v/c}} \cdot f_s ,$$

where V is the *relative* velocity of the source and observer, and C is the velocity of light. In astronomy the measurement of the frequency shift of light received from distant galaxies, the *redshift*, enables their recession velocities to be found.

dose General term for quantity of radiation. See **absorbed dose, dose equivalent, collective dose equivalent, effective dose equivalent, genetically significant dose**.

dose commitment See **committed dose equivalent**.

dose equivalent The quantity obtained by multiplying the absorbed dose by a factor to allow for the different effectiveness of the various ionizing radiations in causing harm to tissues. Unit Sievert (*Sv*).

dose rate The **absorbed dose**, or other dose, received per unit time.

dose reduction factor A factor giving the reduction in radiation sensitivity for a cell or organism which results from some chemical protective agent.

dosimeter Instrument for measuring **dose** and used in radiation surveys, hospitals, laboratories and civil defence. It gives a measure of the radiation field and dosage experienced. Also *dosemeter*. » p. 78.

double beta decay Energetically possible process involving the emission of two β-particles simultaneously. Not to be confused with **dual beta decay**.

doubling time (1) Time required for the neutron flux in a reactor to double. (2) In a breeder reactor, time required for the amount of fissile material to double.

drift chamber A particle detector used in high-energy physics experiments whereby the tracks of charged particles in interactions may be recorded.

dryout margin In a reactor in which steam is directly boiled, the ratio between the heat generated in the fuel at full power to that which would cause steam to boil and cover all the fuel pins and thus markedly reduce the cooling available.

dual beta decay **Branching** where a radioactive nuclide may decay by either electron or positron emission. Not to be confused with **double beta decay**.

Duane and Hunt's law The maximum photon energy in an X-ray spectrum is equal to the kinetic energy of the electrons producing the X-rays, so that the maximum frequency, as deduced from quantum mechanics, is eV/h, where V = applied voltage, e= electronic charge and h = Planck's constant.

duct waveguide Layer in the atmosphere which, because of its refractive properties, keeps electromagnetic radiated (or acoustic) energy within its confines. It is *surface-* or *ground-based* when the surface of the Earth is one confining plane.

ductility Ability of metals and alloys to retain strength and freedom from cracks when shape is altered. See **work-hardening**.

dump condenser Condenser or water-filled tank into which steam destined for the turbines in a nuclear power station can be diverted if the electrical load is suddenly removed.

dust monitor Instrument which separates airborne dust and tests for radioactive contamination.

dye laser A laser using an organic dye and excited by a separate laser. It can be tuned over a significant fraction of the visible spectrum by using a reflection grating as one of the cavity mirrors.

dynamical theory of X-ray and electron diffraction A theoretical approach that takes into account the dynamical equilibrium between the incident and diffracted beams in a crystal, e.g. the effect of interference between the incident beam and multiply diffracted beams.

dyne The unit of force in the CGS system of units. A force of one dyne acting on a mass of 1 g, imparts to it an acceleration of 1 cm s^{-2}. Approximately 981 dynes are equal to 1 g weight. 10^5 dynes = 1 newton.

e Symbol for the electron (e^-) or positron (e^+).

e Symbol for the elementary charge; 1.6022×10^{-19}coulomb.

ε Symbol for (1) **emissivity**; (2) linear strain; (3) permittivity.

η Symbol for **coefficient of viscosity**.

E Symbol for (1) electromotive force; (2) **electric field strength**; (3) **energy**; (4) **illumination**; (5) irradiance.

ebullator A heated surface used to impart heat to a fluid in contact.

ebullition See **boiling**.

ecology The scientific study of (1) the interrelations between living organisms and their environment, including both the physical and biotic factors, and emphasizing both interspecific and intraspecific relations; (2) the distribution and abundance of living organisms (i.e. exactly where they occur and precisely how many there are), and any regular or irregular variations in distribution and abundance, followed by explanation of these phenomena in terms of the physical and biotic factors of the environment.

ECRH *Electron Cyclotron Resonance Heating*. See **cyclotron resonance heating**.

eddy An interruption in the steady flow of a fluid, caused by an obstacle situated in the line of flow; the *vortex* so formed.

effective dose equivalent The quantity obtained by multiplying the dose equivalents to various tissues and organs by the risk weighting factor appropriate to each organ and summing the products. Unit sievert (*Sv*).

effective energy The quantum energy (or wavelength) of a monochromatic beam of X-rays or γ-rays with the same penetrating power as a given heterogeneous beam. Its value depends upon the nature of the absorbing medium. Also *effective wavelength*.

effective half-life The time required for the activity of a radioactive nuclide in the body to fall to half its original value as a result of both biological elimination and radioactive decay. Its value is given by

$$\frac{\tau(b\tfrac{1}{2}) \times \tau(r\tfrac{1}{2})}{\tau(b\tfrac{1}{2}) + \tau(r\tfrac{1}{2})},$$

where $\tau(b\frac{1}{2})$ and $\tau(r\frac{1}{2})$ are the biological and radioactive half-lives respectively.

effective value The value of a simple parameter which has the same effect as a more complex one, e.g. r.m.s. value of ac = dc value for many purposes.

effective wavelength See **effective energy**.

efficiency A non-dimensional measure of the performance of a piece of apparatus, e.g. an engine, obtained from the ratio of the output of a quantity, e.g. power, energy, to its input, often expressed as a percentage. The power efficiency of an *IC engine* is the ratio of the shaft- or brake-horsepower to the rate of intake of fuel, expressed in units of energy content per unit time. It must always be less than the 100% which would imply perpetual motion. Not to be confused with efficacy, which takes account only of the output of the apparatus, and is not given an exact quantitative definition.

effluent The liquid or gaseous waste from a chemical or other plant.

effluent monitor Instrument for measuring level of radioactivity in fluid effluent.

effusion The flow of gases through larger holes than those to which diffusion is strictly applicable. See **Graham's law**. The rate of flow is approximately proportional to the square-root of the pressure difference.

EIA Abbrev. for **Environmental Impact Assessment**.

eigenfunction A solution of an equation which satisfies a set of boundary conditions for that equation. In quantum mechanics, a possible solution for the Schrödinger equation for a given system.

eigentones The natural frequencies of vibration of a system.

eigenvalues Possible values for a parameter of an equation for which the solutions will be compatible with the boundary conditions. In quantum mechanics, the energy eigenvalues for the **Schrödinger equation** are possible *energy levels* for the system. See **energy band**.

Einstein-de Haas effect When a magnetic field is applied to a body the precessional motion of the electrons produce a mechanical moment that is transferred to the body as a whole.

Einstein energy See **mass-energy equation**.

Einstein theory of specific heat of solids Based on the assumption that the thermal vibrations of the atoms in a solid can be represented by harmonic oscillators of one frequency, whose energy is quantized.

EIS Abbrev. for **Environmental Impact Statement**.

elastic collision A collision between two bodies in which, in addition to the total momentum being conserved, the total kinetic energy of the bodies is conserved.

elasticity The tendency of a body to return to its original size or shape, after having been stretched, compressed, or deformed. The ratio of the stress called into play in the body by the action of the deforming forces to the strain or change in dimensions or shape is called the *coefficient (modulus) of elasticity*. See the following definitions and **Hooke's law**.

elasticity of gases If the volume V of a gas is changed by δV when the pressure is changed by δp, the modulus of elasticity is given by

$$-V\frac{\delta p}{\delta V}.$$

This may be shown to be numerically equal to the pressure p for isothermal changes, and equal to γp for adiabatic changes, γ being the ratio of the specific heats of the gas.

elastic limit The limiting value of the deforming force beyond which a body does not return to its original shape or dimensions when the force is removed.

elastic scattering See **scattering**.

electric Said of any phenomena which depend essentially on a peculiarity of electric charges.

electrical Descriptive of means related to, pertaining to, or associated with electricity, but not inherently functional.

electrical conductivity Ratio of current density to applied electric field. Expressed in siemens per metre (S m^{-1}) or ohm^{-1} m^{-1} in SI units. Conductivity of metals at high temperatures varies as T^{-1}, where T is absolute temperature. At very low temperatures, variation is complicated but it increases rapidly (at one stage proportional to T^{-5}), until it is finally limited by material defects of structure.

electrical engineering That branch of engineering chiefly concerned with the design and construction of all electrical machinery and devices, power transmission etc.

electrical resistivity Reciprocal of **electrical conductivity**. Unit is ohm metre.

electric circuit Series of conductors forming a partial, branched or complete path around which either a direct or alternating current can flow.

electric dipole moment Product of the magnitude of the electric charges and the distance between it and its opposite charge in an electric dipole.

electric discharge See **field discharge**.

electric doublet System with a definite electric moment, mathematically equivalent to two equal charges of opposite sign at a very small distance apart.

electric field Region in which forces are exerted on any electric charge present.

electric field strength The strength of an electric field is measured by the force exerted on a unit charge at a given point. Expressed in volts per metre (V m^{-1}). Symbol E.

electric flux Surface integral of the electric field intensity normal to the surface. The electric flux is conceived as emanating from a positive charge and ending on a negative charge without loss. Symbol Ψ.

electrodisintegration Disintegration of nucleus under electron bombardment.

electrofluorescence See **electroluminescence**.

electrokinetics Science of electric charges in motion, without reference to the accompanying magnetic field.

electroluminescence Luminescence produced by the application of an electric field to a dielectric phosphor. Also *electrofluorescence*. See **Gudden-Pohl effect**.

electromagnetic field theory Theory based on *Maxwell's field equations* and concerned with electromagnetic interactions. It predicts the existence of a wave comprising interdependent transverse waves of electric and magnetic fields, **electromagnetic waves**. The scope of the theory is enormous including as it does the operating principles of such electromagnetic devices as motors, television and radar. » p. 82.

electromagnetic interaction An interaction between charged elementary particles and mediated by **photons**; completed in about 10^{-18}s. Intermediate in strength between *strong* and *weak* interactions.

electromagnetic radiation The emission and propagation of electromagnetic energy from a source including long (radio) waves, heat rays, light, X-rays and γ-rays.

electromagnetic separation Isotope separation by *electromagnetic focusing*, as in a mass spectrometer.

electromagnetic spectrum See **electromagnetic wave**.

electromagnetic theory See **electromagnetic field theory**.

electromagnetic units Any system of units based on assigning an arbitrary value to μ_0, the permeability of free space.

$$\mu_0 = 4\pi \times 10^{-7}\ \mathrm{H\ m^{-1}}$$

in the SI system; μ_0 is unity in the CGS electromagnetic system.

electromagnetic wave A wave comprising two interdependent mutually perpendicular transverse waves of electric and magnetic fields. The velocity of propagation in free space for all such waves is that of the velocity of light, 2.99792458×10^8 m s^{-1}. The electromagnetic spectrum ranges from wavelengths of 10^{-15} m to 10^3 m, i.e. from γ-rays through X-rays, ultraviolet, visible light, infrared, microwave, short-, medium- and long-wave radio waves. Electromagnetic waves undergo reflection and refraction, exhibit interference and diffraction effects, and can be polarized. The waves can be channelled, e.g. by waveguides for microwaves or fibre optics for light. » p. 80.

electromotive force Difference of potential produced by sources of electrical energy which can be used to drive currents through external circuits. Abbrev. *emf*. Unit *volt*.

electron A fundamental particle with negative electric charge of 1.602×10^{-19} coulombs and mass 9.109×10^{-31} kg. Electrons are a basic constituent of the atom; they are distributed around the nucleus in *shells* and the electronic structure is responsible for the chemical properties of the atom. Electrons also exist independently and are responsible for many electric effects in materials. Due to their small mass, the wave properties and relativistic effects of electrons are marked. The *positron*, the antiparticle of the electron, is an equivalent particle but with a positive charge. Either electrons or positrons may be emitted in β-decay. Electrons, muons and neutrinos form a group of fundamental particles called **leptons**.

electron affinity See **work function**.

electron attachment Formation of negative ion by attachment of free electron to neutral atom or molecule.

electron binding energy See **ionization potential**.

electron capture That of shell electron (K or L) by the nucleus of its own atom, decreasing the atomic number of the atom without change of mass. The capture is accompanied by the emission of a *neutrino*.

electron density The number of electrons per gram of a material. Approx. 3×10^{23} for most light elements. In an ionized gas the equivalent electron density is the product of the ionic density and the ratio of the mass of an electron to that of a gas ion.

electron dispersion curve A curve showing the electron energy as a function of the wave vector under the influence of the periodic potential of a crystal lattice. Experiments and calculations which determine such curves give important information about the energy gaps, the electron velocities and the density of states.

electronegative Carrying a negative charge of electricity. Tending to form negative ions, i.e. having a relatively positive electrode potential.

electron-electron scattering A possible process that contributes to the electrical resistivity of metals. Important at low temperatures in transition metals.

electronic Pertaining to devices or systems which depend on the flow of electrons; the term covers most branches of electrical science other than electric power generation and distribution. Telecommunications, radar and computers all use electronic components and techniques. Electronic engineering is a field which encompasses the application of electronic devices, as opposed to physical electronics which is the study of electronic phenomena in vacuum, in gases, or in solids.

electronic charge The unit in which all nuclear charges are expressed. It is equal to 1.602×10^{-19} coulombs.

electron mass A result of relativity theory, that mass can be ascribed to kinetic energy, is that the effective mass (m) of the electron should vary with its velocity according to the experimentally confirmed expression:

$$m = \frac{m_0}{\sqrt{1 - (v/c)^2}},$$

where m_0 is the mass for small velocities, c is the velocity of light, and v that of the electron.

electron octet The (up to) eight valency electrons in an outer shell of an atom or molecule. Characterized by great stability, in so far as the complete shell round an atom makes it chemically inert, and round a molecule (by sharing) makes a stable chemical compound.

electron optics Control of free electrons by curved electric and magnetic fields, leading to focusing and formation of images.

electron paramagnetic resonance See **electron spin resonance**.

electron-phonon scattering An important process that contributes to the cause of electrical resistivity; the electrons are scattered by the thermal vibrations of the crystal lattice.

electron radius The classical theoretical value is 2.82×10^{-15} m, but experimentally the effective value varies greatly with the interaction concerned.

electron shell See **atomic structure** p. 121.

electron spectroscopy See **photoelectron spectroscopy**.

electron spin resonance A branch of microwave spectroscopy in which there is resonant absorption of radiation by a paramagnetic substance, possessing unpaired electrons, when the energy levels are split by the application of a strong magnetic field. The difference in energy levels is modified by the environment of the atoms. Information on impurity centres in crystals, the nature of the chemical bond and the effect of radiation damage can be found. Abbrev. *ESR*. Also *electron paramagnetic resonance*.

electron-volt General unit of energy of moving particles, equal to the kinetic energy acquired by an electron losing one volt of potential, equal to 1.602×10^{-19} J. Used in nuclear physics and radiology with the units expressed in thousands (keV) or millions (MeV) of electron-volts. Abbrev. *eV*.

electropositive Carrying a positive charge of electricity. Tending to form positive ions, i.e. having a relatively negative electrode potential.

electroradiescence Emission of ultraviolet or infrared radiation from dielectric phosphors on the application of an electric field.

electrostatics Section of science of electricity which deals with the phenomena of electric charges substantially at rest.

electrothermoluminescence Changes in electroluminescent radiation resulting from changes of dielectric temperature. (Some dielectrics show a series of maxima and minima when heated.) The complementary arrangement of observing changes in thermoluminescent radiation when an electric field is applied is termed **thermoelectroluminescence**.

electroweak theory **Weinberg and Salam's theory** unifying electromagnetic and weak interactions between particles. See **lepton-quark symmetry**.

elementary particle Particle believed to be incapable of subdivision; the term **fundamental particle** is now more generally used.

e/m The ratio of the electric charge to mass of an elementary particle. For slow moving electrons $e/m = 1.759 \times 10^{11}$ C kg^{-1}. This value decreases with increasing velocity because of the relativistic increase in mass. Also *specific charge*.

emanations Obsolescent term for the heavy isotopic inert gases resulting from decay of natural radioactive elements. They are radioisotopes 222, 220, 219 of element 86 or **radon**; these gases are short-lived and decay to other radioactive elements.

emanometer Meter for measuring **radon**.

emergency diesel supply Provision of essential electrical power to a nuclear power plant if power from the grid is lost.

emergency shutdown Rapid shutdown of a reactor to forestall or remedy a dangerous situation.

emergency shutdown system System of shutting down reactor if other methods fail, e.g. injection of boron spheres which absorb neutrons strongly and quickly make reactor subcritical. See **secondary shutdown system**.

emergent ray point See **exit portal**.

emf Abbrev. for **ElectroMotive Force**.

emission Release of electrons from parent atoms on absorption of energy in excess of normal average. This can arise from (1) *thermal* (thermionic) agitation, as in valves, Coolidge X-ray tubes, cathode-ray tubes; (2) *secondary* emission of electrons, which are ejected by impact of higher energy primary electrons; (3) *photoelectric* release on absorption of quanta above a certain energy level; (4) *field* emission by actual stripping from parent atoms by high electric field.

emission spectrum Wavelength distribution of electromagnetic radiation emitted by a self-luminous source.

emission tax Fiscal tax, proportional to the quantity and toxicity of the pollutant emitted, levied to encourage environmentally sounder industrial practices.

emission tomography See **tomography, emission**.

emissive power The total emissive power of a surface is the energy radiated at all wavelengths per unit area per unit time. It depends on the nature of the surface and on its temperature. See **emissivity, emission**.

emissivity The ratio of emissive power of a surface at a given temperature to that of a black body at the same temperature and with the same surroundings. Values range from 1.0 for lampblack down to 0.02 for polished silver. See **Stefan-Boltzmann law**.

empirical formula A formula founded on experience or experimental data only, not deduced in form from purely theoretical considerations.

empiricism The regular scientific procedure whereby scientific laws are induced by inductive reasoning from relevant observations. Critical phenomena are deduced from such laws for experimental observation, as a check on the assumptions or hypotheses inherent in the theory correlating such laws. Scientific procedure, described by empiricism, is not complete without the experimental checking of deductions from theory.

empty band See **energy band**.

emulsion technique Study of nuclear particles, by means of tracks formed in photographic emulsion.

encephalogram X-ray plate produced in **encephalography**.

encephalography Radiography of the

brain after its cavities and spaces have been filled with air, or dye, previously injected into the space round the spinal cord.

endo-ergic process Nuclear process in which energy is consumed. Also *endo-energetic*. The equivalent thermodynamic term is **endothermic**.

endothermic Accompanied by the absorption of heat; ΔH positive.

end product The stable nuclide forming the final member of a radioactive decay series.

end window counter Geiger-Müller counter designed so that radiation of low penetrating power can enter one end. (This is usually covered with a thin mica sheet.)

energy The capacity of a body for doing work. Mechanical energy may be of two kinds: *potential energy*, by virtue of the position of the body, and *kinetic energy*, by virtue of its motion. Energy can take a wide variety of forms. Both *mechanical* and *electrical* energy can be converted into *heat* which is itself a form of energy. Electrical energy can be stored in a capacitor to be recovered at discharge of the capacitor. *Elastic potential energy* is stored when a body is deformed or changes its configuration, e.g. in a compressed spring. All forms of wave motion have energy; in electromagnetic waves it is stored in the electric and magnetic fields. In any closed system, the total energy is constant – the *conservation of energy. Units of energy*: SI unit is the **joule** (symbol J) and is the work done by a force of 1 newton moving through a distance of 1 metre in the direction of the force. The CGS unit, the *erg* is equal to 10^{-7} joules and is the work done by a force of 1 dyne moving through 1 cm in the direction of the force. See **electron-volt, kinetic energy, mechanical equivalent of heat, potential energy**.

energy balance (1) The detailed study of the energy flow in an industrial process. (2) The collection and publication of detailed information on the energy production and demand of a country, as compiled by e.g. the International Energy Agency.

energy band In a solid the energy levels of the individual atoms combine to form bands of allowed energies separated by forbidden regions. The individual electrons are considered to belong to the crystal as a whole rather than to a particular atom. The energy bands are the consequence of the motion of the electron in the periodic potential of the crystal lattice. A solid for which a number of the bands are completely filled and the others empty, is an insulator provided the energy gaps are large. If one band is incompletely filled or the bands overlap then metallic conduction is possible. For semi-conductors, there is a small energy gap between the filled and empty band and *intrinsic* conduction occurs when some electrons acquire sufficient energy to surmount the gap.

energy confinement time In fusion, the ratio of the total energy of a confined plasma to the rate of energy loss from it. » p. 40.

energy fluence The radiation intensity integrated over a short pulse.

energy gap Range of forbidden energy levels between two permitted bands. See **energy band**.

energy-mass equation See **mass-energy equation**.

energy policy A statement of a country's energy production and demand and future intentions.

engineered safeguards The special features built into a nuclear reactor to cope with accidents and malfunctions. » p. 46.

engineered storage Facilities specially designed to store highly radioactive waste, e.g. the dry storage bunkers for spent fuel rods, which must be cooled and kept safe until their contents can be disposed of or processed.

enriched uranium Uranium in which the proportion of the fissile isotope uranium-235 has been increased above its natural abundance.

enrichment (1) Raising the proportion of uranium-235 fissile nuclei above that for natural uranium in reactor fuel. (2) Raising the proportion of the desired isotope in an element above that present initially by isotope separation. » p. 56.

enrichment factor (1) In the UK, the abundance ratio of a product divided by that of the raw material. The *enrichment* is the enrichment factor less unity. (2) US **separation factor**. » p. 57.

enthalpy Thermodynamic property of a working substance defined as $H = U + PV$ where U is the internal energy, P the pressure and V the volume of a system. Associated with the study of heat of reaction, heat capacity and flow processes. SI unit is the joule.

entropy In thermal processes, a quantity which measures the extent to which the energy of a system is available for conversion to work. If a system undergoing an infinitesimal reversible change takes in a quantity of heat dQ at absolute temperature T, its entropy is increased by $dS = dQ/T$. The area under the absolute temperature-entropy graph for a reversible process represents the heat transferred in the process. For an adiabatic process, there is no heat transfer and the temperature-entropy graph is a straight line, the entropy remaining constant during the process. When a thermodynamic system is considered on the microscopic scale, equilibrium is associated with the distribution of molecules that has the greatest probability of occurring, i.e. the state with the greatest degree of disorder. *Statistical mechanics* interprets the increase in entropy in a closed system to a maximum at equilibrium as the consequence of the trend from a less probable to a more probable state. Any process in which no change in entropy occurs is said to be *isentropic*.

entropy of fusion A measure of the increased randomness that accompanies the transition from solid to liquid or liquid to gas; equal to the latent heat divided by the absolute temperature. Also *vaporization*.

entry portal Area through which a beam of radiation enters the body.

environmental impact assessment The European Economic Community's equivalent of the US **environmental impact statement**. Abbrev. *EIA*.

environmental impact statement In the US, a detailed analysis, required of any Agency undertaking a major federal project, of its effects on the environment. Abbrev. *EIS*.

environmental pathway Route by which a radionuclide in the environment can reach Man, e.g. from radioactivity in rain to grass, to cows, to milk, to people.

enzyme A protein with catalytic activity, which is restricted to a limited set of reactions, defining the specificity of the enzyme. They are designated by the suffix *-ase*, frequently attached to the type of reaction catalysed. Thus enzymes catalysing hydrolytic reactions are termed *hydrolases*. Virtually all metabolic reactions in living cells are dependent on and controlled by enzymes.

epithermal neutrons Neutrons having energies just above thermal, comparable to chemical bond energies. See **neutron**.

epithermal reactor See **intermediate reactor**.

equal energy source An electromagnetic or acoustic source whose radiated energy is distributed equally over its whole frequency spectrum.

equi- Prefix from L. *aequus*, equal.

equilibrium The state of a body at rest or moving with constant velocity. A body on which forces are acting can be in equilibrium only if the resultant force is zero and the resultant torque is zero.

erbium laser Laser using erbium in YAG (*Yttrium-Aluminium-Garnet*) glass. It has the advantage of operating between 1.53 and 1.64 μm, a range in which there is a high attenuation in water. This feature is of particular importance, medically, in laser applications to the eye, since a great deal of energy absorption will now occur in the cornea and aqueous humour before reaching the delicate retina.

erg Unit of work or energy in CGS system. One erg of work is done when a force of 1 dyne moves its point of application 1 cm in the direction of the force. See **energy**.

eta (η) In reactor theory, one of the factors in the **four factor formula** which represents average number of fission neutrons produced per neutron absorbed in the fuel. » p. 22.

ether, aether A hypothetical, non-material entity supposed to fill all space whether 'empty' or occupied by matter. The theory that electromagnetic waves need such a medium for propagation is no longer tenable.

eu- Prefix from Gk. *eu*, well, good.

Euratom Abbrev. for *European Atomic Energy Community*.

eV Abbrev. for **electron-volt**.

evaporation The conversion of a liquid into vapour, at temperatures below the boiling point. The rate of evaporation increases with rise of temperature, since it depends on the saturated vapour pressure of the liquid, which rises until it is equal to the atmospheric pressure at the boiling point. Evaporation is used to concentrate a solution.

even-even nuclei Nuclei for which the numbers of protons and neutrons are both even and normally stable.

even-odd nuclei Nuclei with an even number of protons and an odd number of neutrons.

event tree A method of investigating a real or simulated accident which, starting from an initial event, plots the alternative ways in which the accident can proceed depending on whether or not particular safety features function.

eversafe Used to describe a nuclear processing plant which is designed so that a critical amount of fissile material can never accumulate. » p. 60.

ex-, e- Prefixes from L. *ex*, *e*, out of.

exchange (1) The interchange of one particle between two others (e.g. a pion between two nucleons), leading to establishment of exchange forces. (2) Possible interchange of state between two indistinguishable particles, involving no change in the wave function of the system.

exchange force A force acting between particles due to the exchange of some property. In quantum mechanics, such forces can arise when two particles interact. In the ion of the hydrogen molecule, the forces responsible for the binding can be regarded as the continual exchange of the single electron between the two protons. Exchange forces are an important concept in the understanding of nuclear forces. The strong force between nucleons is the exchange of a *pion* (π-meson) between the two interacting nucleons. See **particle exchange**.

excision repair Enzymatic DNA repair process in which a mismatching DNA sequence is removed and the gap filled by synthesis of a new sequence complementary to the remaining strand. » p. 93.

excitation Addition of energy to a system, such as an atom or nucleus, raising it above the **ground state**.

excited atom An atom with more energy than in the normal or ground state. The excess may be associated with the nucleus or an orbital electron.

excited ion An ion resulting from the loss of a valence electron, and the transition of another valence electron to a higher energy level.

excited nucleus A nucleus raised to an excited state with an excess of energy over its ground state. Nuclear reactions frequently leave the product nucleus in an excited state. It returns to its ground state with the emission of γ-rays.

exclusion principle See **Pauli exclusion principle, atomic structure** p. 121.

excursion The rapid increase of reactor power above the predetermined operation level, either deliberately caused for experimental reasons or accidental.

exit portal The area through which a beam of radiation leaves the body. The centre of the exit portal is sometimes called the *emergent ray point*.

exo-ergic process Nuclear process in which energy is liberated. Also *exo-energetic*. The equivalent thermodynamic term is *exothermic*.

exon That part of the transcribed nuclear RNA of eukaryotes which forms the mRNA after the excision of the **introns**. » p. 92.

exothermic Accompanied by the evolution of heat; ΔH negative. Cf. **endothermic**.

expansion See **adiabatic change**.

expansion of gases All gases have very nearly the same coefficient of expansion, namely 0.003 66 per kelvin when kept at constant pressure. See **absolute temperature**, **gas laws**.

exponential reactor A reactor with insufficient fuel to make it diverge; it needs excitation by an external source of neutrons for the determination of its properties. See **divergence**.

exposure dose See **dose**. » p. 103.

extraction column In the nuclear industry, the large vertical columns in which an organic solvent is used to extract uranium and plutonium from nitric acid solution. » p. 61.

f Symbol for **thermal utilization factor**.

F Symbol (1) for **fluorine**; (2) used following a temperature (e.g. 41°F) to indicate the **Fahrenheit scale**.

Fahrenheit scale The method of graduating a thermometer in which freezing point of water is marked 32° and boiling point 212°, the fundamental interval being therefore 180°. Fahrenheit has been largely replaced by the Celsius (Centigrade) and Kelvin scales. To convert °F to °C subtract 32 and multiply by 5/9. For the **Rankine scale** equivalent add 459.67 to °F; this total multiplied by 5/9 gives the Kelvin equivalent.

fail safe A design in which the power supply, control or structure is able to return to a safe condition in the event of failure or maloperation, by the automatic operation of protective devices or otherwise.

fallout Particulate matter in the atmosphere, which is transported by natural turbulence, but which will eventually reach the ground by sedimentation or dry or wet deposition. Applied esp. to airborne radioactive contamination resulting from nuclear explosion, inadequately filtered reactor coolant, failure of reactor containment after an accident etc.

false curvature Curvature of particle tracks (e.g. in cloud chambers, bubble chambers, spark chambers or photographic emulsions) which results from undetected interactions and not from an applied magnetic field.

family The group of radioactive nuclides which form a decay series.

faraday Quantity of electric charge carried by one mole of singly charged ions, i.e. 9.6487×10^4 coulombs. Symbol *F*.

Faraday's law of induction The electromotive force (emf) induced in any circuit is proportional to the rate of change of magnetic flux linked with the circuit. Principle used in every practical electrical machine. *Maxwell's field equations* involve a more general mathematical statement of this law.

Faraday's laws of electrolysis (1) The amount of chemical change produced by a current is proportional to the quantity of electricity passed. (2) The amounts of different substances liberated or deposited by a given quantity of electricity are proportional to the chemical equivalent weights of those substances.

-farious Suffix meaning arranged in so many rows.

fast effect See **fast fission**.

fast fission Uranium-238 has a fission threshold for neutrons with energy of about 1 MeV, and the fission cross-section increases rapidly with energy. Fission of this isotope by fast neutrons may cause a substantial increase in the reactivity of a thermal reactor (*fast effect*). » p. 20.

fast fission factor Ratio of the total number of fast neutrons produced by fissions due to neutrons of all energies (fast and thermal) to the number resulting from thermal/neutron fissions. Symbol ε. » p. 22.

fast neutron See **neutron**. » p. 18.

fast reaction Nuclear reaction involving strong interaction and occurring in a time of the order of 10^{-23} seconds. Due to **strong interaction** forces.

fast reactor A reactor without a moderator in which the chain reaction is maintained almost entirely by fast fission. It may also be a **breeder reactor**. » p. 36.

fathom A unit of measurement. Generally, a nautical measurement of depth = 6 ft.

fault condition A departure from normal operating conditions which might lead directly or indirectly to an accident, damage or a shutdown.

fault tree A representation of the different initial events and the possible successive malfunctions which would lead to an accident.

FBR Abbrev. for *Fast-Breeder Reactor*. See **breeder reactor**, **fast reactor**.

Feather analysis An approximate method of determining the range of β-rays forming part of a combined β-γ spectrum, by comparison of the absorption curve with that for a pure β-emitter.

Fe Symbol for **iron**.

feed The loaded solution introduced into the next stage of a reprocessing plant.

feedwater The de-aerated and chemically treated water fed into a boiler for evaporation.

fermi Femtometre (fm). A very small length unit, i.e. 10^{-15} m. Used in nuclear physics, being of the order of the radius of the proton, 1.2 fm.

Fermi age Slowing down area for neutrons calculated from *Fermi age theory* which assumes that neutrons, on being slowed down, lose energy continuously in an infinite homogeneous medium. Has the dimensions of length^2. Also *neutron age*. See **age theory**.

Fermi constant A universal constant which indicates the coupling between a nucleon and a lepton field. Its value is 1.4×10^{-50} J m^{-3}, and it is important in β-decay theory.

Fermi-Dirac gas An assembly of particles which obey Fermi-Dirac statistics and the Pauli exclusion principle. For an extremely dense Fermi gas, such as electrons in a metal, all energy levels up to a value E_F, the Fermi energy, are occupied at absolute zero.

Fermi-Dirac statistics Statistical mechanics laws obeyed by a system of particles whose wave function changes sign when two particles are interchanged, i.e. the Pauli exclusion principle applies.

fermion A particle which obeys Fermi-Dirac statistics. Fermions have total spin angular momentum of $(n+\frac{1}{2})$ where $n = 0,1,2,...$, and is the Dirac constant. Baryons and leptons are fermions and are subject to the **Pauli exclusion principle**. See **atomic structure** p. 121.

Fermi plot See **Kurie plot**.

Fermi potential The equivalence of the energy of the Fermi level as an electric potential.

Fermi selection rules See **nuclear selection rules**.

Fermi surface A constant energy surface in *k*-space which encloses all occupied electron states at absolute zero in a crystal.

Fermi temperature The degeneracy temperature of a Fermi-Dirac gas which is defined by E_F/k, where E_F is the energy of the Fermi level and k is Boltzmann's constant. This temperature is of the order of tens of thousands of kelvins for the free electrons in a metal.

ferri-, ferro- Prefixes from L. *ferrum*, iron.

fertile An isotope in a nuclear reactor which can be converted by the capture of a neutron into a *fissile* isotope, e.g. uranium-238 is converted by a series of reactions into plutonium-239.

Feynman diagram A diagram which shows the contributions to the rate of an elementary particle reaction. A powerful method of finding the physical properties of a system of interacting particles.

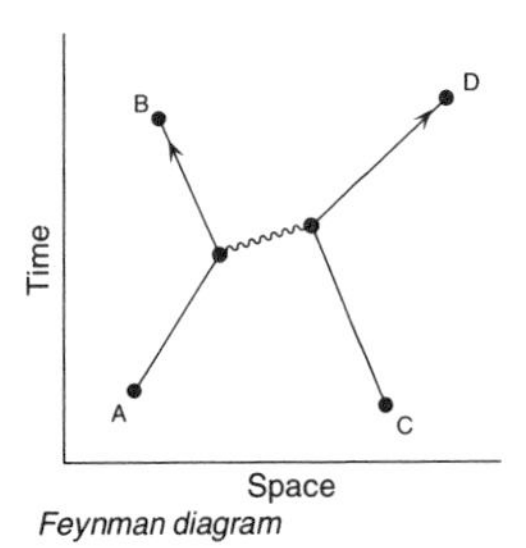

Feynman diagram

The figure represents two electrons, one moving from A to B and emitting a photon which is captured by the second electron moving from C to D.

f-factor The ratio of absorbed dose to exposure dose for a given material and X-ray energy.

fibre camera Instrument for measuring the X-ray diffraction pattern of fibrous materials.

Fick's law of diffusion The rate of diffusion in a given direction is proportional to the negative of the concentration gradient, i.e.

$$\text{molar flux} = -D\frac{\partial C}{\partial x},$$

where D is the diffusion coefficient.

field (1) The interaction between bodies or particles is explained in terms of fields. For example, the *potential energy* of a body may depend on its position and then is represented by a *scalar field* with magnitude only. Other physical quantities carry direction as well as magnitude and they are represented by *vector fields*, e.g. electric, magnetic, gravitational fields. (2) Space in which there are electromagnetic oscillations associated with a radiator; the *induction* field which represents the interchange of energy between the radiator and space is within a few wavelengths of the radiator; *radiation field* represents the energy lost from the radiator to space. Where components radiated by antenna elements are parallel is the *Fraunhofer region* and , where not, is the *Fresnel region*. The latter will exist between the antenna and the Fraunhofer region and is usually taken to extend a distance $2D^2/\lambda$, where λ is the wavelength of the radiation and D is the aerial aperture in a given aspect.

field density The number of lines of force passing normally through unit area of an electric or magnetic field.

field discharge The passage of electricity through a gas as a result of ionization of the gas; it takes the form of a brush discharge, an arc or a spark. Also *electric discharge*.

field intensity See **field strength**.

field of force Principle of *action at a distance*, i.e. mechanical forces experienced by an electric charge, a magnet or a mass, at a distance from an independent electric charge, magnet or mass, because of fields established by these and described by uniform laws.

field strength Vector representing the quotient of a force and the *charge* (or *pole*) in an electric (or magnetic) field, with the direction of the force. Also *field intensity*.

field theory As yet unverified attempt to link the properties of all fields (electric, magnetic, gravitational, nuclear) into a unified system.

film badge Small photographic film used as

radiation monitor and dosimeter. Normally worn on lapel, wrist or finger and sometimes partly covered by cadmium and tin screens so that exposure to neutrons, and to beta and gamma rays, can be estimated separately. » p. 79.

filtration (1) Specifically, in radiology, the removal of longer wavelengths in a composite beam of X-rays by the interposition of thin metal, e.g. copper or aluminium. Similarly for energy of other wavelengths. (2) The physical retention of particles from a solution by a porous membrane or other *filter*.

fine structure analysis The detailed analysis of the neutron flux through the moderator, coolant, structures and fuel in the design of a nuclear reactor.

FINGAL Abbrev. for *Fixation In Glass of Active Liquid*. A method for the long-term storage of active wastes. » p. 62.

fissile Capable of nuclear fission, i.e. the breakdown into lighter elements of certain heavy isotopes (uranium-233, uranium-235, plutonium-239) when they capture neutrons of suitable energy. Also *fissionable*. » p. 18.

fission The spontaneous or induced disintegration of a heavy atomic nucleus into two or more lighter fragments. The energy released in the process is referred to as *nuclear energy*. See **binding energy of a nucleus**. » p. 13.

fissionable See **fissile**.

fission bomb See **atomic bomb**.

fission chain Atoms formed by uranium or plutonium fission have too high a neutron-proton ratio for stability. This is corrected either by neutron emission (the delayed neutrons) or more usually by the emission of a series of beta-particles, so forming a short radioactive decay chain.

fission chamber Ionization chamber lined with a thin layer of uranium. This can experience fission by slow neutrons, which are thereby counted by the consequent ionization. Also *fission counter*. » p. 74.

fission counter See **fission chamber**.

fission neutrons Those released by nuclear fission, having a continuous spectrum of energy with a maximum of about 10^6 eV.

fission parameter The square root of the atomic number of a fissile element divided by its relative atomic mass.

fission poisons Fission products with abnormally high thermal neutron absorption cross-sections, which reduce the reactivity of nuclear reactors. Principally xenon-135 and samarium-149. » p. 19.

fission products Atoms, often radioactive, resulting from nuclear fission. They have masses of roughly half of that of the fissioning nucleus, e.g. strontium-90, a major contributor to radiation in *fallout* from nuclear explosions. » p. 60.

fission spectrum The energy distribution of neutrons released by nuclear fission.

fission track dating Certain glassy minerals may contain uranium-238 which undergoes spontaneous fission at a known rate. Each fission results in the production of two heavy nuclei which travel through and disrupt the molecular lattice of the material, forming microscopically visible tracks. If the amount of uranium is known and the tracks counted per unit area, an estimate can be made of the time since the mineral was last heated to the temperature needed to eliminate any previous tracks. See **radiometric dating** p. 222.

fission yield The percentage of fissions for which one of the products has a specific mass number. Fission yield curves show two peaks of approximately 6% for mass numbers of about 97 and 138. The probability of fission dividing into equal mass products falls to about 0.01%.

fixed points Temperature which can be accurately reproduced and used to define a temperature scale and for the calibration of thermometers. The temperature of pure melting ice and that of steam from pure boiling water at a pressure of one atmosphere define the Celsius and Fahrenheit scales. The *International Practical Temperature Scale* defined 10 fixed points ranging from the triple point of hydrogen (13.81 K) to the freezing point of gold (1337.58 K). See **Kelvin thermodynamic scale of temperature, triple point**.

flame photometry Like **atomic absorption spectroscopy**, except that the flame spectrum is viewed or measured directly.

flashing-off The production of steam by reducing the pressure on super-heated water.

flash radiography High-intensity, short duration X-ray exposure.

flask Lead case for storing or transporting multicurie radioactive sources or a container for the transport of irradiated nuclear fuel. Also *cask, casket, coffin*.

flat region Portion of reactor core over which neutron flux (and hence power level) is approximately uniform.

flattening material Neutron absorber or depleted fuel rod used in centre of reactor core to give larger flat region.

flavour An index which denotes different types of quarks. The six types are: up(u), down(d), strange(s), charm(c), bottom (or beauty) (b), top (or truth) (t). Also US *flavor*.

F-layer Upper ionized layer in the ionosphere resulting from ultraviolet radiation from the Sun and capable of reflecting radio waves back to Earth at frequencies up to 50 MHz. At a regular height of 300 km during the night, it falls to about 200 km during the day. During some seasons, this remains as the F_1 layer while an extra F_2 layer rises to a maximum of 400 km at noon. Considerable variations are possible during particle bombardment from the Sun, the layer rising to great heights or vanishing. Also *Appleton layer*.

flocculation Coagulation of particles by use of reagents which promote formation of flocs, as a preliminary to settlement and removal of excess water by thickening and/or filtration.

flow counter See **gas-flow counter**.

flowmeter Device for measuring, or giving an output signal proportional to, the rate of flow of a fluid in a pipe.

fluid A substance which flows. It differs from a solid in that it can offer no permanent resistance to change of shape. See **gas**, **liquid**.

fluidity The inverse of **viscosity**.

fluidized-bed reactor A reactor in which the active material is supported in a finely divided form by an upwardly moving gas or liquid, as in certain designs of nuclear reactor.

fluorescence Emission of radiation, generally light, from a material during illumination by radiation of usually higher frequency, or from the impact of electrons. Cf. **phosphorescence**.

fluorescent yield Probability of a specific excited atom emitting a photon in preference to an Auger electron.

fluorimeter An instrument used for measuring the intensity of fluorescent radiation.

fluorine Pale greenish-yellow gas, the most electronegative (non-metallic) of the elements and the first of the halogens. Chemically highly corrosive and never found free. Symbol F, at.no. 9, r.a.m. 18.9984, valency 1, mp −223°C, bp −187°C. Its abundance in the Earth's crust is 544 ppm and in seawater 1.3 ppm. It has seven isotopes:

A	Abundance %	half-life	decay mode
17		64.5 s	ε
18		110 m	ε
19	100		
20		11 s	β^-
21		4.3 s	β^-
22		4.2 s	β^-
23		2.2 s	β^-

It is used in the nuclear industry as uranium hexafluoride to enrich uranium-235 from natural uranium. It is a whitish crystalline substance which sublimes at 60°C. Enrichment relies on there being only one stable isotope of fluorine. » p. 56.

fluorography The photography of fluoroscopic images.

fluoroscope Measurement system for examining fluorescence optically. Fluorescent screen assembly used in fluoroscopy.

flux The rate of flow of mass, volume, or energy per unit cross-section normal to the direction of flow.

flux density The number of photons (or particles) passing through unit area normal to the beam, or the energy of the radiation passing through this area.

focus A point to which rays converge after having passed through an optical system, or a point from which such rays appear to diverge. In the first case the focus is said to be *real*; in the second case, *virtual*. The *principal focus* is the focus for a beam of light rays parallel to the principal axis of a lens or spherical mirror.

focusing The convergence to a point of (a) beams of electromagnetic radiation, (b) charged particle beams, (c) sound or ultrasonic beams.

focus-skin distance The distance from the focus of an X-ray tube to the surface of incidence on a patient, usually measured along the beam axis. Abbrev. *FSD*.

food irradiation See panel on p. 160.

foot-pound (force) Unit of work in the ft-lb-sec system of units. The work done in raising a mass of one pound through a vertical distance of one foot against gravity, i.e. 1.3558 J. See **fundamental dynamical units** p. 161.

foot-ton (force) 2240 foot-pounds (force).

forbidden band The gap between two bands of allowed energy levels in a crystalline solid. See **energy band**, **energy gap**.

forbidden lines Spectral lines which cannot be reproduced under laboratory conditions. Such lines correspond to transitions from a metastable state, and occur in extremely rarefied gases, e.g. in the solar corona and in gaseous nebulae.

forbidden transition Transition of electrons between energy states, which, according to Pauli selection rules, have a very low probability in relation to those which have a high probability, and are *allowed*.

force That which, when acting on a body which is free to move, produces an acceleration in the motion of the body, measured by rate of change of momentum of body. The unit of force is that which produces unit acceleration in unit mass. See **newton**,

food irradiation

Gamma rays are able to kill bacteria and yeast and high energy sources, using cobalt-60, have been used for a long time to sterilize medical apparatus like syringes, tubing and dialysis equipment. It was therefore natural to consider whether similar methods could be used to treat certain foodstuffs like spices, which have had to be disinfested with ethylene oxide gas in the past, particularly because irradiation by gamma-ray photons cannot induce radioactivity in the food. The radiation dosage required depends on the desired effect and is expressed in *grays* in the following table. One gray is the *absorbed dose* and is equivalent to one joule absorbed by one kilogram of tissue.

Purpose of irradiation	Dose in kilograys
Inhibition of sprouting	0.05 – 0.15
Delaying ripening	0.5 – 1
Insect disinfestation	0.15 – 0.5
Extension of shelf life	1 – 3
Eliminating spoilage and pathogens	1 – 7
Improving appearance and odour	1 – 7

A major problem for the public acceptability of food irradiation has been the difficulty of finding a reliable test for the presence of irradiated food. It has proved very difficult to detect in the food itself consistent changes due to irradiation which last during the lifetime of the food. However many foods are covered with minute quantities of mineral matter deposited from the soil or by the wind and these have proved very sensitive indicators of any previous irradiation.

A few micrograms of common feldspars and clay minerals are, after irradiation, sufficient to contribute reproducible signals when investigated by two related techniques called thermoluminescence and photoluminescence. Radiation, absorbed by these minerals, causes electrons and ions to become trapped at various points in the mineral structure, where they remain in a *metastable state* until released either by heating or by a strong beam of light at defined wavelengths. The stored energy in the crystals is emitted as light of a wavelength different from the incident light, in a process called *luminescence.* In the laboratory the mineral deposits are carefully removed from the sample and then tested for luminescence. A high signal indicates irradiation but a low signal might mean that the mineral was insensitive, because of its particular mineral form. It is then necessary to irradiate part of the sample kept back for the purpose and then to see if this shows luminescence. If it does, the food stuff was definitely not irradiated previously.

An objective method of detecting any past history of irradiation should allow a more direct comparison of radiation compared to chemical, heat or other methods of preserving and decontaminating food.

dyne, **poundal**. Extended to denote loosely any operating agency. Electromotive force, magnetomotive force, magnetizing force etc. are strictly misnomers.

force on a moving charge If a charge q is moving with a velocity $\mathbf{v}$ in a magnetic field of intensity $\mathbf{B}$, then the force on the charge is $\mathbf{F} = q(\mathbf{v} \times \mathbf{B})$. If the moving charges are within a conductor, the force of a short length $\mathbf{l}$ is $\mathbf{F} = i(\mathbf{l} \times \mathbf{B})$, where i is the current. See **Lorentz force**.

form factor The ratio of the effective value of an alternating quantity to its average value over a half-period.

formula A fixed rule or set form.

forward scatter Scattering of particles through an angle of less than 90° to the original direction of the beam. Cf. **back scatter**.

fundamental dynamical units

The basic equations of dynamics are such as to be the same for any system of fundamental units. Unit force acting on unit mass produces unit acceleration; unit force moved through unit distance does unit work; unit work done in unit time is unit power. Four systems are, or have been, in general use, the SI system now being the only one employed in scientific work. See **SI units** p. 237.

	System				dimensions
	ft lb sec	gravitational	CGS	SI	
length	foot (ft)	foot	centimeter (cm)	metre (m)	L
mass	pound (lb)	slug	gram (g)	kilogram (kg)	M
time	second (s)	second	second	second	T
velocity	ft s^{-1}	ft s^{-1}	cm s^{-1}	m s^{-1}	LT^{-1}
acceleration	ft s^{-2}	ft s^{-2}	cm s^{-2}	m s^{-2}	LT^{-2}
force	poundal (pdl)	pound force (lbf)	dyne	newton (N)	MLT^{-2}
work	ft pdl	ft lbf	erg	joule (J)	$ML^{2}T^{-2}$
power	ft pdl s^{-1}	ft lbf s^{-1}	erg s^{-1}	watt (W)	$ML^{2}T^{-3}$

Notes: (1) There is no name for the unit of power except in the SI system. It is possible to express power in the ft lb s system and in the gravitational system by the horsepower (550 ft lbf s^{-1}) and in the CGS system by the watt (10^7 erg s^{-1}). (2) The unit of force (lbf) in the gravitational system is also known as the

fossil fuel Coal, oil or natural gas, which are derived from fossilized organic matter.

four factor formula That giving the multiplication factor of an infinite thermal reactor as the product of **fast fission factor, resonance escape probability, thermal utilization factor** and the number of neutrons absorbed per neutron absorbed in the fuel (η). » p. 22.

Fourier analysis The determination of the harmonic components of a complex waveform (i.e. the terms of a Fourier series that represents the waveform) either mathematically or by a wave-analyser device.

Fourier integral The limiting form of the *Fourier series* when the period of the waveform becomes infinitely long. It is the Fourier representation of a non-repeated waveform, i.e. pulses, wave-packets and wave trains of limited extent.

Fourier optics The application of Fourier analysis and the use of Fourier transforms to problems in optics, in particular to image formation. The *Fraunhofer diffraction* pattern is the Fourier transform of the distribution of amplitude of light across the diffracting object. The distribution of amplitude in the Fraunhofer pattern is modified by the optical system and the image formed is the transform of this modified distribution. The same principle is used in X-ray crystal structure analysis where an 'image' of the atomic arrangement is constructed mathematically from the X-ray diffraction pattern.

Fourier principle That which shows all repeating waveforms can be resolved into sine wave components consisting of a fundamental and a series of harmonics at multiples of this frequency. It can be extended to prove that non-repeating waveforms occupy a continuous frequency spectrum.

Fourier transform A mathematical relation between the energy in a transient and that in a continuous energy spectrum of adjacent component frequencies. The Fourier transform $F(u)$ of the function $f(x)$ is defined by

$$F(u) = \int_{-\infty}^{+\infty} e^{-iut} f(t)\, dt\ .$$

Some writers use e^{+iut} instead of e^{-iut}.

Fourier transform spectroscopy The production of a spectrum by taking the **Fourier transform** of a two-beam interference pattern.

fp Abbrev. for *freezing-point.*

FPS The system of measuring in *feet, pounds* and *seconds.*

fractionation A system of treatment commonly used in radiotherapy in which doses are given daily or at longer intervals over a period of 3 to 6 weeks.

fracture mechanics The study of the effects on a loaded structure of cracks or other flaws which might occur at specific places.

Franck-Condon principle An electronic transition takes place so fast that a vibrating molecule does not change its internuclear distance appreciably during the transition. Applied to the interpretation of molecular spectra.

Fraunhofer region See **field**.

free-air dose A dose of radiation, measured in air, from which secondary radiation (apart from that arising from air, or associated with the source) is excluded.

free energy The capacity of a system to perform work, a change in free energy being measured by the maximum work obtainable from a given process.

free path See **mean free path**.

frequency Rate of repetition of a periodic disturbance, measured in hertz (cycles per second). Also *periodicity*.

fresnel A unit of optical frequency, equal to 10^{12}Hz =1THz (terahertz).

Fresnel region See **field**.

fretting corrosion Corrosion due to slight movements of unprotected metal surfaces, left in contact either in a corroding atmosphere or under heavy stress.

friction The resistance to motion which is called into play when it is attempted to slide one surface over another with which it is in contact. The frictional force opposing the motion is equal to the moving force up to a value known as the *limiting friction*. Any increase in the moving force will then cause slipping. *Static friction* is the value of the limiting friction just before slipping occurs. *Kinetic friction* is the value of the limiting friction after slipping has occurred. This is slightly less than the static friction. The *coefficient of friction* is the ratio of the limiting friction to the normal reaction between the sliding surfaces. It is constant for a given pair of surfaces.

frisking Searching for radioactive radiation by contamination meter, usually a portable ionization chamber.

FSD Abbrev. for *Full-Scale Deflection*.

fuel *Nuclear fuel*. Fissile material inserted in or passed through a reactor; the source of the chain reaction of neutrons, and so of the energy released.

fuel assembly A group of nuclear fuel elements forming a single unit for purposes of charging or discharging a reactor. The term includes bundles, clusters, stringers etc.

fuel cycle The stages involved in the supply and use of fuel in nuclear power generation. The main steps are mining, milling, extraction, purification, enrichment (if required), fuel fabrication, irradiation in the reactor, cooling, reprocessing, recycling and waste management and disposal. » Chapter 6.

fuel element Unit of nuclear fuel which may consist of a single cartridge, or a cluster of fuel **pins**.

fuelling machine, refuelling machine See **charge (-discharge) machine**.

fuel rating The ratio of total energy released to initial weight of heavy atoms (U, Th, Pu) for reactor fuel. Usually expressed in megawatts per tonne. US *specific power*.

fuel reprocessing The processing of nuclear fuel after use to remove fission products etc. and to recover fissile and fertile materials for further use.

fuel rod Unit of nuclear fuel in rod form for use in a reactor. Short rods are sometimes termed *slugs*.

fundamental dynamical units See panel on p. 161.

fundamental particle A particle that is incapable of subdivision. There are believed to be three kinds of such particle: *leptons*, *quarks* and *gauge bosons*.

fusion (1) The process of forming new atomic nuclei by the fusion of lighter ones; principally the formation of helium nuclei by the fusion of hydrogen and its isotopes. The energy released in the process is referred to as *nuclear energy* or *fusion energy*. See **binding energy of a nucleus**. » p. 38. (2) The conversion of a solid into a liquid state; the reverse of *freezing*. Fusion of a substance takes place at a definite temperature, the melting point, and is accompanied by the absorption of latent heat of fusion.

fusion bomb See **hydrogen bomb**. » p. 68.

fusion energy Energy released by the process of nuclear *fusion* usually in the formation of helium from lighter nuclei. The energy released in stars is by fusion processes. Fusion is the source of energy in the hydrogen bomb. See **proton-proton chain**.

fusion-fission hybrid reactor Proposed reactor system in which neutrons from fusion produce fissile material from a U or Th blanket and electricity. See **blanket**, **breeder reactor**.

fusion reactor Reactor in which **nuclear fusion** is used to produce useful energy. A field of very active current research. » p. 41.

G

g Symbol for **gram**.

g Symbol for (1) acceleration due to gravity; (2) specified efficiency.

γ Symbol for (1) ratio of specific heats of a gas; (2) **surface tension**; (3) propagation coefficient; (4) Gruneisen constant; (5) molar activity coefficient; (6) coefficient of cubic thermal expansion; (7) the greatest refractive index in a biaxial crystal; (8) **electrical conductivity.**

G Symbol for **giga**, i.e. 10^9.

Γ Symbol for surface concentration excess.

gadolinium A rare metallic element, member of the rare earth group. Symbol Gd, at.no. 64, r.a.m. 157.25, valency 3. It has 13 isotopes:

A	Abundance %	half-life	decay mode
149		9.4 d	ε
150		1.8 My	α
151		120 d	ε
152	0.20		
153		242 d	ε
154	2.1		
155	14.8		
156	20.6		
158	15.7		
159		18.6 h	β^-
160	21.8		
161		3.7 m	β^-

Natural gadolinium has a high neutron capture cross-section of 49 000 barns and is expensive, but gadolinium oxide has been used as a **burnable poison** during the start-up of reactors and in nitrate form as a deliberate poison during the shutdown of **CANDU** reactors.

gallon Liquid measure. One imperial gallon is the volume occupied by 10 lb avoirdupois of water. 1 Imp. gallon = 4.54609 litres, = 6/5 US gallon. US gallon = 3.785 43 litres, = 5/6 Imp. gallon.

gamma See **microgram**.

gamma camera See **scintillation camera**.

gamma detector A radiation detector specially designed to record or monitor gamma-radiation. » p. 76.

gamma radiation Electromagnetic radiation of high quantum energy emitted after nuclear reactions or by radioactive atoms when nucleus is left in excited state after emission of α- or β-particle. Also *gamma rays*.

gamma-ray capsule Usually metal, sealed and of sufficient thickness to reduce γ-ray transmission to a safe value.

gamma-ray energy Energy of a gamma-ray photon given by $h\nu$ where ν is the frequency and h is Planck's constant. The energy may be determined by diffraction by a crystal or by the maximum energy of photoelectrons ejected by the γ-rays. The depth of penetration into a material is determined by the energy.

gamma-ray imaging In medical diagnosis, the commonest way of showing the distribution of radioactive isotopes in the body.

gamma-ray photon A quantum of gamma-radiation energy given by $h\nu$ where ν is the frequency and h is Planck's constant.

gamma rays See **gamma radiation**.

gamma-ray source A quantity of matter emitting γ-radiation in a form convenient for radiology.

gamma-ray spectrometer Instrument for investigation of energy distribution of γ-ray quanta. Usually a scintillation or germanium counter followed by a *pulse-height analyser*.

Gamow-Teller selection rules See **nuclear selection rules**.

gangue The portion of an ore which contains no metal; valueless minerals in a lode.

gas (1) A state of matter in which the molecules move freely, thereby causing the matter to expand indefinitely, occupying the total volume of any vessel in which it is contained. (2) The term is sometimes reserved for a gas at a temperature above the critical value. Also defined as a definitely compressible fluid. See **gas laws, density of gases, expansion of gases.**

gas amplification Increase in sensitivity of a Geiger or proportional counter compared with a corresponding ionization chamber.

gas-cooled reactor A reactor in which the cooling medium is gaseous, usually carbon dioxide, air or helium. » p. 31.

gas counter (1) A gas-filled counter operating in the proportional or Geiger-Müller modes. (2) A Geiger counter into which radioactive gases can be introduced. » p. 75.

gas counting That of radioactive materials in gaseous form. The natural radioactive gases (radon isotopes) and carbon dioxide (carbon-14) are common examples. See **gas-flow counter**.

gaseous diffusion Isotope separation process based on principle of molecular diffusion.

gaseous diffusion enrichment The enrichment of uranium isotopes using gaseous uranium hexafluoride passing through a porous barrier. The **separation factor** is 1.0043 and therefore more than 1000 stages are required to produce fuel sufficiently enriched for a commercial nuclear reactor. » p. 56.

gases See **density of gases**, **expansion of gases**.

gas-flow counter (1) Counter tube through which gas is passed to measure its radioactivity, (2) Counter used to measure low-intensity α or β sources. These are introduced into the interior of the counter, and to prevent the ingress of air, the counting gas flows through it at a pressure slightly above atmospheric. Also *flow counter*.

gas laws Boyle's law, Charles's law and the pressure law which are combined in the equation $pV = RT$, where p is the pressure, V the volume, T the absolute temperature and R the gas constant for 1 mole. A gas which obeys the gas laws perfectly is known as an *ideal* or *perfect* gas.

gauge bosons A type of particle that mediates the interaction between two fundamental particles. There are four types: *photons* for electromagnetic interactions, *gluons* for strong interactions, *intermediate vector bosons* for weak interactions and *gravitons* for gravitational interactions.

gauge theory Theories in particle physics which attempt to describe the various types of interaction between fundamental particles. *Quantum electrodynamics* describes relativistic quantum fields. The *Weinberg and Salam* theory unifies the weak and electromagnetic interactions. *Quantum chromodynamics* is designed to explain the binding together of *quarks* to form *hadrons*. Grand unified theories are gauge theories which set out to unify the three interactions, electromagnetic, weak and strong.

gauss CGS electromagnetic unit of magnetic flux density; equal to 1 maxwell cm^{-2}, each unit magnetic pole terminating 4π lines. Now replaced by the SI unit of magnetic flux density, the tesla (T). $1T = 10^4$ gauss.

Gaussian units Formerly widely used system of electric units where quantities associated with electric field are measured in e.s.u. and those associated with magnetic field in e.m.u. This involves introducing a constant c (the free space velocity of electromagnetic waves) into Maxwell's field equations.

Geiger characteristic Plot of recorded count rate against operating potential for a Geiger or proportional counter detecting a beam of radiation of constant intensity.

Geiger-Müller counter An instrument for measuring ionizing radiations, with a tube carrying a high voltage wire in an atmosphere containing argon plus halogen or organic vapour at low pressure, and an electronic circuit which quenches the discharge and passes on an impulse to record the event. Also *Geiger counter*, *G-M counter*. See **Townsend avalanche**. » p. 75.

Geiger-Müller tube The detector of a Geiger-Müller counter, i.e. without associated electronic circuits.

Geiger-Nuttall relationship An empirical rule relating the half-life T of radioactive materials emitting α-particles to the range R of the particles emitted, i.e.

$$\log (1/T) = a \log R + b$$

where a and b are constants.

Geiger region That part of the characteristic of a counting tube, where the charge becomes independent of the nature of the radioactivity intercepted. Also *Geiger plateau*, since in this region the efficiency of counting varies only slowly with voltage on the tube. » p. 75.

Geiger threshold Lowest applied potential for which Geiger tube will operate in Geiger region.

generation time Average life of fission neutron before absorption by fissile nucleus.

genetically significant dose The dose that, if given to every member of a population prior to conception of children, would produce the same genetic or hereditary harm as the actual doses received by the various individuals.

genetic code The rules which relate the four bases of the DNA or RNA with the 20 amino acids found in proteins. » p. 91.

geometrical attenuation Reduction in intensity of radiation on account of the distribution of energy in space, e.g. due to inverse-square law.

geometrical cross-section Area subtended by a particle or nucleus. This does not usually resemble the interaction cross-section.

geometry factor $1/(4\pi)$ of the solid angle subtended by the window or sensitive volume of a radiation detector at the source.

germanium radiation detector A semiconductor detector with relatively large sensitive volume for γ-ray spectrometry. It has a much higher resolution but (in general) less sensitivity than a scintillation spectrometer. » p. 76.

germ line The cells whose descendants give rise to the *gametes*. » p. 110.

GeV Abbrev. for *giga-electron-volt*; unit of particle energy, 10^9 electron-volts, 1.602 ×

10^{-10} J. US sometimes *BeV*.

G-gas Gaseous mixture (based on helium and isobutane) used in low-energy β-counting (e.g. of tritium) by gas-flow proportional counter. See **gas-flow counter**.

giga- Prefix used to denote 10^9 times, e.g. a gigawatt is 10^9 watts.

giga-electron-volt See **GeV**.

gillion Rarely used for 10^9; preferred term is **giga-** or G.

glancing angle The complement of the **angle of incidence**.

glass A hard, amorphous, brittle substance, made by fusing together one or more of the oxides of silicon, boron, or phosphorus, with certain basic oxides (e.g. sodium, magnesium, calcium, potassium), and cooling the product rapidly to prevent crystallization or devitrification. The melting point varies between 800°C and 950°C. The tensile strength of glass resides almost entirely in the outer skin; if this is scratched or corroded, the glass is much more easily broken.

glassification See **vitrification**.

glassy state See **vitreous state**.

glove box An enclosure in which radioactive or toxic material can be handled by use of special rubber gloves attached to the sides of the box, as in the figure, thus preventing contamination of the operator. Normally operated at a slightly reduced pressure so that any leakage is inward.

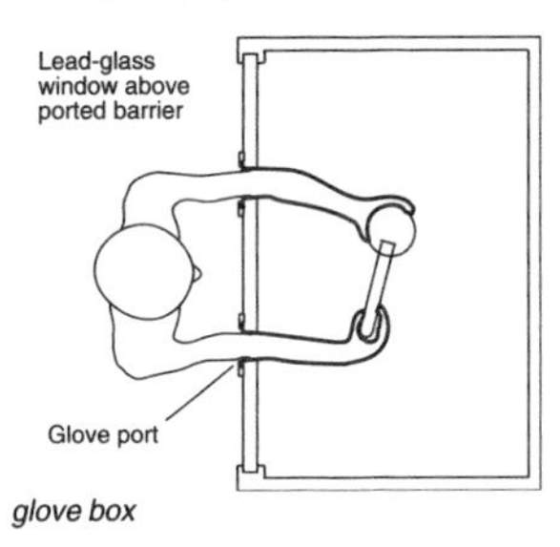

glove box

gluon A *gauge boson* that mediates the strong interaction between quarks (and antiquarks). According to quantum chromodynamic theory, there should be eight different gluons each with zero mass and zero charge.

G-M counter See **Geiger-Müller counter**.

GMT Abbrev. for *Greenwich Mean Time*.

gold grains Small lengths of activated gold wire (half-life 2.70 days), used in a similar manner to **radon seeds.**

gradient The rate of change of a quantity with distance, e.g. the temperature *gradient* in a metal bar is the rate of change of temperature along the bar.

Graham's law Law stating that the velocity of effusion of a gas is inversely proportional to the square root of its density.

gram(me) The unit of *mass* in the CGS system. It was originally intended to be the mass of 1 cm^3 of water at 4°C but was later defined as one-thousandth of the mass of the International Prototype Kilogramme, a cylinder of platinum-iridium kept at Sèvres.

gram(me)-ion Mass in grams of an ion, numerically equal to that of the molecules or atoms constituting the ion.

gram(me) calorie See **calorie**.

gram(me)-röntgen The real conversion of energy when 1 röntgen is delivered to 1 g of air (approx. 8.4 μJ). A convenient multiple is the **meg**

Grand Unified Theories See panel on p. 166.

graphite One of the two naturally occurring forms of crystalline carbon, the other being diamond. It occurs as black, soft masses and, rarely, as shiny crystals (of flaky structure and apparently hexagonal) in igneous rocks; in larger quantities in schists particularly in metamorphosed carbonaceous clays and shales, and in marbles; also in contact metamorphosed coals and in meteorites. Graphite has numerous applications in trade and industry now much overshadowing its use in 'lead' pencils. Much graphite is now produced artificially in electric furnaces using petroleum as a starting material.

graphite reactor A reactor in which fission is produced principally or substantially by slow neutrons moderated by graphite. See **moderator**. » p. 31.

graveyard See **burial site**.

graviton A *gauge boson* that is the agent for gravitational interactions between particles. The graviton has not been detected but is believed to have zero mass and zero charge. It bears the same relationship to the gravitational field as the photon does to the electromagnetic field. Gravitational interactions between particles are so weak that no quantified effects have been observed.

gravity water system A system in which flow occurs under the natural pressure due to gravity, incorporated as part of the safety procedures in nuclear reactors for flooding.

gray SI unit of **absorbed dose**. Symbol Gy.

grazing angle A very small **glancing angle**.

grenz rays X-rays produced by electron beams accelerated through potentials of 25 kV or less. These are generated in many types of electronic equipment; low penetrating power.

Grand Unified Theories

Modern theoretical physics aims to describe the nature of the physical universe with as few assumptions and laws as possible. In a complete framework of physical theory parameters, which we currently have to measure experimentally, such as the speed of light or the fine structure constant, would emerge naturally from the equations. Just as Newton was able to explain Kepler's laws of planetary motion through his more powerful theory of gravitation, so in modern physics there is a desire to find all-embracing physical laws.

Physics recognizes four distinct interactions that affect matter. These are the *electromagnetic* interaction, the *strong* and *weak* interactions affecting particles and nuclei, and the gravitational interaction. Although gravitation is by far the weakest of these, it is always an attractive force and it acts over immense distances; that is why it is so important in cosmology. Each of these interactions has its own theoretical formalism. *Gravitation* is now described through the *general theory of relativity*, for example. Electromagnetic, weak and strong interactions are also described through different *gauge theories.* (Gauge group theory is a powerful branch of mathematics.)

In the 1960s and 1970s, physicists such as S. Weinberg, A. Salam and S. Glashow found ways of achieving partial unification of these different classes of interaction. *Electroweak theory* could explain most features of electromagnetism and the weak interaction. The more elaborate quantum chromodynamics described the interactions between quarks and heavy particles such as neutrons. The unified theories achieved well-publicized successes at the time by predicting new elementary particles that were eventually found by the CERN accelerator.

Grand unified theories (or GUTs) are particularly associated with H.M. Georgi and they aim to fold the electromagnetic, weak and strong interactions into a single gauge group. This theory predicts that the *proton* is not stable, although the half-life is extremely long: 10^{29} years. Experiments to measure the proton decay are being attempted as an important test of the theory. The theory also attempts to describe how elementary particles with energies of 10^{21} electron volts will behave. This energy is so colossal, with enough to run a light bulb for a minute residing on a single particle, that they will never be made in accelerators. They were however abundant in the first 10^{-32} seconds of the Big Bang and it is possible that they could be found in cosmic rays. It is because of these astrophysical connections that astronomers, as well as physicists, are interested in GUTs. A more distant goal is the unification of gravitation also, to produce a viable theory of **quantum gravity**.

grid therapy A method of treatment in radiotherapy, in which radiation is given through a grid of holes in a suitable absorber, e.g. lead, rubber. By careful positioning, a greater depth dose can be given, and skin breakdown reduced to a minimum.

ground state State of nuclear system, atoms etc, when at their lowest energy, i.e. not *excited.* Also *normal state.*

group theory Approximate method for the study of neutron diffusion in a reactor core, in which neutrons are divided into a number of velocity groups in which they are retained before transfer to the next group. See **one-**, **two-**, **multigroup theory**.

group velocity The velocity of energy propagation for a wave in a dispersive medium. Given by

$$v_g = \left(\frac{d\omega}{dk}\right) = v - \lambda\frac{dv}{d\lambda},$$

where $k = 2\pi/\lambda$, $\omega = 2\pi \times$ frequency, v is the phase velocity and λ is the wavelength.

growth (1) Elongation of fuel rods in reactor under irradiation. (2) Build-up of artificial radioactivity in a material under irradiation, or of activity of a daughter product as a result of decay of the parent.

guanine Purine base which occurs in DNA and RNA, pairing with cytosine. See **genetic**

code. » p. 91.

Gudden-Pohl effect A form of electroluminescence which follows metastable excitation of a phosphor by ultraviolet light.

guided wave Electromagnetic or acoustic wave which is constrained within certain boundaries as in a wave guide.

GUT Generic acronym for **Grand Unified Theories**.

gyromagnetic ratio The ratio γ of the magnetic moment of a system to its angular momentum. For orbiting electrons $\gamma = e/2m$ where e is the electronic charge and m is the mass of the electron. γ for electron spin is twice this value.

gyrotron In fusion studies, very high-frequency power generator for microwave heating at the electron cyclotron resonance.

h Symbol for (1) **Planck's constant**; see **Planck's law**; (2) specific enthalpy; (3) height.

H Symbol for **hydrogen**.

hadron An elementary particle that interacts *strongly* with other particles. Hadrons include *baryons* and *mesons*.

hafnium A metallic element in the fourth group of the periodic system. Symbol Hf, at. no. 72, r.a.m. 178.49, mp c. 2200°C, rel.d. 12.1. It has 13 isotopes:

A	Abundance %	half-life	decay mode
171		12.1 h	ε
172		1.87 h	ε
173		24.0 h	ε
174	0.16		
175		70 d	ε
176	5.2		
177	18.6		
178	27.1		
179	13.7		
180	35.2		
181		42.4 d	β^-
182		9 My	β^-
183		64 m	β^-

It occurs in minerals containing zirconium, to which it is chemically similar, but with a much higher neutron capture cross-section. This makes it a troublesome impurity in the zirconium alloys used as fuel cladding. See **zircaloy**.

half-life (1) Time in which half of the atoms of a given quantity of radioactive nuclide undergo at least one disintegration.The half-life T is related to the *decay constant* λ by

$$T_{1/2} = \frac{0.693}{\lambda} .$$

Also *half-value period, half period.*

(2) See **biological half-life**, **effective half-life**.

half-residence time Time in which half the radioactive debris deposited in the stratosphere by a nuclear explosion would be carried down to the troposphere.

half-thickness See **half-value thickness**.

half-value layer See **half-value thickness**.

half-value period, half-period See **half-life** (1).

half-value thickness The thickness of a specified substance which must be placed in the path of a beam of radiation in order to reduce the transmitted intensity by one-half. Also *half-thickness, half-value layer.*

half-width A measure of sharpness on any function $y = f(x)$ which has a maximum value y_m at x_0 and also falls off steeply on either side of the maximum. The half-width is the difference between x_0 and the value of x for which $y = y_m/2$. Used particularly to measure the width of spectral lines or of a response curve.

hali- Prefix from Gk. *hals*, salt. Also *halo-*.

halides Fluorides, chlorides, bromides, iodides and astatides.

halogen-quench Geiger tube Low-voltage tube for which halogen gas (normally bromine) absorbs residual electrons after a current pulse, and so quenches the discharge in preparation for a subsequent count. » p. 75.

hammer track Highly characteristic track resembling a hammer, formed by the decay of a lithium-8 nucleus into two α-particles emitted in opposite directions at right angles to the lithium track.

hand monitor Radiation monitor designed to measure radioactive contamination on the hands of an operator, or to be held in the hand.

haploid Of the reduced number of chromosomes characteristic of the germ cells of a species, equal to half the number in the somatic cells. Cf. **diploid**.

hard radiation Qualitatively, the more penetrating types of X-, beta and gamma rays.

harmonic Sinusoidal component of repetitive complex waveform with frequency which is an exact multiple of basic repetition frequency (the fundamental). The full set of harmonics forms a Fourier series which completely represents the original complex wave. In acoustics, harmonics are often termed overtones, and these are counted in order of frequency above, but excluding, the lowest of the detectable frequencies in the note; the label of the harmonic is always its

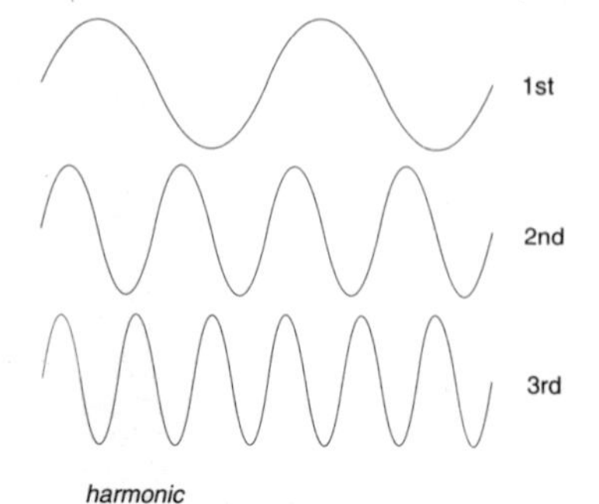

harmonic

frequency divided by the fundamental. The nth overtone is the $(n + 1)^{th}$ harmonic.

harmonic analysis Process of measuring or calculating the relative amplitudes of all the significant harmonic components present in a given complex waveform. The result is frequently presented in the form of a Fourier series, e.g. $A = A_0 \sin \omega\tau + A_1 \sin (2\omega\tau + \vartheta_1) + A_2 \sin(3\omega\tau + \vartheta_2) + ...$etc, where ω = pulsatance and ϑ = phase angle.

harmonic components Any term (but the first) in a Fourier series which represents a complex wave.

harmonic wave See **sinusoidal wave**.

h-bar Symbol for **Dirac's constant**. Equal to Einstein's constant *h* divided by 2π.

He Symbol for **helium**.

head end That part of a reprocessing plant which precedes solvent extraction; it therefore includes facilities for storing and handling fuel assemblies, breaking down and dissolving them. » p. 61.

health physics Branch of radiology concerned with health hazards associated with ionizing radiations, and protection measures required to minimize these. Personnel employed for this work are *health physicists* or *radiological safety officers*.

heat Heat is energy in the process of transfer between a system and its surroundings as a result of temperature differences. However, the term is still used also to refer to the energy contained in a sample of matter. Also for **temperature**, e.g. forging or welding *heat*. For some of the chief branches in the study of heat, see **calorimetry**, **heat units**, **internal energy**, **latent heat**, **mechanical equivalent of heat**, **radiation**, **specific heat capacity**, **temperature**, **thermal conductivity**, **thermometry**.

heat exchanger Any device for the exchange of heat between two substances without intermixing. Typically those in which the gaseous or liquid coolant from a nuclear reactor heats water to provide steam for the turbines in an *indirect-cycle* reactor. » p. 24.

heat transfer See **conduction of heat**, **convection of heat**, **radiation**.

heat units See **joule**, **calorie**, **British thermal unit**. Also *calorific value*.

Heaviside-Lorentz units Rationalized CGS Gaussian system of units, for which corresponding electric and magnetic laws are always similar in form. See **rationalized units**.

heavy-aggregate concrete Concrete containing very dense aggregate material such as lead, barytes or iron nodules in place of some or all of the usual gravel, so increasing its gamma-ray absorption coefficient. See **loaded concrete**.

heavy particle See **hyperon**.

heavy water Deuterium oxide, or water containing a substantial proportion of deuterium atoms (D_2O or HDO).

heavy-water reactor A reactor using **heavy water** as moderator, e.g. **CANDU** and **SGHWR** types. » p. 29.

hecto- Metric prefix meaning ×100.

Heisenberg principle See **uncertainty principle**.

helium Inert gaseous element. Symbol He, at.no. 2, r.a.m. 4.0026, with extremely stable nucleus identical to α-particle. It has two isotopes:

A	Abundance %	half-life
3	1.4×10^{-4}	stable
4	99.9998	stable

It liquefies at temperatures below 4 K, and undergoes a phase change to a form known as *liquid helium II* at 2.2 K. The latter form has many unusual properties believed to be due to a substantial proportion of the molecules existing in the lowest possible quantum energy state. Liquid helium is the standard coolant for devices working at cryogenic temperatures. The abundance of helium in the Earth's crust is 0.003 ppm and in the atmosphere 5.2 ppm (vol). Almost all of it is of radiogenic origin and obtained from gas wells. It has been used as a coolant in gas-cooled reactors where its chemical stability and very low neutron capture cross-sections are advantageous, but it is expensive and only finds wide use in low-volume applications like surrounding the fuel in fuel pins.

helium-neon laser A laser using a mixture of helium and neon, energized electrically. Its output is on the visible region at 632.8 nm and can have a continuous power of 1 W or a pulsed output with peak powers up to 100 W. Also *He-Ne laser*. See **laser** p. 182.

helix A line, thread, wire, or other structure curved into a shape such as it would assume if wound in a single layer round a cylinder; a form like a screw-thread or coiled spring which is very common in biological macromolecules, e.g. DNA.

helix Path like a loosely wound spring which is followed by charged particle in a magnetic field.

hemi- Prefix from Gk. *hemi*, half.

hemisphere The half of a sphere, obtained by cutting it by a plane passing through the centre. As applied to the Earth, the term usually refers to the *northern* or the *southern hemisphere*, the division being by the equatorial plane.

He-Ne laser See **helium-neon laser**.

henry SI unit of self- and mutual inductance. (1) A circuit has an inductance of 1 henry if an e.m.f. of 1 volt is induced in the circuit by a current variation of 1 ampere per second. (2) A coil has a self-inductance of 1 henry when the magnetic flux linked with it is 1 weber per ampere. (3) The mutual inductance of two circuits is 1 henry when the flux linked with one circuit is 1 weber per ampere of current in the other. Symbol H.

hertz SI unit of frequency, indicating number of cycles per second (c/s). Symbol Hz.

Hertzian waves Electromagnetic waves from, e.g. 10^4 to 10^{10} Hz, used for communication through space, covering the range from very low to ultra-high frequencies, i.e. from audio reproduction, through radio broadcasting and television to radar. » p. 81.

hetero- Prefix from Gk. *heteros*, other, different.

heterogeneous radiation Radiation comprising a range of wavelengths or particle energies.

heterogeneous reactor A reactor in which the fuel is present as rods spread in an array or lattice within (but separate from) the moderator. Cf. **homogeneous reactor**.

hex Abbrev. for **uranium hexafluoride**.

hex- Prefix from Gk. *hex*, six.

Hf Symbol for **hafnium**.

high-energy physics See **particle physics**.

high-flux reactor A reactor designed to operate with a greater neutron flux than normal for testing materials for radiation effects and for experiments requiring intense beams of neutrons. Also *materials testing reactor*.

high-frequency heating Heating (induction or dielectric) in which the frequency of the current is above mains frequency; from rotary generators up to ~3000 Hz and from electronic generators 1–100 MHz. Also *radio heating*. See **microwave heating**.

high-level waste Nuclear waste requiring continuous cooling to remove the heat produced by radioactive decay. » p. 60.

high spot A small volume so situated that the dose therein is significantly above the general dose level in the region treated.

high-temperature reactor A reactor designed to attain core temperatures above 660°C. Usually uses coated uranium dioxide or carbide pellets, cooled by helium gas. » p. 35.

hodoscope Apparatus (e.g. an array of radiation detectors) which is used for tracing paths of charged particles in a magnetic field.

holdback Agent for reducing an effect, e.g. a large quantity of inactive isotope reduces the coprecipitation or absorption of a radioactive isotope of the same element.

holo-, hol- Prefixes from Gk. *holos*, whole.

homo- Prefix from Gk. *homos*, same.

homogeneous ionization chamber An ionization chamber in which both walls and gas have similar atomic composition, and hence similar energy absorption per unit mass. » p. 74.

homogeneous light See **monochromatic light**.

homogeneous radiation That of constant wavelength (monochromatic), or constant particle energy.

homogeneous reactor A reactor in which the fuel and moderator are finely divided and mixed (or the fuel may be dissolved in a liquid moderator) so as to produce an effectively homogeneous core material. See **slurry reactor**.

homogenizer A device in which coarse and polydisperse emulsions are transformed into nearly monodisperse systems. The liquid is subjected to an energetic shear.

Hooke's law For an elastic material the *strain* is proportional to the applied *stress*. The value of the stress at which a material ceases to obey Hooke's law is the *limit of proportionality*. See **elasticity**.

hormesis The stimulating effect of a small dose of a substance which at larger doses is toxic. Might apply to radiation by e.g. stimulating the production of repair enzymes.

hot Colloquial reference to high levels of radioactivity; hence *hot laboratory*, a designated area for handling radioactive substances where extra precautions against irradiation of staff are taken.

hot cell See **cave**.

hot spot (1) The position of highest temperature in a reactor fuel pin. (2) A highly radioactive region in a plant or reactor, needing protective screening.

HTR Abbrev. for **High-temperature Reactor**.

hulls Small pieces of fuel rod cladding left after fuel pins have been dissolved in a reprocessing plant.

hyal-, hyalo- Prefixes from Gk. *hyalos*, clearstone, glass.

hydr-, hydro- Prefixes from Gk. *hydro*, water.

hydraulics The science relating to the flow of fluids. Adj. *hydraulic*.

hydrostatics Branch of statics which studies the forces arising from the presence of fluids.

hybrid electromagnetic wave A wave having longitudinal components of both the electric and magnetic field vectors. Abbrev. *HEM*.

hydrogen The least dense element, forming diatomic molecules H_2. Symbol H, at.no. 1, r.a.m. 1.00797, valency 1, mp −259.14°C, bp −252.7°C, density 0.089 88 g dm^{-3} at stp. It is a colourless, odourless, diatomic gas, water being formed when it is burnt, occurring as three isotopes; those of mass number 2 and 3 being deuterium and tritium.

A	Abundance %	half-life	decay mode
1	99.98		
2	0.015		
3		12.3 y	β^-

It is cosmically the most abundant of all elements but in its elemental form is not of major importance in the Earth's crust (abundance 1520 ppm; 0.53 ppm by volume in the atmosphere).

hydrogen bomb Bomb using the nuclear fusion process to release vast amounts of energy. Extremely high temperatures are required for the process to occur and these temperatures are obtained by a *fission bomb* near which the fusion material is arranged. Lithium deuteride can initiate a number of fusion processes involving the hydrogen isotopes, deuterium and tritium. These reactions also produce high-energy neutrons capable of causing *fission* in a surrounding layer of the most abundant isotope of uranium, uranium-238 so that further energy is released. » p. 69.

hydrogenous Said of a substance rich in hydrogen and therefore suitable for use as a moderator of neutrons in a nuclear reactor.

hydrometer An instrument by which the relative density of a liquid may be determined by measuring the length of the stem of the hydrometer immersed, when it floats in the liquid with its stem vertical.

hyper- Prefix from Gk. *hyper*, above.

hyperbaric chamber A chamber containing oxygen at high pressures, in which patients are placed to undergo radiotherapy or treatment for certain forms of poisoning.

hypercharge See **strangeness**.

hyperon Elementary particles with masses greater than that of a neutron and less than that of a deuteron, and having a lifetime of the order of 10^{-10} s.

hypo- Prefix from Gk. *hypo*, under.

hypothesis A prediction based on theory, an educated guess derived from various assumptions, which can be tested using a range of methods, but is most often associated with experimental procedure; a proposition put forward for proof or discussion.

hypso- Prefix from Gk. *hypsos*, height.

hysteresis The retardation or lagging of an effect behind the cause of the effect, e.g. dielectric hysteresis, magnetic hysteresis etc.

Hz Symbol for **hertz**.

I

I Symbol for **iodine**.

I Symbol for (1) electric current; (2) luminous intensity.

IA Abbrev. for *International Ångström*.

IAEA Abbrev. for **International Atomic Energy Agency**.

ICRH Abbrev. for *Ion Cyclotron Resonance Heating*. See **cyclotron resonance heating**.

idio- Prefix from Gk. *idios*, peculiar, distinct.

ignition temperature In fusion, the point at which alpha-particle heating can sustain the fusion reaction.

illuminance See **illumination**.

illumination The quantity of light or luminous flux falling on a unit area of a surface. Illumination is inversely proportional to the square of the distance of the surface from the source of light, and proportional to the cosine of the angle made by the normal to the surface with the direction of the light rays. The unit of illumination is the *lux*, which is an illumination of 1 lumen m^{-2}. Symbol E. Also *illuminance*.

image intensifier An electronic device screen for enhancing the brightness of an image. Used in low light or in, e.g. fluoroscopy, to reduce patient dose. See **intensifying screen**.

impact For the direct impact of two elastic bodies, the ratio of the relative velocity after impact to that before impact is constant and is called the *coefficient of restitution* for the materials of which the bodies are composed. This constant has the value 0.95 for glass/glass and 0.2 for lead/lead, the values for most other solids lying between these two figures.

impact parameter The distance at which two particles which collide would have passed if no interaction had occurred between them.

imperfect dielectric A dielectric in which there is a loss element resulting in part of the electric energy of the applied field being used to heat the medium.

implant The radioactive material, in an appropriate container, which is to be imbedded in a tissue for therapeutic use, e.g. needle or seed.

imploding linear system Fusion device in which a cylindrical plasma is formed by the implosion of material lining the reactor vessel. » p. 69.

impulse When two bodies collide, over the period of impact there is a large reaction between them. Such a force can only be measured by its time integral ($\int F\,dt$) which is defined as the impulse of the force, and which equals the change of momentum produced in either body.

In Symbol for **indium**.

in- Prefix from L. *in*, in(to), not.

incident beam Any wave or particle beam, the path of which intercepts a surface of discontinuity.

incoherent Said of radiation of the same frequency emitted from discrete sources with random phase relationships. All light sources except the *laser* emit incoherent radiation.

independent particle model of nucleus Model in which each nucleon is assumed to act quite separately in a common field to which they all contribute.

indeterminacy principle See **uncertainty principle**.

indirect cycle Nuclear power plant in which the core coolant passes through a heat exchanger in which the secondary circuit of water produces steam for the turbines, as in a Pressurized Water Reactor. » p. 24.

indium A silvery metallic element in the third group of the periodic system. Symbol In, at.no. 49, r.a.m. 114.82, mp 155°C, bp 2100°C, rel.d. 7.28 at 13°C, electrical resistivity 9×10^{-8} ohm metres. Found in traces in zinc ores. The metal is soft and marks paper like lead; it forms compounds with carbon compounds. It has a high capture cross-section for slow neutrons and so is readily activated. It has eight isotopes:

A	Abundance %	half-life	decay mode
110		69.1 m	ε
111		2.83 d	ε
112		14.4 m	ε
113	4.3		
114		71.9 s	β^-
115	95.7		
116		14.1 s	β^-
117		43.8 m	β^-

individual risk The probability of radiation damage to an individual.

induced emf That which appears in a circuit as a result of changes in the interlinkages of magnetic flux with part of the circuit, like the emf in the secondary of a transformer. Discovered by Faraday in 1831. See **Faraday's law of induction**.

induced radioactivity The radioactivity induced in non-radioactive elements by neutrons in a reactor, or protons or deuterons in a cyclotron or linear accelerator. X-rays or gamma rays do not induce radioactivity unless the gamma-ray energy is exceptionally high.

inductance (1) That property of a component or circuit which, when carrying a current, is characterized by the formation of a magnetic field and the storage of magnetic energy. (2) The magnitude of such capability.

induction field See **field**.

inelastic collision In atomic or nuclear physics, a collision in which there is a change in the total energies of the particles concerned resulting from the excitation or de-excitation of one or both of the particles. See **collision**.

inelastic scattering See **scattering**.

inertia The property of a body, proportional to its mass, which opposes a change in the motion of the body.

inertial confinement In fusion studies, short-term plasma confinement arising from inertial resistance to outward forces (mainly by the compression and heating of deuterium or mixed deuterium-tritium pellets by a powerful laser). See **containment**, **inertial fusion system**, **magnetic confinement**.

inertial fusion system System in which small capsules (*pellets*) containing deuterium and tritium are injected into a reaction chamber and ignited by high-energy laser or ion beams. See **magnetic confinement fusion system**. » p. 38.

inert gas See **noble gases**.

inert metal Alloy (usually Ti-Zr) for which scattering of neutrons by nuclei is negligible.

I neutrons Neutrons possessing the energy required to undergo resonance absorption by iodine.

infra- Prefix from L. *infra*, below.

infrared radiation Electromagnetic radiation in the wavelength range from 0.75 to 1000 μm approximately, i.e. between the visible and microwave regions of the spectrum. The *near* infrared is from 0.75 to 1.5 μm, the *intermediate* from 1.5 to 20 μm, and the *far* from 20 to 1000 μm. » p. 82.

infrared spectrometer An instrument similar to an optical spectrometer but employing non-visual detection and designed for use with infrared radiation. The infrared spectrum of a molecule gives information as to the functional groups present in the molecule and is very useful in the identification of unknown compounds.

inherent filtration Filtration introduced by the wall of the X-ray tube as distinct from added primary or secondary filters.

inhour Unit of reactivity equal to reciprocal of period of nuclear reactor in hours, e.g. a reactivity of two inhours will result in a **period** of half an hour. From *inverse hour*.

inhour equation An equation relating the decay constant, ω, to the excess multiplication δk of the overall neutron flux in a reactor.

$$\delta k = l\omega + k\omega \sum_{i=1}^{I} \frac{\beta_i}{\omega + \lambda_i},$$

where l is the neutron lifetime (typically 10^{-3} s), k is the multiplication constant of the reactor, i refers to the **delayed neutron groups** (conventionally 6), λ the decay constant of group i and β the fractional yield of the group. For small values of δk, when ω can be neglected, the right-hand term in the equation is 0.083 s for uranium-235 fissions, which gives a reactor period of 850 s. This shows the important effect of the delayed neutrons on the stability of a reactor. » p. 20.

in-pile test A test in which the effects of irradiation are measured while the specimen is subjected to radiation and neutrons in a reactor.

insensitive time See **dead time**.

instability In a *plasma*, various instabilities of which the principal are: bending of plasma, *kink instability*, and bead-like instability, *sausage instability*, in both of which the magnetic field change magnifies the variations; acceleration-driven *Rayleigh-Taylor instability* when the magnetic fields and plasma behave like a light fluid in contact with and accelerated against a heavier one, causing spiky irregularities; *flute instability* in mirror machines. See **unstable equilibrium**.

instantaneous specific heat capacity Specific heat capacity at any one temperature level; *true s.h.c.* to distinguish from *mean s.h.c.*.

instantaneous value A term used to indicate the value of a varying quantity at a particular instant. More correctly it is the average value of that quantity over an infinitesimally small time interval.

instrument range The intermediate range of reaction rate in a nuclear reactor, when the neutron flux can be measured by permanently installed control instruments, e.g. ion chambers. See **start-up procedure**.

insulation (1) Any means for confining as far as possible a transmissible phenomenon (e.g. electricity, heat, sound, vibration) to a particular channel or location in order to obviate or minimize loss, damage or annoyance. (2) Any material (also *insulant*) or means suitable for such a purpose in given conditions, e.g. dry air suitably enclosed, polystyrene and polyurethane foam slab, glass fibre, rubber, porcelain, mica, asbestos, hydrated magnesium carbonate, cork, kapok, crumpled aluminium foil etc. See **shielding**.

insulator Material having a high electrical resistivity in the range $10^6 - 10^{15}$ ohm metres. The electrons in the material are not free to move under the influence of an electric field.

integral dose See **dose**.

intensifying screen (1) Layer or screen of fluorescent material adjacent to a photographic surface, so that registration by incident X-rays is augmented by local fluorescence. (2) Thin layer of lead which performs a similar function for high-energy X-rays or gamma rays, as a result of ionization produced by secondary electrons.

intensitometer Instrument for measuring intensities of X-rays during exposures.

intensity of radiation Energy flux, i.e. of photons or particles, per unit area normal to the direction of propagation.

interaction (1) Transfer of energy between two particles. (2) Interchange of energy between particles and a wave motion. (3) Between waves. See **interference**.

intercavitary X-ray therapy X-ray therapy in which the appropriate part of suitable X-ray apparatus is placed in a body cavity.

interlock A mechanical and/or electrical device to prevent hazardous operation of a reactor, e.g. to prevent withdrawal of control rods before coolant flow has been established.

intermediate-level waste Nuclear waste not included in the categories **high-level waste** or **low-level waste**.

intermediate neutrons See **neutron**.

intermediate reactor A reactor designed so that the majority of fissions will be produced by the absorption of intermediate **neutrons**. Also *epithermal reactor*.

intermolecular forces Term referring to the forces binding one molecule to another. They are very much weaker than the bonding forces holding together the atoms of a molecule. See **van der Waals' forces**.

internal conversion Nuclear transition where energy released is given to orbital electron which is usually ejected from the atom (conversion electron), instead of appearing in the form of a γ-ray photon. The *conversion coefficient* for a given transition is given by the ratio of conversion electrons to photons.

internal energy For a thermodynamic system, the difference between the heat absorbed by the system and the external work done by the system, is the change in its internal energy (the *first law of thermodynamics*). The internal energy takes the form of the kinetic energy of the constituent molecules and their potential energies due to the molecular interactions. The internal energy is manifest as the temperature of the system, latent heat, as shown by a change of state, or the repulsive forces between molecules, seen as expansion. Symbol U. SI unit is the joule.

internal pair production Production of electron-positron pair in the coulomb field of a nucleus. For transitions where the excitation energy released exceeds 1.02 MeV, this process will be competitive with both internal conversion and γ-ray emission. It occurs most readily in nuclei of low atomic number when the excitation energy is several MeV.

international Ångström A unit which, although very nearly equal to the ångström unit (10^{-10} m), is defined in a different way. It is such that the red cadmium line at 15°C and 760 mm Hg pressure would have a wavelength of 6438.469 6 IÅ. Formerly the reference standard for metrology and spectroscopy. Abbrev. *IÅ*.

International Atomic Energy Agency Autonomous intergovernmental body for promoting the peaceful uses of nuclear energy and ensuring as far as possible that it is not used to further military objectives. Abbrev. *IAEA*.

international electrical units Units (amp, volt, ohm, watt) for expressing magnitudes of electrical quantities, adopted internationally until 1947; replaced first by **MKSA**, later by **SI units** p. 237.

international practical temperature scale A practical scale of temperature which is defined to conform as closely as possible with the thermodynamic scale. Various fixed points were defined initially using the gas thermometer, and intermediate temperatures are measured with a stated form of thermometer according to the temperature range involved. The majority of temperature measurements of this scale are now made with platinum resistance thermometers.

international system of units See **SI units** p. 237.

interventional radiology A term used to describe radiological procedures undertaken for therapeutic rather than diagnostic purposes, e.g. *angioplasty*, where a balloon on a catheter is dilated in a narrowed artery.

intracavitary therapy Treatment applied within cavities of the body; said, e.g. of radium placed in the cavity of the uterus, also of irradiation of part of the body through natural or artificial body cavities.

intranuclear forces Those forces which operate between nucleons at close range comprising **short-range forces** and **coulomb forces**. According to the hypothesis of charge independence, the former are always the same for two nucleons of corresponding angular momentum and spin regardless of whether they are protons or neutrons. See **isotopic spin**.

intrinsic angular momentum The total spin of atom, nucleus or particle as an idealized point, or arising from orbital motion. When quantized, the former and latter are, respectively,

$$\frac{1}{2}\frac{h}{2\pi} \text{ and } \frac{h}{2\pi},$$

where h = Planck's constant.

intron Genes in eukaryotes are organized in such a way that while the whole sequence is transcribed, only part of it forms the messenger RNA. *Introns* or *intervening sequences*, which are often long, are excised during the maturation of the RNA. Cf. **exons**, the part which is expressed in the protein product. » p. 92.

inventory Total quantity of fissile material in a reactor.

inverse square law The law stating that the intensity of a field of radiation is inversely proportional to the square of the distance from the source. Applies to any system with a spherical wavefront and a negligible energy absorption.

iodine A black lustrous non-metallic halogen solid that sublimes easily. Symbol I, at.no. 53, r.a.m. 126.905, mp 113.5°C, bp 184°C, rel.d. 4.98. Abundance in the Earth's crust 0.46 ppm, in seawater 0.05 ppm. Concentrated by biological processes in marine plants and animals. It has 11 isotopes:

A	Abundance %	half-life	decay mode
123		13.2 h	ε
124		4.18 d	ε
125		60.2 d	ε
126		13.0 d	ε
127	100		
128		25.0 m	β^-
129		16 My	β^-
130		12.4 h	β^-
131		8.04 d	β^-
132		2.30 h	β^-
135		6.7 h	β^-

Iodine-135 with its 6.7 h half-life is an important fission product in nuclear reactors because it is the major source, by β-decay, of the nuclear poison, xenon-135. It is therefore the reason for the long delay before xenon reaches its maximum after power is reduced in a reactor.

ion Strictly, any atom or molecule which has resultant electric charge due to loss or gain of valency electrons. Free electrons are sometimes loosely classified as *negative ions*. Ionic crystals are formed of ionized atoms and in solution exhibit ionic conduction. In gases, ions are normally molecular and cases of double or treble ionization may be encountered. When almost completely ionized, gases form a fourth state of matter, known as a **plasma**. Since matter is electrically neutral, ions are normally produced in pairs. » p. 59.

ion beam A beam of ions moving in the same direction with similar speeds, esp. when produced by some form of accelerating machine or mass spectrograph. Also *ionic beam*.

ion concentration The number of ions of either sign, or of ion pairs, per unit volume. Also *ionization density*.

ion cyclotron frequency An ion in a uniform magnetic field, perpendicular to its motion, follows a circular path similar to that in a *cyclotron*. The critical frequency depends on the magnitude of the field and on the charge and mass of the ion. See **particle accelerators** p. 206.

ionic Appertaining to or associated with gaseous or electrolytic ions. NB *Ion* is frequently used interchangeably with *ionic* as an adjective, e.g. in *ion(ic)* conduction.

ionic beam See **ion beam**.

ionization Formation of ions by separating atoms, molecules or radicals, or by adding or subtracting electrons from atoms by strong electric fields in a gas, or by weakening the electric attractions in a liquid, particularly water.

ionization by collision The removal of one or more electrons from an atom by its collision with another particle such as another electron or an α-particle. Very prominent in electrical discharge through a rarefied gas.

ionization chamber Instrument used in study of ionized gases and/or ionizing radiations. It comprises a gas-filled enclosure with parallel plate or co-axial electrodes, in which ionization of the gas occurs. For fairly large applied voltages, the current through the chamber is dependent only upon the rate of ion production. For very large voltages, additional ionization by collision enhances the current. The system is then known as a **proportional counter** or **Geiger-Müller**

counter. » p. 74.

ionization continuum The energy spectrum above the ionization threshold of an atom.

ionization cross-section Effective geometrical cross-section offered by an atom or molecule to an ionizing collision.

ionization current (1) The current passing through an ionization chamber. (2) The current passed by an ionization gauge, when used for measuring low gas pressures.

ionization density See **ion concentration**.

ionization potential Energy, in electron-volts (eV), required to detach an electron from a neutral atom. For hydrogen, the value is 13.6 eV. Atoms, other than hydrogen, may lose more than one electron and can be multiply ionized. Also *electron binding energy, radiation potential.* See **photoelectron spectroscopy**.

ionization temperature A critical temperature, different for different elements, at which the constituent electrons of an atom will become dissociated from the nucleus; hence a factor in deducing stellar temperatures from observed spectral lines indicating any known stage of ionization.

ionized (1) Electrolytically dissociated. (2) Converted into an ion by the loss or gain of an electron.

ionized atom An atom with a resultant charge arising from capture or loss of electrons; an *ion* in gas or liquid.

ionizing collision Interaction between atoms or elementary particles in which an ion pair is produced.

ionizing energy The energy required to produce an ion pair in a gas under specified conditions. Measured in eV. For air it is about 32 eV.

ionizing event Any interaction which leads to the production of ions.

ionizing particle Charged particle which produces considerable ionization on passing through a medium. Neutrons, neutrinos and photons are not ionizing particles although they may produce some ions.

ionizing radiation Any electromagnetic or particulate radiation which produces ion pairs when passing through a medium. » p. 86.

ion pair Positive and negative ions produced together by transfer of electron from one atom or molecule to another.

ion source Device for releasing ions as these are required in a particle accelerator. A minute jet of gas or vapour of the required compound is ionized by heating at a filament or with an electron beam. The water-cooled magnet concentrates the ions near the filament and focuses them as they are accelerated into the main beam by the high-voltage electrode.

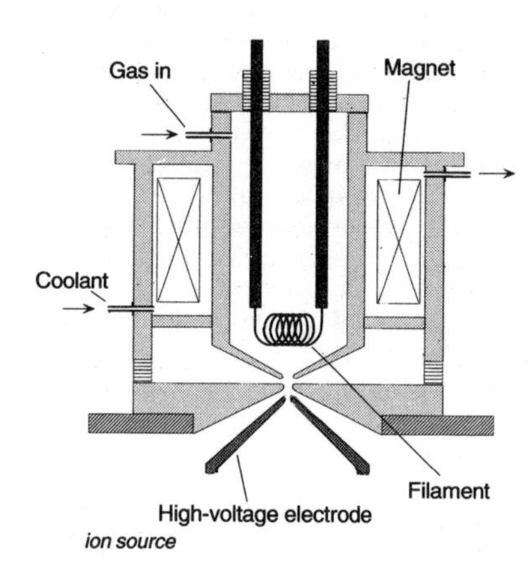

ion source

ion yield The average number of ion pairs produced by each incident particle or photon.

iron A metallic element in the eighth group of the periodic system. Symbol Fe, at.no. 26, r.a.m. 55.847, mp 1525°C, bp 2800°C, rel.d. at 20°C 7.86, electrical resistivity $9.8{\times}10^{-8}$ ohm metres. As the basis metal in steel and cast iron, it is the most widely used of all metals. It is the fourth commonest element of the Earth's crust, with an abundance of 6.2%, and it is thought to make up 80% of the core of the Earth. It has 12 isotopes:

A	Abundance %	half-life	decay mode
51		0.25 s	ε
52		8.27 h	ε
53		8.51 m	ε
54	5.8		
55		2.7 y	ε
56	91.8		
57	2.15		
58	0.29		
59		44.6 d	β^-
60		1.5 My	β^-
61		6.0 m	β^-
62		68 s	β^-

As the major component in stainless and other steels used in nuclear reactors, iron's behaviour on neutron irradiation is important in determining the lifetime and the difficulty of decommissioning nuclear reactors.

irradiance See **radiant-flux density**.

irradiation Exposure of a body to X-rays, gamma-rays or other ionizing radiations.

irradiation of food See **food irradiation**, p. 160.

irradiation swelling Changes in density and volume of materials due to neutron irradiation.

irreversibility Physical systems have a tendency to change spontaneously from one state to another but not to change in the reverse direction. *Entropy* provides an indication of irreversibility.

isenthalpic Of a process carried out at constant enthalpy, or heat function H.

isentropic See **entropy**.

iso- Prefix from Gk. *isos*, equal.

isobar One of a set of nuclides having the same total of protons and neutrons with the same *mass number* and approximately the same *atomic mass*, e.g. the isotopes of hydrogen and helium, $^3_1$H and $^3_2$He.

isobaric spin See **isotopic spin**.

isocount contours The curves formed by the intersection of a series of **isocount surfaces** with a specified surface.

isocount surface A surface on which the counting rate is everywhere the same.

isodiametric, isodiametrical Having the same length vertically and horizontally.

isodiapheres Two or more nuclides having the same difference between the number of neutrons and the number of protons.

isodose chart A graphical representation of a number of isodose contours in a given plane, which usually contains the central ray.

isodose curve, isodose contour The curve obtained at the intersection of a particular **isodose surface** with a given plane.

isodose surface A surface on which the dose received is everywhere the same as in the figure.

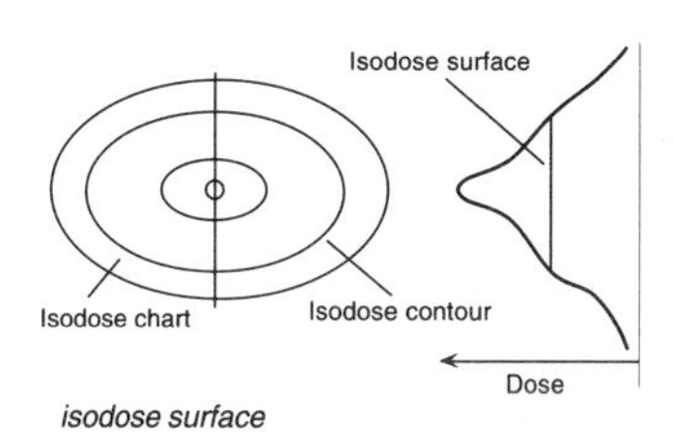

isodose surface

isomerism The existence of nuclides which have the same atomic number and the same mass number, but are distinguishable by their energy states; that having the lowest energy is stable, the others having varying lifetimes. If the lifetimes are measurable the nuclides are said to be in *isomeric states* and undergo *isomeric transitions* to the ground state.

isomer separation The chemical separation of the lower energy member of a pair of nuclear isomers.

isospin Contraction of **isotopic spin**.

isothermal (1) Occurring at constant temperature. (2) A curve relating quantities measured at constant temperature.

isothermal change A change in the volume and pressure of a substance which takes place at constant temperature. For gases, Boyle's law applies to isothermal changes.

isothermal lines, curves Curves obtained by plotting pressure against volume for a gas kept at constant temperature. For a gas sufficiently above its critical temperature for Boyle's law to be obeyed, such curves are rectangular hyperbolas.

isothermal temperature coefficient In a reactor in which the temperature is uniform over the core, the change in reactivity for a given temperature alteration.

isotones Nuclei with the same neutron number but different atomic numbers, i.e. those lying in a vertical column of a **Segrè chart**.

isotope One of a set of chemically identical species of atom which have the same **atomic number** but different **mass numbers**. A few elements have only one natural isotope, but all elements have artificially produced radioisotopes. (Gk. *isos*, same; *topos*, place). » p. 10.

isotope separation Process of altering the relative abundance of isotopes in a mixture. The separation may be virtually complete as in a mass spectrograph, or may give slight enrichment only as in each stage of a diffusion plant. » p. 56.

isotope structure Hyperfine structure of spectrum lines resulting from a mixture of isotopes in a source material. The wavelength difference is termed the *isotope shift*.

isotope therapy Radiotherapy by means of radioisotopes.

isotopic abundance See **abundance.**

isotopic dilution The mixing of a particular nuclide with one or more of its isotopes.

isotopic dilution analysis A method of determining the amount of an element in a specimen by observing the change in isotopic composition produced by the addition of a known amount of radioactive allobar.

isotopic number See **neutron excess**.

isotopic spin A quantum number assigned to members of a group of elementary particles differing only in electric charge; the particle groups are known as *multiplets*. Thus it is convenient to regard protons and neutrons as two manifestations of the nucleon, with isotopic spin either parallel or antiparallel to some preferred direction, i.e.

they have isotopic spin +½ and −½. The nucleon is then a doublet. This can be extended to all baryons and mesons. For example, the triplet π-meson consists of three pions. The small mass differences between the members of a multiplet is associated with their differing charges. The number of members of a multiplet set is $2I + 1$ where I is the isotopic spin, 0 for a singlet, ½ for a doublet, 1 for a triplet etc. The justification for the classification of particles is that all the members of a multiplet respond identically to strong nuclear interactions, the charges affecting only electromagnetic interactions. Isotopic spin is conserved in all strong interactions and never changes by more than one in a weak interaction. This classification is introduced by *analogy* with the spin or intrinsic angular momentum of atomic spectroscopy; isotopic spin has nothing to do with the nuclear spin of the particles. Also *isobaric spin*, *isospin*, *i-spin*.

isotron A device for the separation of isotopes. Pulses from a source of ions are synchronized with a deflecting field. The ions undergo deflections according to their mass.

isotropic Medium, having physical properties, e.g. magnetic susceptibility or elastic constants, which do not vary with direction.

iterated fission expectation Limiting value, after a long time, of the number of fissions per generation in the chain reaction initiated by a specified neutron to which this term applies.

JK

J Symbol for **joule**.

J Symbol for (1) electric current density; (2) **magnetic polarization**.

jacket See **can**.

jet (1) A fluid stream issuing from an orifice or nozzle. (2) A small nozzle, as the *jet* of a carburettor.

JET Abbrev. for *Joint European Torus*. Large **tokamak** experiment designed to use deuterium and tritium to produce energy by a fusion process. » p. 41.

jet nozzle process Process whereby isotope separation, based on the mass dependence of centrifugal force, is obtained by the fast flow of uranium hexafluoride in a curved duct. » p. 58.

Joshi effect Change of current in a gas because of light irradiation.

joule SI unit of work, energy and heat. One joule is the work done when a force of 1 newton moves its point of application 1 metre in the direction of the force. Symbol *J*. 1 erg = 10^{-7}J, 1 kW h = 3.6×10^{6}J, 1 eV = 1.602×10^{-19} J, 1 calorie = 4.18 J, 1 BTU = 1055 J.

Joule's equivalent See **mechanical equivalent of heat**.

Joule-Kelvin effect See **Joule-Thomson effect**.

Joule-Thomson effect When a gas is subjected to an *adiabatic expansion* through a porous plug or similar device, the temperature of the gas generally decreases. This effect is due to energy being used to overcome the cohesion of the molecules of the gas. The liquefaction of gases by the *Lindé process* depends on this effect. Also *Joule-Kelvin effect*.

J/Ψ particle A *meson* with zero strangeness and zero *charm*, but with a very high mass, 3.1 GeV, and an exceptionally long life. Discovered in 1974 at Brookhaven, by high-energy proton beam experiments (J-particle) and independently at the Stanford Linear Accelerator Centre by electron-positron colliding beam experiments (ψ-particle). The existence of this particle necessitated the postulation of the charm and the anti-charm quarks: the J/ψ particle is composed of one charm and one anti-charm quark. Also ψ/J *particle*.

JT60 Large **tokamak** experiment at the Japan Atomic Energy Research Institute. » p. 38.

k Symbol for (1) **Boltzmann's constant**; (2) radius of gyration.

κ – Symbol for (1) compressibility; (2) magnetic susceptibility.

K Symbol for **kelvin**.

kaon See **meson**.

kata- Prefix from Gk. *kata*, down. Also *cata-*.

K-capture Absorption of an electron from the innermost (K) shell of an atom into its nucleus. An alternative to ejection of a *positron* from the nucleus of a radioisotope. Also *K-electron capture*. See **X-rays**.

kCi Abbrev. for *kiloCurie*, i.e. radioactivity equivalent to 1000 curies.

K-electron capture See **K-capture**.

kelvin The SI unit of thermodynamic temperature. It is 1/273.15 of the temperature of the **triple point** of water above **absolute zero**. The temperature interval of 1 kelvin equals that of 1°C (degree Celsius). Symbol K. See **Kelvin thermodynamic scale of temperature**.

Kelvin thermodynamic scale of temperature A scale of temperature based on the thermodynamic principle of the performance of a reversible heat engine. The scale cannot have negative values so *absolute zero* is a well defined thermodynamic temperature. The temperature of the **triple point** of water is assigned the value 273.16 K. The temperature interval corresponds to that of the Celsius scale so that the freezing point of water (0°C) is 273.15 K. Unit is the **kelvin**. Symbol K.

kerma The initial energy of all the charged ionizing particles released by an uncharged ionizing particle in a given mass of tissue. Measured in gray (J kg^{-1}). » p. 103.

keV Abbrev. for *kilo-electron-volt*; unit of particle energy, 10^3 **electron volts**.

killer See **poison**.

kilo- Prefix denoting 1000; used in the metric system, e.g. 1 kilogram = 1000 grams.

kilocalorie See **calorie**.

kilocurie source Powerful radioactive source, usually in the form of cobalt-60.

kilocycles per second See **kilohertz**.

kilo-electron-volt See **keV**.

kilogram(me) Unit of mass in the MKSA (SI) system, being the mass of the *International prototype kilogram*, a cylinder of platinum-iridium alloy kept at Sèvres.

kilohertz One thousand **hertz** or cycles per second. A multiple of the SI unit of frequency. Abbrev. *kHz*.

kiloton Unit of explosive power for nuclear weapons equal to that of 10^3 tons of TNT.

kinematic viscosity The coefficient of

viscosity of a fluid divided by its density. Symbol ν. Thus $\nu = \eta/\rho$. Unit in the CGS system is the stokes ($cm^2\ s^{-1}$); in SI $m^2\ s^{-1}$.

kinetic energy The energy arising from motion. For a particle of mass m moving with a velocity v it is $\frac{1}{2} m v^2$, and for a body of mass M, moment of inertia I_g, velocity of centre of gravity v_γ and angular velocity ω, it is

$$\frac{1}{2} M v_\gamma^2 + \frac{1}{2} I_g \omega^2 .$$

kinetic friction See **friction**.

kink instability See **instability**.

Kirchhoff's law The ratio of the coefficient of absorption to the coefficient of emission is the same for all substances and depends only on the temperature. The law holds for the total emission and also for the emission of any particular frequency.

Klein-Gordon equation The equation describing the motion of a spinless charged particle in an electromagnetic field.

Klein-Nishina formula Theoretical expression for the cross-section of free electrons for the scattering of photons. See **Compton effect**.

***K*-lines** Characteristic X-ray frequencies from the atoms due to excitation of electrons in the K-shell. Denoted by *K*a, *K*b, *K*c; the lines are all doublets.

K- ,L- ,M- , ...Q-shells Imaginary spherical shells surrounding the nucleus of a many-electron atom in which groups of electrons are arranged according to the *Pauli exclusion principle*. Starting with the innermost shell the shells are called K, L, M, ..., Q corresponding to the principal quantum numbers n of 1,2,3,....7. Each shell contains $2n^2$ electrons. See **atomic structure** p. 121.

Knight shift Shift in nuclear magnetic resonance frequency in metals from that of the same isotope in chemical compounds in the same magnetic field. It is due to the paramagnetism of the conduction electrons.

Kr Symbol for **krypton**.

Kronig-Penny model A relatively simple model for a one-dimensional lattice from which the essential features of the behaviour of electrons in a periodic potential may be illustrated.

krypton A zero-valent element, one of the noble gases. Symbol Kr, at.no. 36, r.a.m. 83.80, mp −169C, bp −151.7C, density 3.743 g dm^{-3} at stp. It has 15 isotopes:

A	Abundance %	half-life	decay mode
75		4.3 m	ε
76		14.8 h	ε
77		75 m	ε
78	0.356		
79		35.0 h	ε
80	2.27		
81		0.21 My	ε
82	11.6		
83	11.5		
84	57.0		
85		10.7 y	β^-
86	17.3		
87		76 m	β^-
88		2.84 y	β^-
89		3.18 m	β^-

It is a colourless and odourless monatomic gas, and constitutes about 1-millionth by volume of the atmosphere, from which it is obtained by liquefaction. Krypton-85 is discharged into the atmosphere by nuclear plants, but not at a level considered to be a problem at present. Various methods for storing it over long periods have been investigated.

K-shell The innermost *electron shell* in an atom corresponding to a principal quantum number of unity. The shell can contain two electrons. See **atomic structure** p. 121.

Kurie plot Plot used for determining the energy limit of a β-ray spectrum from the intercept of a straight line graph. Prepared by plotting a function of the observed intensity against energy, the intercept on the axis being the energy limit for the spectrum. Also *Fermi plot*.

kV Abbrev. for *kiloVolt*.

kVp Abbrev. for *kiloVolts, peak*. Voltage applied across an X-ray tube, hence designating the maximum energy of emitted X-ray photons.

kymography A method of recording in a single radiograph the excursions of moving organs in the body, by use of a *kymograph*, which records physiological muscular waves, pulse beats or respirations.

λ Symbol for (1) **wavelength**; (2) linear coefficient of thermal expansion; (3) **mean free path**; (4) radioactive decay constant; (5) **thermal conductivity**.

L Symbol for (1) **angular momentum**; (2) luminance; (3) self-inductance.

labelled atom That atomic position in a molecular formula which is occupied by an isotopic tracer. Also *tagged atom*. See **radioactive tracer**.

labelled molecules Molecules containing a radioactive isotopic tracer.

labyrinth See **radiation trap** (2).

laevo- Prefix from L. *laevus*, left.

lambda point (1) Transition temperature of helium I to helium II. (2) Temperatures characteristic of second order phase change, e.g. ferromagnetic Curie point.

lambert Unit of luminance or surface brightness of a diffuse reflector emitting 1 lumen cm^{-2} or $10^4/\pi$ cd m^{-2}. The millilambert is used for low illuminations. See **lumen**.

Lamb shift A small difference in the $2S_{1/2}$ and $2P_{1/2}$ energy levels in the hydrogen atom, which, according to the Dirac theory, should be the same. The effect is explained by the theory of **quantum electrodynamics**.

lamell-, lamelli- Prefixes from L. *lamella*, thin plate.

laminar flow A type of fluid flow in which adjacent layers do not mix except on the molecular scale.

laminography See **tomography, emission**.

Landé splitting factor The factor employed in the calculation of the splitting of atomic energy levels by a magnetic field. See **Zeeman effect**.

large calorie See **calorie**.

Larmor radius The radius of the circular or helical path followed by a charged particle in a uniform magnetic field.

laser See panel on p. 182.

laser compression See **inertial confinement**.

laser enrichment The enrichment of uranium isotopes using a powerful laser to ionize atoms of the selected isotope of molecules bearing the uranium isotope. The separation can then be done by chemical or physical means. » p. 59.

laser fusion The initiation of the nuclear fusion process by directing energy from a laser beam on to the fusion fuel contained in pellets. The laser both heats and compresses the material. » p. 42.

laser fusion reactor A proposed type of fusion reactor in which pellets of deuterium and tritium are contained in a small target sphere and bombarded by a pulsed laser beam. Also *inertial fusion system*. » p. 42.

laser threshold The minimum pumping power (or energy) required to operate a laser.

latent heat More correctly, *specific latent heat*. The heat which is required to change the state of unit mass of a substance from solid to liquid, or from liquid to gas, without change of temperature. Most substances have a latent heat of fusion and a latent heat of vaporization. The specific latent heat is the difference in *enthalpies* of the substance in its two states.

latent neutrons In reactor theory, the delayed neutrons due from (but not yet emitted by) fission products. » p. 18.

latent period The time between exposure to radiation and its effect.

lateral Situated on or at, or pertaining to, a side.

lattice Regular geometrical pattern of discrete bodies of fissionable and non-fissionable material in a nuclear reactor. The arrangement must be subcritical or just critical if it is desired to study the properties of the system. » p. 26.

lattice constants The constants effecting the neutron flux in a reactor, chiefly those found in the **four factor formula**. » p. 22.

law A scientific law is a rule or generalization which describes specified natural phenomena within the limits of experimental observation. An apparent exception to a law tests the validity of the law under the specified conditions. A true scientific law admits of no exception. A law is of no scientific value unless it can be related to other laws comprehending relevant phenomena.

Lawson criterion The minimum physical conditions of plasma temperature (T), plasma density (n) and confinement time (τ) needed for the production of net power in fusion. It is expressed as $n\,\tau\,f(T)$ where $f(T)$ has a minimum value of around 2×10^{20} s m^{-3} for the deuterium-tritium reaction. » p. 40.

lb Abbrev. for **pound**.

L-capture Absorption of an electron from the L-shell into the nucleus, giving rise to X-rays of characteristic wavelength depending on atomic number of the element. See **K-capture**, **X-rays**.

leaching The removal, usually by water, of a substance from a solid as in the removal of fission products from the vitrified blocks designed to contain them.

laser

Light Amplification by Stimulated Emission of Radiation. A source of intense monochromatic light in the ultraviolet, visible or infrared region of the spectrum. It operates by producing a large population of atoms with their electrons in a certain high energy level. By *stimulated emission*, transitions to a lower level are induced, the emitted photons travelling in the same direction as the stimulating photons. If the beam of inducing light is produced by reflection from mirrors or **Brewster windows** at the ends of a resonant cavity, the emitted radiation from all stimulated atoms is in phase or *coherent*, and the output is a very narrow beam of coherent monochromatic light. Solids, liquids and gases have been used as lasing materials. They can vary in size from those used in printers and compact disk readers or even smaller, to powerful lasers which can cut thick metal, separate uranium isotopes (p. 59) or implode fusion devices (p. 42).

The quantum photolectric theory (p. 84) shows how a photon with the right energy can cause an electron to jump to a more energetic orbit but in addition an electron already in a more energetic orbit will be forced to return to its original orbit if it collides with an electron with the appropriate energy. In doing so it will emit an additional photon with the same energy and the same phase and travelling in the same direction. This is called *stimulated emission.* Normally most electrons will be in their least energetic state but in special conditions it is possible to force a high proportion into a more energetic state so that a *population inversion* occurs. An incoming electron can then cause a burst of coherent and monochromatic light.

There are many kinds of laser depending on the output and wavelength required but the helium-neon laser can illustrate the principle. If helium at low pressure is placed in a gas-discharge tube the electrons passing from the cathode to the anode will excite the helium atoms into a high-energy state (Fig. 1) in which the excited atoms can collide with any other atoms that are present. It so happens that neon also has an excited state at the same energy level as helium and if excited helium atoms collide with rest-state neon atoms, there is a good chance of the neon atoms moving into their excited state. By adjusting the proportions of helium and neon in the tube a high proportion of the neon atoms can be maintained in the high-energy state and a population inversion will occur. An appropriate incoming photon can now force a neon electron into a lower-energy state and emit another photon which then stimulates another neon electron, producing a pulse of highly coherent light. The output can be further enhanced by having a fully reflecting mirror at one end and a semireflecting mirror at the other end of the tube and adjusting its length (see **mode-locking**). This will cause the photons to traverse the tube many times, stimulating further emissions, before they finally escape.

Powerful lasers are often arranged in tandem so that the stimulated emission of one laser can cause the population inversion in a second and so enhance the power of the whole system. See **carbon-dioxide laser**, **maser**, **Q-switching.**

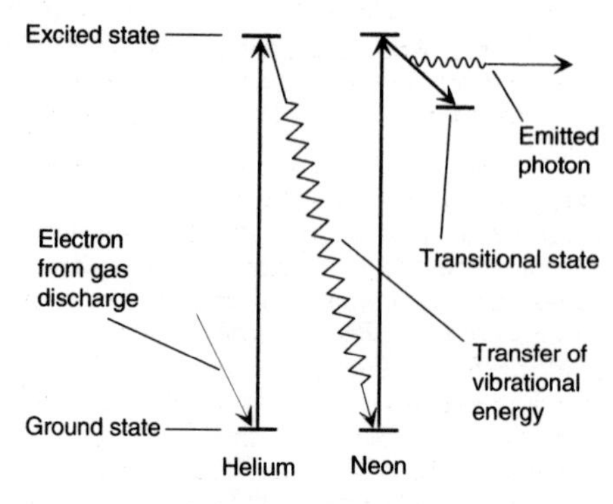

Fig. 1 **Energy states in a helium-neon laser.**

Lifetime Study

A term applied to the long-term study of the survivors of the Hiroshima and Nagasaki bombs, detonated on 6 and 10 August 1945. This research is now supported by the Radiation Effects Research Foundation and has been carried out by Japanese and US scientists, publishing a large number of reports over the years. It is by far the largest study of the biological effects of a single acute radiation dose.

Two methods have been used. For 80 000 of the survivors it has been possible to calculate from position, shielding and symptoms, dose rates varing from nothing to over four grays. Their death rates from cancer were then compared to that of the total Japanese population. In addition, about 68 000 children were divided into three groups depending on their parents' history: those irradiated within 2000 metres of the two epicentres; those placed at more than 2500 metres from them and those who returned to the cities after the bombs. The children were further classified according to the gonadal dose to both their parents.

The absolute risk for different cancers in the survivors is shown in Fig. 2 (the scale is the excess death per million person-years for a dose of 10 milligrays). It shows the expected rise in excess deaths from leukaemia comparatively soon after irradiation, followed by a fall 30 years later. Other cancers, with their longer latent periods, become more significant after 20 years. In general those who received more than two grays had about twice as many cancers during the 33 years following the bomb, except for leukaemias and multiple myeloma, where there were just over 10 and 5 times as many cases.

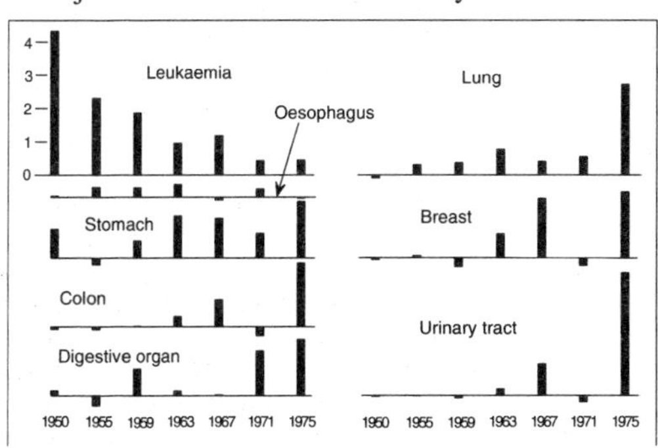

Fig. 2 **Absolute cancer risk.** All sites measured on the same scale. Digestive organ refers to sites other than rectum, pancreas and those recorded here. Dates refer to beginning of each period.

The 68 000 children have been investigated in many ways, new methods being added as they have become available. The number of cancers was recorded but also the number of congenital abnormalities, deaths other than cancer, chromosomal rearrangements, abnormal number of sex chromosomes, sex ratio, abnormal proteins as evidence for genetic recessives, and general growth and development. But there is no statistically significant evidence for any difference between the children of irradiated parents and the matched controls. Despite this and because such a difference has been found in mice, the investigators have tried to derive a figure for the genetic doubling dose. In other words: what is the dose necessary to double the number of mutations compared to the background? They have calculated a figure between two and three sieverts for gonadal exposure compared to 0.3 to 0.4 for male mice.

lead A metallic element in the fourth group of the periodic system. Symbol Pb, at.no. 82, r.a.m. 207.19, valency 2 or 4, mp 327.4°C, bp 1750°C, rel.d. at 20°C 11.35, specific electrical resistivity 20.6×10^{-8} ohm metres. Occurs chiefly as **galena**. It has 12 isotopes:

A	Abundance %	half-life	decay mode
201		9.3 h	ε
202		0.05 My	ε
203		51.9 h	ε
204	1.42		
205		15 My	ε
206	24.1		
207	22.1		
208	52.3		
209		3.25 h	β^-
210		22.3 y	β^-
211		36.1 m	β^-
212		10.6 h	β^-

It is used as shielding in X-ray and nuclear engineering because of its relative cheapness, high density and nuclear properties. Abundance in the Earth's crust 13 ppm. Toxic.

lead equivalent Absorbing power of radiation screen expressed in terms of the thickness of lead which would be equally effective for the same radiation.

lead protection Protection provided by metallic lead against ionizing radiation. Joined with other substances to provide further protection, e.g. lead glass, lead rubber.

lead rubber Rubber containing high properties of lead compounds. Used as flexible protective material for, e.g. gloves and aprons.

leakage Net loss of particles from a region or across a boundary, e.g. neutrons from the core of a reactor, often split up for calculation purposes into *fast non-leakage probability* (P_f); *resonance non-leakage probability* (P_r); *thermal non-leakage probability* (P_t). » p. 20.

leak detector (1) Device for indicating points of ingress of gas into a high vacuum system. (2) See **burst-can detector**.

least energy principle A system is only in stable equilibrium under those conditions for which its potential energy is a minimum.

lepto- Prefix from Gk. *leptos*, slender.

lepton A fundamental particle that does not interact strongly with other particles. There are several different types of lepton; the electron, the negative muon, the tau-minus particle and their associated neutrinos. See **antilepton**.

lepton number The number of **leptons** less the number of corresponding antiparticles taking part in a process. The number appears to be conserved in any process.

lepton-quark symmetry The number of *pairs* of leptons (Charge $-e$ and charge 0) is equal to the number of *pairs* of quarks (charge $+2e/3$ and charge $-e/3$). The number of types of leptons is thus equal to the number of types of quarks. Required by the theory unifying electromagnetic and weak interactions between particles. See **Weinberg and Salam's theory**.

lethargy of neutrons The natural logarithm of the ratio of initial and actual energies of neutrons during the moderation process. The lethargy change per collision is defined similarly in terms of the energy values before and after the collision. Also *logarithmic energy decrement.*

leuco-, leuko- Prefixes from Gk. *leukos*, white.

level Possible energy value of electron or nuclear particle.

levitron Toroidal fusion system in which the poloidal magnetic field is provided by a current flowing in a solid ring situated at the circular axis of the torus. The ring may be supercooled so it can be levitated indefinitely, avoiding the use of material supports.

Li Symbol for **lithium**.

lifetime The mean period between the birth and death of a charge carrier in a semi-conductor. See **half-life, mean life**.

Lifetime Study See panel on p. 183.

light **Electromagnetic radiation** capable of inducing visual sensation, with wavelengths between about 400 and 800 nm. See **illumination, speed of light**.

light efficiency The ratio of total luminous flux over total power input, expressed in lumens per watt in e.g. an electric lamp.

light flux The measure of the quantity of light passing through an area, e.g. through a lens system. Light flux is measured in lumens. Illumination is light flux per unit area.

light quanta When light interacts with matter, the energy appears to be concentrated in discrete packets called *photons*. The energy of each photon $E = h\nu$ where ν is the frequency and h is Planck's constant. See **gamma radiation, quantum, X-rays**. » p. 80.

light sensitive Thin surfaces for which the electrical resistance, emission of electrons or generation of a current, depends on the incidence of light.

light water Normal water (H_2O) as distinct from **heavy water**.

light-water reactor A reactor using ordinary water as moderator and coolant. See **boiling water reactor, pressurized water**

reactor. Abbrev. *LWR*. » p. 24.

light watt The photometric radiation equivalent, the radiation being measured in lumens or watts. The ratio of lumens to watts depends on the wavelength. At the wavelength of greatest sensitivity of the eye λ = 555 nm, the value of the light watt is 682 lumens.

limiter An aperture which defines the boundary of a plasma and protects the vacuum vessel from damage by contact with hot plasma.

limit values A European Community term specifying environmental quality standards and emission standards, which defines the limits that member states must set, e.g. on concentrations of smoke and sulphur dioxide at ground level.

line Single frequency of radiation as in a luminous, X-ray or neutron spectrum.

linear Said of any device or motion where the effect is exactly proportional to the cause, as rotation and progression of a screw, current and voltage in a wire resistor (Ohm's law) at constant temperature, output versus input of a modulator (or demodulator).

linear absorption coefficient See **absorption coefficient**.

linear energy transfer The linear rate of energy dissipation by particulate or electromagnetic radiation while penetrating absorbing media. Abbrev. *LET*.

linear stopping power See **stopping power**.

line spectrum A spectrum consisting of relatively sharp lines, as distinct from a **band spectrum** or a **continuous spectrum**. Line spectra originate in the atoms of incandescent gases or vapours. See **Bohr theory**, **spectrum**.

lingering period The time interval during which an electron remains in its orbit of highest excitation before jumping to the energy level of a lower orbit.

liquefaction temperature The temperature at which a gas changes state to liquid. Physically the same as the boiling point.

liquid A state of matter between a solid and a gas, in which the shape of a given mass depends on the containing vessel, the volume being independent. A liquid is practically incompressible.

liquid counter A counter for measuring the radioactivity of a liquid, usually designed to measure β-, as well as γ-rays. Also *liquid scintillation counter*. » p. 76.

liquid-drop model A model of the atomic nucleus using the analogy of a liquid drop in which the various concepts of surface tension, heat of evaporation etc. are employed. A semi-empirical mass formula can be developed from the model which describes the masses of many stable and unstable nuclei in terms of a few parameters.

liquid-flow counter A counter for continuous monitoring of radioactivity in flowing liquids.

liquid-metal reactor (1) Normally a reactor designed for liquid-metal (usually sodium) cooling. (2) Occasionally a liquid-metal fuelled reactor. Abbrev. *LMR*. » p. 36.

liquid scintillation counter See **liquid counter**. » p. 76.

lithium An element. Symbol Li, at.no. 3, r.a.m. 6.939, mp 186°C, bp 1360°C, rel.d. 0.585. It is the least dense solid, chemically resembling sodium but less active. Abundance in the Earth's crust 20 ppm. It has three isotopes:

A	Abundance %	half-life	decay mode
6	7.5		
7	92.5		
8		0.84 s	β–

The isotope lithium-6 is an important constituent of thermonuclear bombs. » p. 68. It is also used in alloys and in the production of tritium; also as a basis for lubricant grease with high resistance to moisture and extremes of temperature. In nuclear engineering it will be used as the *chemical blanket* for extracting heat from neutron capture in fusion reactors. » p. 41.

lithium-drifted silicon detector An energy-sensitive solid-state detector for ionizing radiation. Lithium is thermally diffused into almost pure but p-type silicon crystal. Used for low-energy X-ray and gamma ray spectroscopy. The crystal is kept at liquid nitrogen temperature to reduce thermal noise and to avoid the redistribution of lithium ions. » p. 76.

lithium-drift germanium detector One type of **germanium radiation detector**. There is also an *intrinsic germanium detector* which does not contain lithium. » p. 76.

litre Unit of volume equal to 1 cubic decimetre.

L-lines Characteristic X-ray frequencies from atoms due to the excitation of electrons from the L-shell. Denoted by *L*a,*L*b, ...; the lines are multiplets.

LMFBR Abbrev. for *Liquid-Metal-cooled Fast-Breeder Reactor*. See **breeder reactor**, **fast reactor**. » p. 36.

LMR Abbrev. for **Liquid-Metal Reactor**.

load (1) The weight supported by a structure. (2) The mechanical force applied to a body.

loaded concrete Concrete used for shielding nuclear reactors, loaded with elements of high atomic number, e.g. lead, iron shot, barytes.

load factor The ratio of the average load to peak load over a period.

load following See **self regulating.** » p. 24.

loading The introduction of fuel into a reactor.

loading and unloading machine The system for introducing and withdrawing fuel elements from a reactor, with safety provision for personnel. See **charge, discharge machine**.

LOCA Abbrev. for *Loss-Of-Coolant-Accident*. The conditions which might arise in the event of loss of primary or secondary coolant in a reactor. » p. 46.

logarithmic decrement Logarithm to base *e* of the ratio of the amplitude of diminishing successive oscillations.

log-dec Abbrev. for **LOGarithmic DECrement**.

longi- Prefix from L. *longus*, long.

long ton A unit of mass, 2240 lb. See **ton**.

loop-type reactor An **indirect-cycle** reactor in which the heat exchanger is outside the pressure vessel.

Lorentz force The force experienced by a point charge q moving with a velocity $\mathbf{v}$ in a field of magnetic induction $\mathbf{B}$ and in an electric field $\mathbf{E}$. The force $\mathbf{F}$ is given by

$$\mathbf{F} = q\,[(\mathbf{v} \times \mathbf{B}) + \mathbf{E}\,]\,.$$

Lorentz transformations The relations between the co-ordinates of space and time of the same event as measured in two inertial frames of reference moving with a uniform velocity relative to one another. They are derived from the postulates of the special theory of **relativity**.

low-level waste In general, those radioactive wastes which, because of their low activity, do not require shielding during normal handling or transport.

L-shell The *electron shell* in an atom corresponding to a principal quantum number of two. The shell can contain up to 8 electrons. See **atomic structure** p. 121.

lumen Unit of **luminous flux**, being the amount of light emitted in unit solid angle by a small source of output, one **candela**. In other words, the lumen is the amount of light which falls on unit area when the surface area is at unit distance from a source of one candela. Abbrev. *lm*.

luminescence Emission of light (other than from thermal energy causes) such as *bioluminescence* etc. Thermal luminescence in excess of that arising from temperature occurs in certain minerals. See **fluorescence, phosphorescence**.

luminous flux The flux emitted within unit solid angle of one steradian by a point source of uniform intensity of one **candela**. Unit of measurement is the *lumen*.

LWR Abbrev. for **Light-Water Reactor**.

Lysholm grid A type of grid interposed between the patient and film in diagnostic radiography in order to minimize the effect of scattered radiation.

M

m Abbrev. for **milli-**.

m Symbol for (1) electromagnetic moment; (2) **mass**; (3) mass of electron; (4) molality.

μ Symbol for (1) index of refraction; (2) magnetic permeability; (3) **micron** (obsolete); (4) the prefix **micro-**.

machine A device for overcoming a resistance at one point by the application of a force at some other point. Typical simple machines are the inclined plane, the lever, the pulley, and the screw. See **mechanical advantage**.

Mach principle The principle that scientific laws are descriptions of nature, are based on observation, and alone can provide deductions which can be tested by experiment and/or observation.

macr-, macro- Prefixes from Gk. *makros*, large.

macroscopic Visible to the naked eye.

magic number Extra stable atomic nuclei exist which have a *magic number*, 2, 8, 20, 28, 50, 82 or 126, of either protons or neutrons.

magnesium A light metallic element in the second group of the periodic system. Symbol Mg, at.no. 12, r.a.m. 24.312, mp 651°C, bp 1120°C at 1 atm, rel.d. 1.74, electrical resistivity 42×10^{-8} ohm metres, specific latent heat of fusion 377 kJ kg^{-1}. It has nine isotopes:

A	Abundance %	half-life	decay mode
21		0.123 s	ε
22		3.86 s	ε
23		11.3 s	ε
24	78.99		
25	10.00		
26	11.01		
27		9.46 m	β^-
28		21.0 h	β^-
29		1.4 s	β^-

The metal is brilliant white, and magnesium ribbon burns in air, giving an intense white light. The sixth most abundant element of the Earth's crust, which contains 2.76% Mg but the figure is much higher in the Earth as a whole. There is 0.13% in sea-water. Apart from its widespread use as a structural material in the aircraft and other industries, it is also widely used alloyed with 0.8% aluminium and 0.5% beryllium in the form of **Magnox** as a fuel cladding in gas-cooled reactors. » p. 31. The element has the following properties:

σ_{cap} mb	mp °C	Therm. conduct. W $m^{-1}K^{-1}$	density g cm^{-3}
63	650	156	1.85

magnet A mass of iron or other material which possesses the property of attracting or repelling other masses of iron, and which also exerts a force on a current-carrying conductor placed in its vicinity.

magnet core The iron core within the coil of an electromagnet.

magnetic Said of all phenomena essentially depending on magnetism.

magnetic axis A line through the effective centres of the poles of a magnet.

magnetic bottle The containment of a plasma during thermonuclear experiments by applying a specific pattern of magnetic fields. Also *magnetic trap*. See **magnetic confinement.**

magnetic confinement In fusion research, the use of shaped magnetic fields to confine a plasma.

magnetic confinement fusion system Fusion system in which plasma is confined, i.e. not allowed to come in contact with the walls of the chamber, by use of specially shaped magnetic fields. Cf. **inertial fusion system**. » p. 38.

magnetic lens The counterpart of an optical lens in which a magnet or system of magnets is used to produce a field which acts on a beam of electrons in a similar way to a glass lens on light rays. Comprises current-carrying solenoids of suitable design. The figure shows a quadrupole lens of which two, the second rotated by 90° with respect to the first, are needed in tandem to focus a beam of electrons.

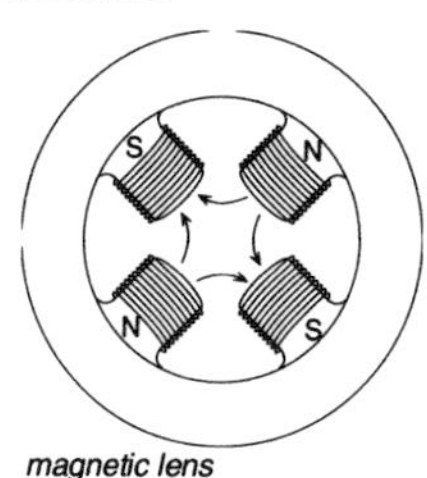

magnetic lens

magnetic levitation A method of opposing the force of gravity using the mutual repulsion between two like magnetic poles.

magnetic mirror Device based on the principle that ions moving in a magnetic field towards a region of considerably higher magnetic field strength are reflected. This principle is used in **mirror machines**. Mirrors may be *simple, minimum* β (more efficient) or *tandem* (solenoid plugged at both ends by minimum β-filters).

magnetic moment Vector such that its product with the magnetic induction gives the torque on a magnet in a homogeneous magnetic field. Also *moment of a magnet.*

magnetic polarization The magnetic polarization **J** is related to the magnetic field **H** and the magnetic induction **B** by

$$\mathbf{B} = \mu_0 \mathbf{H} + \mathbf{J},$$

where μ_0 is the permeability of free space.

magnetic pressure The pressure which a magnetic field is capable of exerting on a plasma.

magnetic pumping Use of radio frequency currents in coils over bulges in the tube of a stellarator or tokamak fusion reactor to modulate the steady axial field and provide heat to the plasma. This process is most efficient when there is resonance between the radiofrequency signals and vibrations of the molecules of plasma. The process is then called *resonance heating*.

magnetic resonance imaging The use of nuclear magnetic resonance of protons to produce proton density maps or images of the human body.

magnetic rigidity A measure of the momentum of a particle. It is given by the product of the magnetic intensity perpendicular to the path of the particle and the resultant radius of curvature of this path.

magnetic saturation The limiting value of the magnetic induction in a medium when its *magnetization* is complete and perfect.

magnetic separator A device for separating, by means of an electromagnet, any magnetic particles in a mixture from the remainder of the mixture, e.g. for separating iron filings from brass filings.

magnetic stability A term used to denote the power of permanent magnets to retain their magnetism in spite of the influence of external magnetic fields, vibration etc.

magnetic trap See **magnetic bottle, magnetic confinement**.

magnetic tube See **trapping region**.

magnetic units Units for electric and magnetic measurements in which μ_0, the permeability of free space, is taken as unity. Now replaced by SI units in which μ_0 is given the value $4\,\pi \times 10^{-7}$ H m^{-1}.

magnetism Science covering magnetic fields and their effect on materials, due to unbalanced spin of electrons in atoms. See **Coulomb's law for magnetism**.

magnetization Orientation from randomness of saturated *domains* in a ferromagnetic material. The magnetization per unit volume **M** is related to the field strength **H** and the magnetic induction **B** by

$$\mathbf{B} = \mu_0 (\mathbf{H} + \mathbf{M}) \text{ tesla},$$

where μ_0 is the permeability of free space. SI unit is ampere per metre.

magnetize (1) To induce magnetization in ferromagnetic material by direct or impulsive current in a coil. (2) To apply alternating voltage to a transformer or choke having ferromagnetic material, generally laminated, in its core.

magnetohydrodynamic instability See **instability**.

magnetohydrodynamics The study of the motions of an electrically conducting fluid in the presence of a magnetic field. The motion of the fluid gives rise to induced electric currents which interact with the magnetic field which in turn modifies the motion. The phenomenon has applications both to magnetic fields in space and to the possibility of generating electricity. If the free electrons in a plasma or high velocity flame are subjected to a strong magnetic field, then the electrons will constitute a current flowing between two electrodes in a flame. Abbrev. *MHD*.

magnetomotive force Line integral of the magnetic field intensity round a closed path. Abbrev. *m.m.f.*

magnetostatics Study of steady-state magnetic fields.

magnifier Any thermionic amplifier, esp. one used for the amplification of audiofrequencies.

Magnox Name of a magnesium-containing fuel cladding for natural uranium metallic fuel used in the first commercial power reactors built in the UK, which are consequently called Magnox reactors. Magnox alloy has the following properties:

σ_{cap} mb	mp °C	Therm. conduct. W m^{-1}K^{-1}	density g cm^{-3}
65	647	156	1.75

» p. 31.

Magnus effect Force experienced by a spinning ball or cylinder in a fluid. The effect is responsible for the swerving of golf and tennis balls when hit with a slice.

malignant (1) Tending to go from bad to worse. (2) Cancerous. See **tumour**.

mamu Abbrev. for **millimass unit**.

manipulator Remote handling device used, e.g. with radioactive materials. The diagram shows a much simplified design, the operator being protected by a lead-glass window while watching the process. In practice a number of additional movements would be necessary.

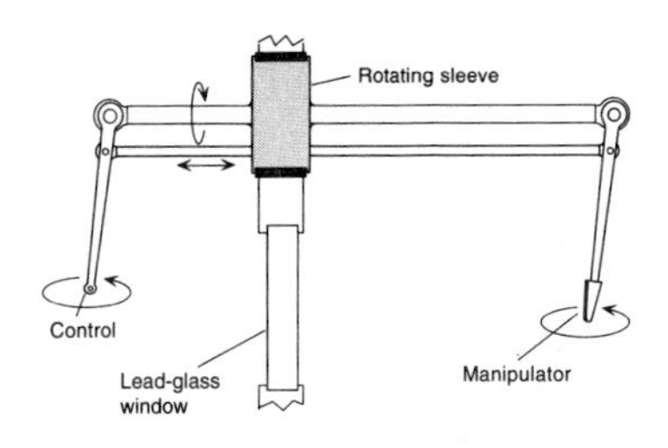

manipulator

maser Abbrev. for *Microwave Amplification by Stimulated Emission of Radiation*. A *microwave* oscillator that operates on the same principle as the **laser**. Maser oscillations produce coherent monochromatic radiation in a very narrow beam. Less noise is generated than in other kinds of microwave oscillators. Materials used are generally solid-state, but masers have been made using gases and liquids.

maser relaxation The process by which excited molecules in the higher energy state revert spontaneously or under external stimulation to their equilibrium or ground state. Maser action arises when energy is released in this process to a stimulating microwave field which is thereby reinforced.

mass The quantity of matter in a body. Two principle properties are concerned with mass: (1) the *inertial mass*, i.e. the mass as a measure of resistance of a body to changes in its motion; (2) the *gravitational mass*, i.e. the mass as a measure of the attraction of one body to another. The general theory of relativity shows that inertial and gravitational mass are equivalent. By a suitable choice of units they can be made numerically equal for a given body, which has been confirmed to a high degree of precision by experiment. Mass is always conserved. Cf. **rest mass**. See **mass-energy equation**, **relativistic mass equation**.

mass absorption co-efficient See **absorption co-efficient**.

mass decrement, mass defect Both terms may be used for (1) the **binding energy of a nucleus** expressed in mass units, and (2) the measured mass of an isotope less its mass number. The BSI recommend the use of *mass decrement* for (2) above. In the USA, it is usually used for (1) and *mass excess* for (2). See **packing fraction**. » p. 13.

mass-energy equation The equation $E = mc^2$. Confirmed deduction from Einstein's special theory of relativity that all energy has mass. If a body gains energy E, its inertia is increased by the amount of mass:

$$m = \frac{E}{c^2},$$

where c is the speed of light. Derived from the assumption that all conservation laws must hold equally in all frames of reference and using the principle of conservation of *momentum*, of *energy* and of *mass*. See **rest-mass energy**.

mass excess See **mass decrement**.

mass limitation The method of **criticality control** in a plant in which the total mass of fissile material is limited.

mass number Total of *protons* and *neutrons* in a nucleus, each being taken as a unit of mass. Also *nuclear-, nucleon number*.

mass of electron See **electron mass**.

mass spectrometer A *mass spectrograph* in which the charged particles are detected electrically instead of photographically.

mass spectrum See **spectrum**.

mass stopping power See **stopping power**.

mass unit See **atomic mass unit**.

materialization Reverse of Einstein energy released with annihilation of mass. A common example is by *pair production* (electron-positron) from gamma rays.

materials testing reactor See **high-flux reactor**.

mathematical modelling The representation by mathematical expressions of a physical event as an aid to understanding the process.

mathematics The study of the logical consequences of sets of axioms. *Pure mathematics*, roughly speaking, comprises those branches studied for their own sake or their relation to other branches. The most important of these are algebra, analysis and topology. The term *applied mathematics* is usually restricted to applications in physics. Applications in other fields, e.g. economics, mainly statistical, are sometimes referred to as *applicable mathematics*.

matter The substances of which the physical universe is composed. Matter is characterized by gravitational properties (on Earth by weight) and by its indestructibility under normal conditions.

maximum permissible concentration The recommended upper limit for the dose which may be received during a specified period by a person exposed to ionizing radiation. Also *permissible dose*.

maximum permissible dose rate, flux That dose rate or flux which, if continued throughout the exposure time, would lead to the absorption of the maximum permissible dose.

maximum permissible level A phrase used loosely to indicate *maximum permissible concentration, dose* or *dose rate*.

maximum value See **peak value**.

Maxwell's circuital theorems The generalized forms of Faraday's law of induction and Ampère's law (modified to incorporate the concept of displacement current). Two of *Maxwell's field equations* are direct developments of the circuital theorems.

Maxwell's distribution law The distribution of numbers of gas molecules which have given speeds, or kinetic energies in a gas of uniform temperature. The law can be deduced from the *kinetic theory of gases*.

Maxwell's field equations Mathematical formulations of the laws of Gauss, Faraday and Ampère from which the theory of electromagnetic waves can be conveniently derived.

Maxwell's thermodynamic relations Four mathematical identities relating the pressure, the volume, the entropy and the thermodynamic temperature for a system in equilibrium. They are expressed in the form of partial derivatives relating the quantities.

maze See **radiation trap** (2).

mean calorie See **calorie**.

mean free path The mean distance travelled by a molecule in a gas between collisions. It is dependent on the molecular cross-section $\pi\sigma^2$ so that

$$\lambda = \frac{1}{\sqrt{2}\pi\, n\sigma^2},$$

where n is the number of molecules per unit volume. According to the kinetic theory of gases it is related to the viscosity η by $\lambda = k\eta/\rho u$ where ρ = density, u = mean molecular velocity and k is a constant between ⅓ and ½ depending on the approximation in the theory.

mean free time Average time between collisions of electrons with impurity atoms in semiconductors; also of intermolecular collision of gas molecules.

mean lethal dose The single dose of whole body irradiation which will cause death, within a certain period, to 50% of those receiving it. Abbrev. *MLD*.

mean life (1) The average time during which an atom or other system exists in a particular form, e.g. for a thermal neutron it will be the average time interval between the instant at which it becomes thermal and the instant of its disappearance in the reactor by leakage or by absorption. Mean life = 1.443 × *half-life*. Also *average life*. (2) The mean time between birth and death of a charge carrier in a semiconductor, a particle (e.g. an ion, a pion) etc.

mean residence time Mean period during which radioactive debris from nuclear weapon tests remains in the stratosphere.

mechanical advantage The ratio of the resistance (or load) to the applied force (or effort) in a **machine**.

mechanical equivalent of heat Originally conceived as a conversion coefficient between mechanical work and heat (4.186 joules = 1 calorie) thereby denying the identity of the concepts. Now recognized simply as the specific heat capacity of water, 4.186 kJ kg^{-1}.

mechanical resonance In a mechanical vibrating system, the enhanced response to a driving force as the frequency of this force is increased through a resonant frequency at which the inertial reactance balances the reactance due to the compliance of the system.

mechanics The study of forces on bodies and of the motions they produce.

mechanomotive force The r.m.s. value of an alternating mechanical force, in newtons, developed in a transducer.

medi-, medio- Prefixes from L. *medius*, middle.

mega- Prefix denoting 1 million, or 10^6, e.g. a frequency of 1 *megahertz* is equal to 10^6 Hz; 1 *megawatt* = 10^6 watts; 1 *megavolt* = 10^6 volts.

mega-electron-volt See **MeV**.

mega-gramme-röntgen See **gram(me)-röntgen**.

megaton Explosive force equivalent to 1 000 000 tons of TNT, taken as 10^{-12} calories. Used as a unit for classifying nuclear weapons. » p. 65.

megavoltage therapy See **supervoltage therapy**.

megawatt days per tonne Unit for energy output from reactor fuel; a measure of burn-up. Expressed as MWd/tonne.

meltdown A type of nuclear reactor accident in which the fuel becomes excessively overheated so that it melts and collapses into or through the fabric of the reactor. Overheating may be caused by loss of *coolant* function or of **moderator**.

Mendeleev's law 'Law of Octaves'. If elements are listed according to increasing atomic weight, their properties vary but show a general similarity at each period of eight true rises.

mesic atom Short-lived atom in which a

negative *muon* has displaced a normal electron.

meso- Prefix from Gk. *mesos*, middle. Also *mes-*.

meson A hadron with a baryon number of 0. Mesons generally have masses intermediate between those of electrons and nucleons and can have negative, zero or positive charges. Mesons are *bosons* and may be created or annihilated freely. There are three groups of mesons: π-mesons (pions), K-mesons (kaons) and η-mesons.

meson field The field concerned with the interchange of protons and neutrons in the nucleus of an atom, mesons transferring the energy.

messenger RNA The RNA whose sequence of nucleotide triplets determines the sequence of a polypeptide.

meta- Prefix from Gk. *meta*, after.

metastasic Said of electrons which move from one shell to another or are absorbed from a shell into the nucleus.

metastasis The transfer, by lymphatic channels or blood vessels, of diseased tissue (esp. cells of malignant tumours) from one part of the body to another; the diseased area arising from such transfer.

meter US for **metre**.

metre The Système Internationale (SI) fundamental unit of length. The metre is defined (1983) in terms of the velocity of light. The metre is the length of path travelled by light in vacuum during a time interval of 1/299 792 458 of a **second**. Originally intended to represent 10^{-7} of the distance on the Earth's surface between the North Pole and the Equator, formerly it was defined in terms of a line on a platinum bar and later (1960) in terms of the wavelength from krypton-86. US *meter*.

metre-kilogram(me)-second-ampere See **MKSA**.

metric system A system of weights and measures based on the principle that each quantity should have one unit whose multiples and submultiples are all derived by multiplying or dividing by powers of 10. This simplifies conversion, and eliminates completely the complicated tables of weights and measures found in the traditional British system. Originally introduced in France, it is the basis for the *Système International* (*SI*) now universally adopted. See **SI units** p. 237.

metrology The science of measuring.

MeV Abbrev. for mega-electron-volt. Unit of particle energy. 10^6 electron-volts.

Mg Symbol for **magnesium**.

mg Abbrev. for *milligram*.

MHD Abbrev. for **magnetohydrodynamics**.

micro- Prefix from Gk. *mikros*, small. When used of units it indicates the basic unit $\times 10^{-6}$, e.g. 1 microampere (μA) = 10^{-6} ampere. Symbol μ.

microbar The unit of pressure = 10^{-6} bar = 10^{-1}N m^{-2} = 1dyne cm^{-2}. Used in acoustics.

microgram(me) Unit of mass equal to one millionth of a gram (10^{-9} kg). Sometimes known as *gamma*.

micrometre One-millionth of a metre. Symbol μm. Also *micron* in past.

micromicro- Prefix for 1-million-millionth, or 10^{-12}. Replaced in SI by *pico-* (p).

micron Obsolete measure of length equal to one millionth of a metre, symbol μ. Replaced in SI by *micrometre*, symbol μm.

microradiography Exposure of small thin objects to **soft X-rays**, with registration on a fine-grain emulsion and subsequent enlargement up to 100 times. Also used to signify the optical reproduction of an image formed, e.g. by an electron microscope.

microwave heating Heating (induction or dielectric) of materials in which the current frequency is in the range 0.3×10^{12} to 10^9 Hz. Extensively used in domestic microwave ovens.

microwave spectrometer An instrument designed to separate a complex microwave signal into its various components and to measure the frequency of each; analogous to an *optical spectrometer*. See **spectrometer**.

microwave spectroscopy The study of atomic and/or molecular resonances in the microwave spectrum.

microwave spectrum The part of the electromagnetic spectrum corresponding to microwave frequencies.

migration area One sixth of the mean square distance covered by a neutron between creation and capture. Its square root is called *migration length*. See **slowing-down area, diffusion area**.

migration length See **migration area**.

mil A unit of length equal to 10^{-3} in, used in measurement of small thicknesses, i.e. thin sheets. Also colloq. *thou*.

milli- Prefix from L. *mille*, thousand. When attached to units, it denotes the basic unit $\times10^{-3}$. Abbrev. *m*.

millicurie One thousandth of a **curie**.

millimass unit Equal to 0.001 of atomic mass unit. Abbrev. *mu*.

millimetre The thousandth part of a metre. Abbrev. *mm*.

million-electron-volt See **MeV**.

milliradian 10^{-3} radian.

minimum ionization The smallest possible value of the specific ionization that a charged particle can produce in passing through a given substance. It occurs for particles having velocities = 0.95 c, where c = velocity of light.

minimum wavelength The shortest wavelength emitted in an X-ray spectrum. It is determined by the maximum voltage applied to the X-ray tube. The emitted X-ray photon acquires all the energy of the electron accelerated by the voltage, V, so the minimum wavelength

$$\lambda_{\mathrm{m}} = \frac{hc}{eV} = \frac{1.2396 \times 10^{-6}}{V} \ \mathrm{m},$$

where h is Planck's constant, e is the electronic charge and c is the speed of light.

minute (1) A 60th part of an hour of time. (2) A 60th part of an angular degree. (3) A 60th part of the lower diameter of a column.

mirror A highly-polished reflecting surface capable of reflecting light rays without appreciable diffusion. The commonest forms are plane, spherical (convex and concave) and paraboloidal (usually concave). The materials used are glass silvered on the back or front, speculum metal, or stainless steel.

mirror machine Fusion machine using the **magnetic mirror** principle to trap high-energy ions in a plasma.

mirror nuclides Nuclides with the same number of nucleons, but with proton and neutron numbers interchanged.

mirror symmetry See **parity**.

missile shield The structure placed over or round a reactor to prevent an explosively ejected fuel or control rod from piercing the containment building.

mixed oxide fuel Oxides of plutonium and uranium mixed together and forming a breeder reactor fuel, neutrons from the fissile plutonium bombarding the fertile uranium. Abbrev *MOX*. » p. 35.

mixed-settler solvent extraction A method in which two immiscible liquids are agitated and then allowed to settle; both phases then being separated and mixed with fresh solvent and the process repeated to form a cascade for separating e.g. plutonium and uranium from fission products.

mixing In gaseous isotope separation, the process of reducing the concentration gradient for the lighter isotope close to the diffusion barrier.

mixing efficiency A measure of the effectiveness of the mixing process in isotope separation.

MKSA Abbrev. for the *Metre-Kilogram(me)-Second-Ampere* system of units, adopted by the International Electrotechnical Commission, in place of all other systems of units. See **SI units** p. 237.

ml Abbrev. for *millilitre*.

MLD Abbrev. for **Mean Lethal Dose**.

M-lines Characteristic X-ray frequencies from atoms due to the excitation of electrons from the M-shell. Only developed in atoms of high atomic number.

mm Abbrev. for *millimetre*.

m.m.f. Abbrev. for *magnetomotive force*.

Mo Symbol for **molybdenum**.

mode (1) One of several electromagnetic wave frequencies which a given oscillator may generate, or to which a given resonator may respond, e.g. magnetron modes, tuned line modes. In a waveguide, the mode gives the number of half-period field variations parallel to the transverse axes of the guide. Similarly, for a cavity resonator, the half-period variations parallel to all three axes must be specified. In all cases, different modes will be characterized by different field configurations. (2) Similarly, one of several frequencies of mechanical vibration which a body may execute or with which it may respond to a forcing signal. (3) A well-defined distribution of the radiation amplitude in a cavity which results in the corresponding distribution pattern in the laser output beam. In a multimodal system the beam will tend to diverge.

mode-locking A technique for producing laser pulses of extremely short duration. Laser cavities have modes with frequency spacing $c/2L$ where c is the speed of light and L is the length of the cavity. Oscillations can occur in any mode as long as its frequency is within the natural line width of the laser transition. If a property of the cavity is modulated at a frequency of $c/2L$, then all the modes become coherently coupled. A train of extremely short pulses is emitted where the time duration is roughly the inverse of the line width. See **laser** p. 182.

moderating ratio A figure of merit balancing a moderator's ability to capture neutrons and to slow them down.

moderation The slowing down of neutrons in a reactor to thermal energies. *Degradation* is the unintentional slowing of neutrons. » p. 19.

moderator Material such as water, heavy water and graphite used to slow down neutrons in a reactor. See **lethargy of neutrons**. » p. 19.

moderator control Control of a reactor by varying the position or quantity of the moderator.

modulus Constant for units conversion between systems.

molal specific heat capacity The *specific heat capacity* of 1 mole of an element or compound. Also *volumetric heat* (for gases).

molar heat capacity The heat required to raise the temperature of a substance by 1 K. The symbol for that measured at constant volume is C_V and for that at constant pressure C_P.

mole The amount of substance that contains as many entities (atoms, molecules, ions, electrons, photons etc) as there are atoms in 12 g of ^{12}C. It replaces in SI the older terms *gram-atom, gram-molecule* etc, and for any chemical compound will correspond to a mass equal to the relative molecular mass in grams. Abbrev. *mol.* See **Avogadro number.**

molecular diffusion The process used in gaseous diffusion plants to separate molecules of gas with slightly different molecular weights by forcing them through very small holes. » p. 56.

molecular stopping power The energy loss per molecule per unit area normal to the motion of the particle in travelling unit distance. It is *approximately* equal to the sum of the atomic stopping powers of the constituent atoms.

molybdenum A metallic element in the sixth group of the periodic system. Symbol Mo, at.no. 42, r.a.m. 95.94, mp 2625°C, bp 3200°C, rel.d. at 20°C 10.2, resistivity *c.* 5×10^{-8} ohm metres, hardness 147 (Brinell). It has 12 isotopes:

A	Abundance %	half-life	decay mode
90		5.67 h	ε
91		15.5 m	ε
92	14.8		
93		3500 y	ε
94	9.3		
95	15.9		
96	16.7		
97	9.6		
98	24.1		
99		66.0 h	β^-
100	9.6		
101		14.6 m	β^-

Its physical properties are similar to those of iron; its chemical properties are like those of a non-metal.

moment of a magnet See **magnetic moment.**

moment of inertia Of a body about an axis: the sum Σmr^2 taken over all particles of the body where m is the mass of a particle and r its perpendicular distance from the specified axis. When expressed in the form Mk^2, where M is the total mass of the body ($M = \Sigma m$), k is called the *radius of gyration* about the specified axis. Also used erroneously for *second moment of area.*

momentum A dynamical quantity, conserved within a closed system. A body of mass M and whose centre of gravity G has a velocity v has a *linear momentum* of Mv. It has an *angular momentum* about a point O defined as the moment of the linear momentum about O. About G this reduces to $I\omega$ where I is the moment of inertia about G and ω the angular velocity of the body.

monazite Monoclinic phosphate of the rare earth metals, $CePO_4$, containing cerium as the principal metallic constituent, and also some thorium. Monazite is exploited from beach sands, where it may be relatively abundant; one of the principal sources of rare earths and thorium.

mon-, mono- Prefixes from Gk. *monos,* alone, single.

monitor Ionization chamber or other radiation detector arranged to give a continuous indication of intensity of radiation, as in radiation laboratories, radiation protection from fallout contamination, industrial operations or X-ray exposure. » p. 78.

monitoring Periodic or continuous determination of the amount of specified substances, e.g. toxic materials or radioactive contamination, present in a region or a person or of a flow rate, pressure etc, in a (continuous) process, as safety measures.

monochromatic By extension from *monochromatic light,* any form of oscillation or radiation characterized by a unique or very narrow band of frequency.

monochromatic light Light containing radiation of a single wavelength only. No source emits truly monochromatic light, but a very narrow band of wavelengths can be obtained, e.g. the cadmium red spectral line, wavelength 643.8 nm with a *half-width* of 0.0013 nm. Light from some lasers have extremely narrow line widths. Also *homogeneous* light.

monochromatic radiation Electromagnetic radiation (originally visible) of one single frequency component. By extension, a beam of particulate radiation comprising particles all of the same type and energy. *Homogeneous* or *monoenergic* is preferable in this sense.

monochromator Device for converting heterogeneous radiation (electromagnetic or particulate) into a homogeneous beam by absorption, refraction or diffraction processes.

monoenergic See **monochromatic radiation.**

Moseley's law For one of the series of characteristic lines in the X-ray spectrum of atoms, the square root of the frequency of the lines is directly proportional to the *atomic number* of the element. This result stresses the importance of atomic number rather than atomic weight.

Mössbauer effect When an atomic nucleus emits a γ-ray photon it must recoil to conserve linear momentum. Consequently there is a change of frequency of the radiation due to the movement of the source (*Doppler effect*). If the atom is firmly bound in a crystal lattice so that it may not recoil, the momentum is taken up by the whole lattice; an effect much used in the study of the structure of solids.

moving-field therapy A form of crossfire technique in which there is relative movement between the beam of radiation and the patient, so that the entry portal of the beam is constantly changing. See **converging-field therapy**.

MOX Abbrev. for **Mixed Oxide Fuel**.

MPD Abbrev. for *Maximum Permissible Dose*.

M-shell The *electron shell* in an atom corresponding to a principal quantum number of three. The shell can contain up to 18 electrons. See **atomic structure** p. 121.

multigroup theory Theoretical reactor model in which presence of several energy groups among the neutrons is taken into account. See **delayed neutron groups**.

multiple decay, multiple disintegration **Branching** in a radioactive decay series.

multiplets (1) Optical spectrum lines showing fine structure with several components, i.e. triplet or more complex structures. Due to spin-orbit interactions in the atom. (2) See **isotopic spin**.

multiplication factor See **neutron reproduction factor**.

mu-meson An elementary particle once thought to be a *meson* but now known to be a *lepton*. See **muon**.

muon Fundamental particle with a rest mass equivalent to 106 MeV. It is one of the *leptons* and has a negative charge and a half-life of about 2 μs. Decays to electron, neutrino and antineutrino. It participates only in *weak* interactions. The *antimuon* has a positive charge and decays to positron, neutrino and antineutrino.

mutation A change, spontaneous or induced, that converts one *allele* into another (*point mutation*). More generally, any change of a gene or of chromosomal structure or number. A *somatic mutation* is one occurring in a somatic cell and not in the germ line.

mutation rate The frequency, per gamete, of mutations of a particular gene or a class of genes. Sometimes, esp. in micro-organisms, the number per unit of time (esp. generation time).

MWE, MWe Abbrev. for *MegaWatt Electric*. Unit for electric power generated by a nuclear reactor.

N

n Symbol for **nano-**.

n Symbol for **neutron**.

ν Symbol for (1) **frequency**; (2) **neutrino**; (3) **Poisson's ratio**; (4) **kinematic viscosity**.

N Symbol for **newton**.

N Symbol for **neutron number**.

N_A Symbol for **Avogadro number**.

NAA Abbrev. for **Neutron Activation Analysis**. See **trace element analysis** p. 251.

nano- Prefix for 10^{-9}, i.e. equivalent to millimicro or one thousand-millionth. Symbol n.

National Bureau of Standards US federal department set up in 1901 to promulgate standards of weights and measures and generally investigate and establish data in all branches of physical and industrial sciences. Abbrev. *NBS*.

National Physical Laboratory The UK authority for establishing basic units of mass, length, time, resistance, frequency, radioactivity etc. Founded by the Royal Society in 1900; now government controlled and engaged in a very wide range of research. Abbrev. *NPL*.

natural abundance See **abundance**.

natural background In the detection of nuclear radiation, the radiation due to *natural radioactivity* and to cosmic rays, enhanced by contamination and fallout.

natural radioactivity Radioactivity which occurs in nature. Such radioactivity indicates that the isotopes involved have a half-life comparable with the age of the Earth or result from the decay of such isotopes. Most of these nuclides can be grouped in one of three **radioactive series**. Natural radioactivity accounts for 86% of all the radiation received by humans and originates from space, from natural radioactive elements, e.g. uranium-238, present in small amounts in rocks and soil, and from food, water and buildings. Uranium decays in a sequence of products (forming lead ultimately), one of which is the *noble gas* **radon**: radioactive radon seeps through soil and may accumulate in poorly ventilated buildings. Radioactive elements naturally present in fossil fuels are released when the fuels are burned. See **ionizing radiation**.

natural uranium reactor A reactor in which natural, i.e. unenriched, uranium is the chief fissionable material. » p. 29.

negative Designation to electric charge, introduced by Franklin, now known to be exhibited by the electron, which, in moving forms the normal electric current.

negative proton See **antiproton**.

net transport The difference between the actual **transport** in an isotope separation plant and that which would be obtained by the same plant with raw material of natural isotopic abundance. » p. 56.

neutral beam A beam of high-energy *atoms* used to heat a plasma. As the atoms are neutral the beam is not affected by magnetic fields. » p. 41.

neutral equilibrium The state of equilibrium of a body when a slight displacement does not alter its potential energy.

neutral injection The additional heating of a plasma by injecting beams of accelerated atoms into it. » p. 41.

neutrino A fundamental particle, a **lepton**, with zero charge and zero mass. A different type of neutrino is associated with each of the four charged leptons. Its existence was predicted by Pauli in 1931 to avoid β-decay infringing the laws of conservation of energy and angular momentum. As they have very weak interactions with matter, neutrinos were not observed experimentally until 1956. See **antineutrino**.

neutron Uncharged subatomic particle, mass approximately equal to that of the proton, which enters into the structure of atomic nuclei. Interacts with matter primarily by collisions. Spin quantum number of neutron = +½, rest mass = 1.008 665 amu, the charge is zero and the magnetic moment −1.9125 nuclear Bohr magnetons. Although stable in nuclei, isolated neutrons decay by β-emission into protons, with a half-life of 10.6 minutes. See **neutron energy**. » p. 10.

neutron absorption cross-section The cross-section for a nuclear reaction initiated by neutrons. This is expressed in *barns*. For many materials this rises to a large value at particular neutron energies due to resonance effects, e.g. a thin sheet of cadmium forms an almost impenetrable barrier to thermal neutrons. » p. 14.

neutron activation analysis See **trace element analysis** p. 251.

neutron age See **Fermi age**.

neutron balance For a constant power level in a reactor, there must be a balance between the rate of production of both prompt and delayed neutrons, and their rate of loss due to both absorption and leakage from the reactor. » p. 20.

neutron current The net rate of flow of neutrons through a surface perpendicular to the direction in which they are migrating.

Neutron current is a *vector* whereas *neutron flux* is a *scalar*.

neutron detection Observation of charged particle recoils following collisions of neutrons with protons in a counter containing hydrogen gas or a compound of hydrogen; or of charged particles produced by interaction of neutrons with atomic nuclei, e.g. in a boron-trifluoride *counter*. Absorption of neutrons by boron gives rise to α-particles which can be detected.

neutron diffraction The coherent elastic scattering of neutrons by the atoms in a crystal. If the scattering is by the nucleus then the atomic structure of the crystal can be deduced from the measurements of the diffraction pattern. If the scattering is by atoms with electron configurations that have a magnetic moment, then details of the magnetic structure of the crystal can be determined.

neutron diffusion The migration of neutrons from regions of high neutron density to those of low density, in a medium in which neutron capture is small compared to neutron scattering. See **age theory**, **group theory**, **transport theory**.

neutron economy The keeping of the losses of neutrons in a reactor to a minimum.

neutron elastic scattering A beam of thermal neutrons, whose electrons are such that their associated de Broglie wavelength is of the same order of magnitude as interatomic distances, will be diffracted by a crystal. Most of the intensity of the beam will be diffracted with no loss of energy, i.e. no wavelength change, and this is said to be *neutron elastic scattering*. See **neutron inelastic scattering.**

neutron energy (1) The binding energy of a neutron in a nucleus, usually several MeV. (2) The energy of a free neutron which in a reactor will be classed in several groups: high-energy neutrons, energy 10 MeV; fast neutrons, energy 10 MeV to 20 keV; intermediate neutrons, energy 20 keV to 100 eV; epithermal neutrons, energy 100 eV to 0.025 eV; thermal or slow neutrons, energy approx. 0.025 eV. See **multigroup theory**.

neutron excess The difference between the neutron number and the proton number for a nuclide. Also *isotopic number*.

neutron flux The number of neutrons passing through a unit area in unit time, *or* the product of number of neutrons per unit volume and their mean speed. In a nuclear reactor the flux is of the order of 10^{16} to 10^{18} m^{-2} s^{-1}.

neutron gun Block of moderating material with a channel through it used for producing a beam of fast neutrons.

neutron hardening The increase of the average energy of a beam of neutrons by passing them through a medium which shows preferential absorption of slow neutrons.

neutron inelastic scattering A beam of thermal neutrons, whose energies are of the same order of magnitude as a quantum of lattice vibrational energy, a *phonon*, will be diffracted by a crystal with exchanges of energy with the excited travelling waves. A detailed examination of the change in direction and energy of the scattered neutrons gives valuable information about the *lattice dynamics* of the crystal.

neutron leakage Escape of neutrons in a reactor from the core containing the fissile material; reduced by using a reflector. » p. 20.

neutron-magnetic scattering The magnetic moment in a crystal has contributions from both electron spin and electron orbital angular momenta. Both of these will interact with the neutron spin magnetic moment and will give rise to the magnetic scattering of neutrons. This is a powerful method for the study of magnetic structures.

neutron-nuclear scattering length A measure of the ability of different *nuclei* to scatter neutrons; it is independent of the scattering angle. The scattering length is different for different isotopes of the same atom, so that, in addition to coherent scattering from a crystal containing more than one isotope, there will be incoherent scattering. Neutron-nuclear scattering is spin dependent, and this leads to further incoherent scattering from a nucleus with non-zero spin.

neutron number The number of neutrons in a nucleus. Equal to the difference between the relative atomic mass (the total number of nucleons) and the atomic number (the number of protons).

neutron poison Any material other than fissionable, which absorbs neutrons; used for the control of nuclear reactors. » p. 19.

neutron radiography Radiography by beam of neutrons from nuclear reactor which then produces an image on a photoelectric image intensifier following neutron absorption. It has theoretical advantages over X-ray radiography because the mass absorption coefficients for neutrons are very different for different parts of the specimen, although the atomic numbers (and hence X-ray absorption) are very similar, e.g. specimens containing hydrogenous material. Clinical trials do not demonstrate much, if any, therapeutic advantage.

neutron reproduction factor The factor,

k, which is the ratio between the number of neutrons inducing fission in the fuel to the number required to maintain the chain reaction. If $k = 1$, the reactor is *critical* and operating at a steady state; if it is <1 it is *subcritical* and in a shutdown state; if it is >1 it is *supercritical* or *divergent*. Also *multiplication factor*.

neutron scatter plug A mechanical device in a fuel assembly designed to reflect those neutrons back into the core which would otherwise have escaped through the coolant pipe.

neutron shield Radiation shield erected to protect personnel from neutron irradiation. In contra-distinction to gamma-ray shields, it must be constructed of very light hydrogenous materials which will quickly moderate the neutrons which can be absorbed, for instance, by boron incorporated in the shield.

neutron source A source giving a high neutron flux, e.g. for neutron activation analysis. Apart from reactors, these are chemical sources such as a radium-beryllium mixture emitting neutrons as a result of the (αv) reaction, and accelerator sources in which deuterium nuclei are usually accelerated to strike a tritium-impregnated titanium target, thus releasing neutrons by the (DT) reaction. The former are continuously active but the latter have the advantage of becoming inert as soon as the accelerating voltage is switched off.

neutron spectrometer Instrument for investigation of energy spectrum of neutrons. See **crystal spectrometer**, **time-of-flight spectrometer**. Other techniques depend on the nuclear reaction of neutrons with helium-3 *(helium-3 spectrometer)*, lithium-6 or hydrogen *(proton recoil spectrometer)*.

neutron spectroscopy Experimental determination of the intensity and change in energy (wavelength) of the neutrons scattered in a particular direction when a beam of mono-energetic neutrons are incident on a crystal. A powerful method of studying lattice vibrations and phonon energies.

neutron velocity selector (1) Device using a rotating **chopper** or a rotating helical device, or a pulsed accelerator to provide a pulse of neutrons, the velocity being selected by time of flight (see **time-of-flight spectrometer**). (2) Device using neutron diffraction. See **crystal spectrometer.**

neutron yield The average number of neutrons emitted at each fission; of the order of 2.5.

newton The unit of force in the SI system, being the force required to impart, to a mass of 1 kg, an acceleration of 1 m sec^{-2}. Symbol *N*. 1 newton = 0.2248 pounds force.

Ni Symbol for **niobium**.

NII Abbrev. for *Nuclear Installations Inspectorate*. A branch of the Health and Safety Executive in the UK responsible for the safety assessment and inspection of nuclear facilities.

nile 1 nile corresponds to a **reactivity** of 0.01. In indicating reactivity changes, it is more usual to use the smaller unit, the millinile, equal to a change in reactivity of 10^{-5}.

niobium A rare metallic element. Symbol Nb, at.no. 41, r.a.m. 92.9064, mp 2500°C. Used in high-temperature engineering products (e.g. gas turbines and nuclear reactors) owing to the strength of its alloys at temperatures above 1200°C. It has nine isotopes:

A	Abundance %	half-life	decay mode
89		2.0 h	ε
90		14.6 h	ε
91		700 y	ε
92		35 My	ε
93	100		
94		20 000 y	β^-
95		35.0 d	β^-
96		23.4 h	β^-
97		72 m	β^-

Combined with tin (Nb_3Sn) it has a high degree of superconductivity. Used alone or as an alloy in nuclear fuel cladding; it has the following properties:

σ_{cap} mb	mp °C	Therm. conduct. W $m^{-1}K^{-1}$	density g cm^{-3}
1150	2470	54	10.2

nm Abbrev. for *nanometre* = 10 ångströms, = 10^{-9} m.

NMR Abbrev. for **Nuclear Magnetic Resonance**.

noble gases The elements helium, neon, argon, krypton, xenon and radon-222, much used (except the last) in gas-discharge tubes. (Radon-222 has short-lived radioactivity, half-life less than 4 days.) Their outer (valence) electron shells are complete, thus rendering them inert to all the usual chemical reactions; a property for which argon, the most abundant, finds increasing industrial use. The heavier elements, Rn, Xe, Kr, are known to form a few unstable compounds, e.g. XeF_4. Also *inert gases*, *rare gases*.

node See **antinode**.

non-ionizing fields and radiation See panel on p. 198.

non-leakage probability For neutrons in a

non-ionizing fields and radiation

A steady electric current will produce a steady magnetic field as well as a steady electric field. The strength of such fields are measured in amperes per metre and volts per metre respectively, with their flux densities measured in teslas (T) and watts per square metre (W m^{-2}). The fields are not propagated and remain static around the source declining linearly with distance from the source. Oscillating electric or magnetic fields are propagated as photons, as was described in Chapter 9. Neither the static fields nor the long wavelength photons from electromagnetic sources have sufficient energy to ionize an atom. The kind of effects found may well therefore be different from those produced by, for example, free radicals (see p. 102).

Static fields

Typical natural static electrical fields are about 150 volts per metre but may reach thousands of volts per metre in a thunderstorm. Even higher electrical fields are generated near television sets or even by friction. The natural static magnetic field, that detected by a compass, is between 10 and a 100 microtesla (μT), but small magnets found in the home may have a field a hundred times higher and the large superconducting magnets in magnetic resonance imaging machines will have flux densities of 2.5 tesla or higher.

Static electric fields do not penetrate the body but cause an electric charge to build up on the surface that can sometimes be felt by the movement of bodily hair. With very high voltages the surrounding air can become ionized and the surface voltage will discharge to earth as a spark and may be fatal. There is very little evidence that such charges can cause any internal effect; volunteers have stayed in 600 volts per metre charged compartments for several weeks without ill effect. Static magnetic fields, on the other hand, penetrate living tissue and any molecules which themselves behave as small magnets will become aligned in the field, this being the basis for the medical imaging procedures that exploit magnetic resonance.

All of the data from people exposed for short times in magnetic resonance imaging machines and most of the data from animal experiments show that magnetic fields up to several tesla fail to induce any behavioural difference, genetic or somatic chromosomal change or a wide range of other effects. However, vertigo and nausea have been reported in a few individuals exposed to fields of about 4 tesla near magnetic resonance machines. It has been suggested that average exposures should be restricted to 200 millitesla per day for those working near large magnets and to less than this figure for the general public.

Extremely low-frequency radiation below 3 kHz

People under high-voltage AC power lines are subject to an electric field of about 10 kilovolts per metre and a magnetic field of 40 microtesla (40 μT), but industrial welders and similar workers may receive over a thousand times as much, while in the home the magnetic field is probably 50 times less than the power line figure. The electrical effects of these low-frequency fields is very like that for static fields, but electric fields will also occur within the body, typically of the order of a million times less than that outside. Low-frequency magnetic fields induce electrical currents within the body which will depend on how the body is placed relative to the magnetic field and the radii of the conduction loops. Currents of 5 amperes per square metre could be induced by a magnetic field of 1 tesla .

continued on next page

non-ionizing fields and radiation (contd)

The effect of an electric current passing through body fluids is to alter the electrical potentials which exist across membranes and so change their properties which, for nerve cells, include the transmission of electrical impulses. One problem is that the currents are so small that they are less than the electrical noise generated by the thermal agitation of the atoms present (thermal noise), although it is possible to calculate that if the bandwidth of the current is very restricted the effect could be detected above background noise for even very small currents.

A considerable amount of animal research has been done on whether low doses of electromagnetic radiation have any immediate effect on the foetus or longer term genetic effect. Unfortunately they do not provide clear-cut results. Some defects in animals have been noted but other laboratories have often failed to repeat their results. It has been suggested that high electrical (20 kV m^{-1}) and magnetic (5 mT) fields which can be detected by the individual and may cause irritability should be avoided. It is thought that more subtle effects on nervous tissue will be avoided if external magnetic fields are kept below that needed to induce a secondary current of 10 mA m^{-2}. Further more definitive research on the problems of genetic and developmental damage is clearly needed.

Microwaves

Above 100 kHz the predominantly electrical effects on living tissue induced by a varying magnetic field give way to the thermal excitation of the atoms and the consequent heating of the tissue. It therefore becomes important to measure the rate at which the energy is deposited for which a *specific energy absorption rate* (SAR) of watts per kilogram (W kg^{-1}) is used. A SAR of 1 watt per kilogram is needed to raise body temperature by 1°C in an hour if cooling is neglected, with higher figures in a normal environment. A SAR of 0.4 W $kg^{-1}h^{-1}$ has been adopted by several authorities for maximum exposure.

Most research indicates that effects other than those associated with temperature rise are unlikely with microwave radiation and that provided the overall temperature rise is kept below 1°C there are no unfavourable biological effects.

reactor, the ratio of the actual multiplication constant to that calculated for a reactor of infinite size. » p. 20.

non-quantized See **classical**.

non-relativistic Said of any procedure in which effects arising from relativity theory are absent or can be disregarded, e.g. properties of particles moving with low velocity, e.g. 1/20th that of light propagation.

non-spectral colour Colour outside the range which contributes to white light, but which affects photocells.

normal frequency See **normal modes**.

normal modes If a vibrating system consists of a number of oscillators coupled together, the resulting motion is in general complicated. However, by choosing the starting conditions correctly, it is possible to make the system vibrate in such a way that every part has the same frequency. Such simple vibrations are called the *normal modes* and the associated frequencies, the *normal* or *natural frequencies*. If there are *N* degrees of freedom in the system then there are *N* normal modes.

normal state See **ground state**.

N-shell The **electron shell** in an atom corresponding to a principal quantum number of four. The shell contains up to 32 electrons and is the last shell to be filled completely by electrons in the naturally occuring elements.

$n\tau$ The **Lawson criterion**. The product of the plasma density n and the plasma confinement time τ for nuclear fusion processes. $n\tau$ 10^{20} s m^{-3} is the value for fusion to produce useful energy. » p. 40.

nuclear battery A battery in which the electric current is produced from the energy of radioactive decay, either directly by collecting beta particles or indirectly, e.g. by using the heat liberated to operate a thermojunction. In general, nuclear batteries have very low outputs (often only microwatts) but long and trouble-free operating lives.

nuclear binding energy The binding energy that holds together the constituent nucleons of the nucleus of an atom. It is the energy equivalence of the mass difference between the masses of the atom and the sum of the individual masses of its constituents; expressed in MeV or amu. » p. 13.

nuclear breeder A nuclear reactor in which in each generation there is more fissionable material produced than is used up in fission. » p. 36.

nuclear charge Positive charge arising in the atomic nucleus because of protons, equal in number to the atomic number.

nuclear conversion ratio The ratio of the fissile atoms produced to the fissile atoms consumed, in a breeder reactor. » p. 36.

nuclear cross-section See **cross-section**.

nuclear disintegration Fission, radioactive decay, internal conversion or isomeric transition. Also *nuclear reaction.* » pp. 10, 18.

nuclear emission Emission of gamma ray or particle from nucleus of atom as distinct from emission associated with orbital phenomena.

nuclear emulsion Thick photographic coating in which the tracks of various fundamental particles are revealed by development as black traces.

nuclear energy In principle, the binding energy of a system of particles forming an atomic nucleus. More usually, the energy released during nuclear reactions involving regrouping of such particles (e.g. fission or fusion processes). The term *atomic energy* is deprecated as it implies rearrangement of atoms rather than of nuclear particles. » p. 13.

nuclear field Postulated short-range field within a nucleus, which holds protons and neutrons together, possibly in shells.

nuclear fission The spontaneous or induced disintegration of the nucleus of a heavy atom into two lighter atoms. The process involves a loss of mass which is converted into nuclear energy. » p. 18.

nuclear force The force which keeps neutrons and protons together in a nucleus, differing in nature from electric and magnetic forces, gravitational forces being negligible. The force is of short range, is practically independent of charge, and arises from the exchange of *pions* between the nucleons (Yukawa). See **meson, short-range forces, exchange force**.

nuclear fuel See **reactor types** p. 225.

nuclear fusion The process of forming atoms of new elements by the fusion of atoms of lighter ones. Usually the formation of helium by the fusion of hydrogen and its isotopes. The process involves a loss of mass which is converted into nuclear energy. The basis of a **fusion reactor**. » p. 38.

nuclear isomer A nuclide existing in an excited metastable state. It has a finite half-life after which it returns to the ground state with the emission of a γ-quantum or by internal conversion. Metastable isomers are indicated by adding *m* to the mass number.

nuclear magnetic resonance Certain atomic nuclei, e.g. hydrogen-1, fluorine-19, phosphorous-31, have a nuclear magnetic moment. When placed in a strong magnetic field, the nuclear moments can only take up certain discrete orientations, each orientation corresponding to a different energy state. Transitions between these energy levels can be induced by the application of radiofrequency radiation. This is known as nuclear magnetic resonance. For example, the protons in water experience this resonance effect in a field of 0.3 T at a radiation frequency of 12.6 MHz. The resonant frequency depends on the magnetic field at the nucleus which in turn depends on the environment of the particular nucleus. Nuclear magnetic resonance gives invaluable information on the structure of molecules. It has also been developed to provide a non-invasive clinical imaging of the human body, *magnetic resonance imaging* (MRI); multiple projections are combined to form images of sections through the body, providing a powerful diagnostic aid. Abbrev. *NMR*.

nuclear model A model giving an explanation of the properties of the atomic nucleus and its interactions with other particles. See **collective model of the nucleus, independent particle model of the nucleus, liquid-drop model, optical model of the nucleus, shell model, unified model of the nucleus**.

nuclear number See **mass number**.

nuclear paramagnetic resonance See **nuclear magnetic resonance**.

nuclear photoeffect See **photodisintegration**.

nuclear potential The potential energy of a nuclear particle in the field of a nucleus as determined by the short-range forces acting on it. Plotted as a function of position it will normally represent some sort of **potential well**.

nuclear power Power generated by the release of energy in a nuclear reaction.

nuclear radius The somewhat indefinite radius of a nucleus within which the density of *nucleons* (protons and neutrons) is experimentally found to be nearly constant. The radius in metres is 1.2×10^{-15} times the cube root of the nuclear number (atomic mass). It is not a precise determinable quantity. » p. 9.

nuclear reaction The interaction of a photon or particle with a target nucleus. An amount of energy, Q-value, is released or absorbed, depending on whether the mass of the reaction products is less than or more than the mass of the reactants. Reactions fall into two broad categories; those in which the reaction proceeds *via* the formation of a *compound nucleus*; the alternative mechanism is by *direct interaction*. At higher energies *spallation* occurs in which the target nucleus splits up into a number of fragments.

nuclear reactor See **reactor types** p. 225.

nuclear reactor oscillator Device producing variations in reactivity by oscillatory movement of sample to measure reactor properties, or neutron capture cross-section of sample.

nuclear selection rules Rules specifying the transitions of electrons or nucleons between different energy levels which may take place (*allowed transitions*). The rules may be derived theoretically through wave mechanics but are not obeyed rigorously; so-called *forbidden transitions* merely being highly improbable.

nucleic acid General term for natural polymers in which *bases* (purines or pyrimidines) are attached to a sugar phosphate backbone. Can be single- or double-stranded. Short molecules are called *oligonucleotides*. Also *polynucleotide*.

nucleogenesis Theoretical process(es) by which nuclei could be created from possible fundamental dense plasma. See **ylem**.

nucleonics The science and technology of nuclear studies.

nucleon number See **mass number**.

nucleons Protons and neutrons in a nucleus of an atom.

nucleor The hypothetical core of a nucleon. It is suggested that it is the same for protons and neutrons.

nucleus (1) In *physics* it is the structure composed of protons (positively charged) and neutrons (no charge), and which constitutes practically all the mass of the atom. Its charge equals the atomic number; its diameter is from 10^{-15} to 10^{-14} m. With protons, equal to the atomic number, and neutrons to make up the atomic mass number, the positive charge of the protons is balanced by the same number of extranuclear electrons. » p. 10. (2) In *biology* it is the compartment within the animal or plant cell bounded by a double membrane and containing the genomic DNA, with its associated functions of *transcription* and processing. » p. 87.

nuclide An atomic nucleus as characterized by its atomic number, its mass number and nuclear energy state.

Nusselt number The significant non-dimensional parameter in convective heat loss problems, defined by $Qd/k\Delta\theta$, where Q is rate of heat lost per unit area from a solid body, $\Delta\theta$ is the temperature difference between the body and its surroundings, k is the thermal conductivity of the surrounding fluid and d is the significant linear dimension of the solid.

O Symbol for **oxygen**.

ω Symbol for (1) angular frequency; (2) angular velocity; (3) **dispersive power**; (4) pulsatance.

Ω Symbol for (1) **ohm**; (2) angular velocity.

oblate Globose, but noticeably wider than long.

obscuration Fraction of incident radiation which is removed in passing through a body or a medium.

odd-even nuclei Nuclei containing an odd number of protons and an even number of neutrons.

odd-odd nuclei Nuclei with an odd number of both protons and neutrons. Very few are stable.

off gas Gas escaping from a reactor, diffusion plant or other installation.

ohm SI unit of electrical resistance, such that 1 ampere through it produces a potential difference across it of 1 volt. Symbol Ω.

ohm cm CGS unit of **resistivity**.

ohm metre SI unit of **resistivity**.

Ohm's law In metallic conductors, at constant temperature and zero magnetic field, the current I flowing through a component is proportional to the potential difference V between its ends, the constant of proportionality being the *conductance* of the component. So $I = V/R$ or $V = IR$, where R is the *resistance* of the component. Law is strictly applicable only to electrical components carrying direct current and for practical purposes to those of negligible reactance carrying alternating current. Extended by analogy to any physical situation where a pressure difference causes a flow through an impedance, e.g. heat through walls, liquid through pipes.

OK Abbrev. for *Odourless Kerosene*.

Oklo natural fission reactor See panel on p. 203.

olig-, oligo- Prefixes from Gk. *oligos*, a few, small.

omega-minus particle The heaviest hyperon (1672 MeV), discovered in 1964. Its existence produces strong evidence for the classification, developed from *Lie algebra* (group theory), of those elementary particles which interact strongly.

once-through fuel cycle A reactor in which the fuel is not reprocessed when spent.

oncogene Genetic locus originally identified in RNA tumour viruses which is capable of giving a host cell some of the properties of a tumour cell. Implicated as cause of certain cancers. » p. 97.

one-group theory Greatly simplified reactor model in which all neutrons are regarded as having the same energy. See **multigroup theory**.

one-particle model of a nucleus A form of **shell model** in which nuclear spin and magnetic moment are regarded as associated with one resident nucleon.

opacity Reciprocal of the optical *transmission ratio*.

opaque Totally absorbent of rays of a specified wavelength, e.g. wood is opaque to visible light but slightly transparent to infrared rays, and completely transparent to X-rays and waves for radio communication.

Oppenheimer-Phillips (OP) process A form of **stripping** in which a deuteron surrenders its neutron to a nucleus without entering it.

optical black A body when it absorbs all radiation falling on its surface. No substance is completely black in this context.

optical constants The refractive index (n) and the absorption coefficient (k) of an absorbing medium. Together these determine the complex refractive index ($n–k$) of the medium.

optical maser A maser in which the stimulating frequency is visible or infrared radiation. See **laser**.

optical model of the nucleus In nuclear reactions the target nucleus can frequently be treated as a sphere that partly absorbs and partly transmits the incidental radiation, i.e. it has, in the optical analogy, a *complex* refractive index. The nucleus can thus be represented by a 'potential well' having real and imaginary components.

optical pumping A mechanism by which an external light source of suitable frequency stimulates the material to produce a *population inversion* for a particular energy transition in a **laser**.

optical pyrometer An instrument which measures the temperatures of furnaces by estimating the colour of the radiation, or by matching it with that of a glowing filament.

optical spectrum The visible radiation emitted from a source separated into its component frequencies.

optics The study of light. *Physical optics* deals with the nature of light and its wave properties; *geometrical optics* ignores the wave nature of light and treats problems of reflection and refraction from the ray aspect.

orbital The properties of each electron in a

Oklo natural fission reactor

One of the places were uranium has been mined is at Oklo in what is now the Republic of Gabon. In 1972 a sample of uranium was found to have a slightly lower abundance of uranium-235 than had been found anywhere else in the world. The normal value for this abundance is 0.720% to a high degree of accuracy and the value found by French scientists at their Atomic Energy Commission was 0.717, and subsequent samples had even less, down to 0.440. The only possible explanation for such a discrepancy is that uranium-235 fission had occurrred.

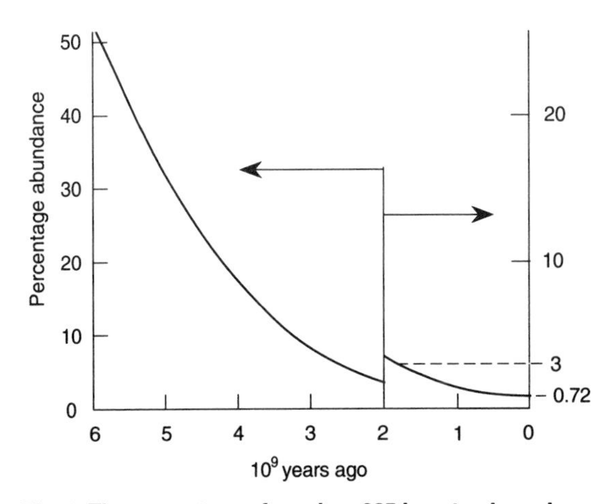

Fig. 1 **The percentage of uranium-235 in natural uranium.** It was 3% just under 2000 million years ago. A 2 × enlarged scale is used on the right.

The half-life of uranium-235 is 0.7×10^9 years compared to 4.5×10^9 for uranium-238. Therefore uranium-235 must have been much more abundant in the geological past. Indeed it is not unreasonable to suppose that the conditions at the formation of the galaxy did not favour one isotope against the other. Knowing the different half-lives of the isotopes, their relative abundance at earlier times can be calculated. This relation is illustrated in Fig. 1. It shows not only that 6×10^9 years ago, when the galaxy was formed, the relative abundance of uranium-235 was about 50%, but also that 2×10^9 years ago it was over 3%. Fission, moderated by natural water, would therefore have been possible.

Sceptics were confounded and fission confirmed by analysing the isotopic abundances of elements like ruthenium and neodymium in the deposit and comparing them with those in spent fission fuel and in nature. The Oklo abundances were much closer to those in reactor fission products and unlike those found elsewhere in nature.

Estimates of the size of the Oklo uranium deposit and the average amount of uranium-235 depletion indicate that perhaps 5000 kg of uranium-235 was involved and that fission could have occurred over a period of perhaps a million years. It would not have occurred rapidly because the water would have boiled and the reaction immediately halted until the rocks cooled enough to let more water seep back in. Nevertheless the energy produced was very large, equivalent in total to one hundred million tons of TNT.

many-electron atom may be reasonably described by its response to the potential due to the nucleus and to the other electrons. The *wave function*, which expresses the probability of finding the electron in a region, is specified by a set of four quantum numbers and defines the *orbital* of the electron. The state of the many-electron atom is given by defining the orbitals of all the electrons subject to the Pauli exclusion principle.

orbit shift coils Coils placed on the magnetic pole faces of a betatron or synchrotron, so that by passing a current pulse through the coils the particles may be momentarily displaced from the stable orbit to strike a target.

organic phosphor Organic chemical used as solid or liquid scintillator in radiation detection.

ortho- Prefix from Gk. *orthos*, straight.

orthodiagraph An X-ray apparatus for recording exactly the size and form of organs and structures inside the body.

oscillations Any motion that repeats itself in equal intervals of time is a *periodic motion*. A particle in periodic motion moves back and forward over the same path and is said to be *oscillating* or *vibrating*. If the oscillations are not precisely repeated due to frictional forces which dissipate the energy of motion, the oscillations are said to be *damped*.

O-shell The *electron shell* in an atom corresponding to a principal quantum number of five. In naturally occuring elements the shell is never completely filled but has most electrons, 21, for the element uranium. See **atomic structure** p. 121.

overall luminous efficiency Ratio of luminous flux of lamp to total energy input. Not to be confused with *luminous efficiency*.

oxide fuel Fuel made from the oxides of fissile elements, which can operate at much higher temperatures than the metal.

oxygen A non-metallic element. Symbol O, at.no. 8, r.a.m. 15.9994, valency 2, mp −218.4°C, bp −183°C, density at stp 1.429 04 g dm^{-3}, formula O_2. It is a colourless, odourless gas which supports combustion and is essential for the respiration of most forms of life. An unstable form is ozone, O_3. Oxygen is the most abundant element, forming 21% by volume of the atmosphere, 89% by weight of water, and nearly 50% by weight of the rocks of the Earth's crust. It has seven isotopes:

A	Abundance %	half-life	decay mode
14		71 s	ε
15		122 s	ε
16	99.76		
17	0.038		
18	0.204		
19		26.9 s	β^-
20		13.5 s	β^-

oxygen-isotope determinations A method of using the oxygen-16 and oxygen-18 isotope ratio measurements from cores taken from Greenland and Antarctic ice sheets or from oxygen-bearing geological materials, e.g. carbonates from the shells of marine organisms. The results may be used to estimate (1) the temperature at which the original snow fell before turning to ice; (2) the sea temperature at the time of deposition of marine fossils and (3) the global ice volume, thus giving a chronology of the ice ages during the Pleistocene. See **radiometric dating** p. 222.

P

p Abbrev. for (1) **pico-**; (2) **proton**.

p Symbol for (1) **electric dipole moment**; (2) impulse; (3) **momentum**; (4) pressure.

π See **pi**.

ϕ Symbol for (1) heat flow rate; (2) **luminous flux**; (3) magnetic flux; (4) **work function**.

P Symbol for **poise**.

P Symbol for (1) electric polarization; (2) **power**; (3) pressure.

Ψ Symbol for (1) **electric flux**; (2) magnetic field strength.

Pa Abbrev. for **pascal**.

packaged Said of a reactor of limited power which can be packaged and erected easily on a remote site.

packing fraction The mass *M* of a nuclide expressed in *atomic mass units* differs slightly from a whole number, the mass number *A*. The packing fraction is defined as

$$\frac{(M-A)}{A}.$$

The *mass defect* or *mass decrement* is $(M-A)$.

pairing energy A component of the binding energy of the nucleus that represents an increase in this energy where the number of neutrons and the number of protons are even. See **even-even nuclei**.

palaeogenetic Originating in the past.

panradiometer Instrument for measuring radiant heat irrespective of the wavelength.

para- Prefix from Gk. *para*, beside.

parallax Generally, the apparent change in the position of an object seen against a more distant background when the viewpoint is changed. Absence of parallax is often used to adjust two objects, or two images, at equal distances from the observer.

paralysis time The time for which a radiation detector is rendered inoperative by an electronic switch in the control circuit. See **dead time**.

parameter Generally, a variable in terms of which it is convenient to express other interrelated variables which may then be regarded as being dependent upon the parameter. More specifically, the quantitative design elements which relate to a specific system.

parasitic capture Neutron capture in reactor not followed by fission.

parent In radioactive particle decay of *A* into *B*, *A* is the parent and *B* the daughter.

parent peak The component of a mass spectrum resulting from the undissociated molecule.

parity The conservation of parity or space-reflection symmetry (*mirror symmetry*) states that no fundamental difference can be made between right and left, or in other words the laws of physics are identical in a right- or in a left-handed system of coordinates. The law is obeyed for all phenomena described by classical physics but was recently shown to be violated by the weak interactions between elementary particles.

particle (1) A useful concept of a small body which has a finite mass but negligible dimensions so that it has no *moment of inertia* about its *centre of mass*. (2) A volume of air or fluid which has dimensions very small compared with the wavelength of a propagated sound wave, but large compared with molecular dimensions.

particle accelerators See panel on p. 206.

particle exchange The interaction between *fundamental particles* by the exchange of another fundamental particle. See **gauge boson**.

particle-induced X-ray emission See **trace element analysis** p. 251.

particle physics Study of the properties of fundamental particles and of fundamental interactions. Often *high-energy physics*.

particle scattering See **scattering**.

partition coefficient The ratio of the equilibrium concentrations of a substance dissolved in two immiscible solvents. If no chemical interaction occurs, it is independent of the actual values of the concentrations. Also *distribution coefficients*.

pascal The SI derived unit of pressure or stress, equals 1 newton per square metre. Abbrev. *Pa*.

Paschen series A series of lines like the **Balmer series**, in the infrared spectrum of hydrogen. » p. 84.

Pauli exclusion principle A fundamental law of quantum mechanics, that no two **fermions** in a system can exist in identical quantum states, i.e. have the same set of quantum numbers. This constraint explains the electronic structure of atoms and also the general nature of the periodic table. As neutrons and protons are also fermions, nuclear structure is also strongly influenced by the exclusion principle. See **atomic structure** p. 121.

Pb Symbol for **lead**.

pd, p.d. Abbrev. for **potential difference**.

peak The *instantaneous* value of the local maximum of a varying quantity. It is $\sqrt{2}$ times the r.m.s. value for a sinusoid.

particle accelerators

Particle accelerators require a source of ions, a high voltage to accelerate the ions and magnetic or electrostatic fields for focusing and redirecting the beams. The main technical difficulty is to provide and maintain the very high voltages needed. In the original Cockcroft-Walton device of 1932, 800 thousand volts (800 kV) were obtained and this was enough to cause the nuclear disintegration of lithium-7 into two helium-4 nuclei by proton bombardment.

Van de Graaff generators

The Cockcroft-Walton design used a multiple voltage doubling arrangement of condensers and diodes which is still found as a proton injector into higher energy accelerators. A more widely used electrostatic device is the Van de Graaff generator which has a moving belt of insulating material to transfer positive charges into a hollow high-voltage terminal (Fig. 1). Corona points, placed at the machine's earthy end, are held at about +20 000 volts. The points ionize the air in their immediate vicinity, repelling the cations which are transferred on to the moving insulated belt by corona discharge. The belt carries the charges to the inside of the positive terminal head, where similar points transfer the charges to the head.

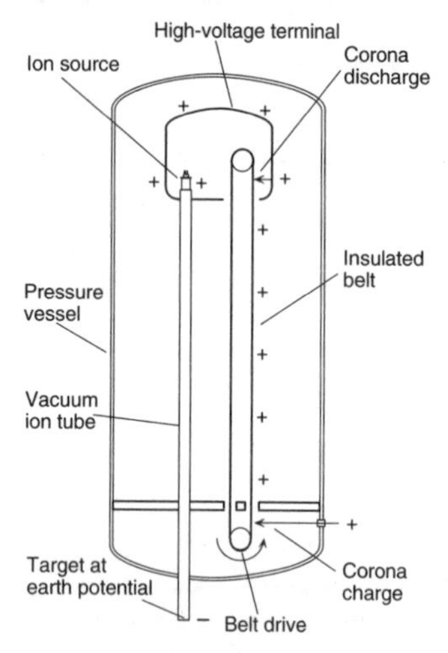

Fig. 1 **Van de Graaff generator**.

The ion source is placed within the head and the ions are accelerated back to ground potential through an insulating vacuum tube. Nowadays the whole machine is placed in a pressure tank containing an insulating gas such as sulphur hexafluoride (SF6) at about 10 atmospheres to suppress arcing. Steady voltages of over 10 million volts can be obtained in the mA range of current. Various improvements including a *tandem design* and means of converting negative into positive ions (see **stripping foil**) have ensured the continued use of these machines.

Cyclotrons

Unlike the two previous devices, which are single-stage accelerators, the cyclotron is a multi-stage device in which ions are progressively accelerated a step at a time up to energies of several hundred million volts. Originally built by Lawrence and Livingston in Berkeley, California, USA in 1931, their essential feature is that the particles travel in an outward spiral path within hollow conductors, called *dees* (see Fig. 2), and that each time they cross the gap between the dees, they are accelerated by an alternating voltage

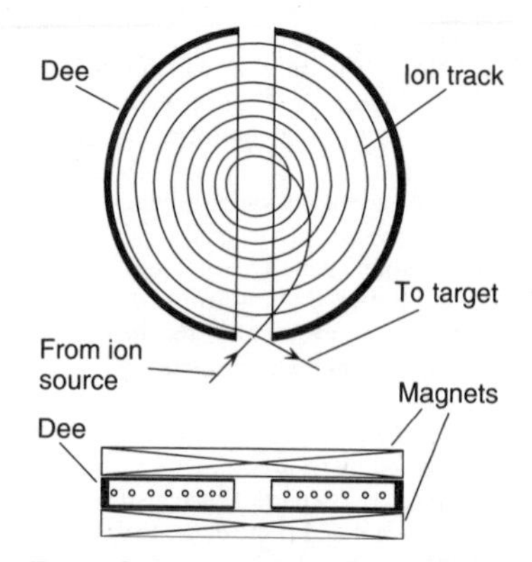

Fig. 2 **Cross-sections of a cyclotron**.

continued on next page

applied to the dees. Further, the accelerated and higher-energy particles, travelling at a larger radius, arrive at the next gap at the same time as lower-energy particles at smaller radii. The particles therefore move round the spiral in exact synchrony with the AC voltage of many megahertz. There are also two large electromagnets, placed above and below the dees which keep the ions positioned and focused within the dees.

The major problem with the fixed-frequency cyclotron is that, as the velocity of the particles increases, the momentum of the particles starts to behave in a relativistic way and they become progressively defocused above about 40 MeV. The only way to overcome this problem is to vary the frequency of the voltage applied to the dees and to reduce the acceleration applied as the particles cross the gaps. The result is to progressively bunch the particles, slowing the faster and accelerating the slower, which now have to traverse many more turns of the spiral. If now the frequency is also *decreased* the particles will slowly move out to the exit position.

Synchrocyclotrons

Such machines are called *synchrocyclotrons* and the largest, completed at Berkeley in 1946, can produce protons of 740 MeV. Various other improvements were made to the design of accelerators including the so-called *azimuthally varying field* cyclotron in which the magnetic field is varied by placing magnets with complex shapes around the ion path so that high and low magnetic fields alternate. Particles no longer move in a circular path but change to a smaller radius as they pass through the lower field. The net effect is to give acceptable accelerations at a higher current than in a synchrocyclotron.

Synchrotrons

Building synchrocyclotrons of higher energy means larger diameter paths and, because the magnets have to cover the complete diameter, larger more expensive magnets, even magnets of 4 m diameter such as those at Berkeley are very expensive. The solution is to vary both the frequency of the accelerating voltage and the strength of the magnetic field in order to keep the particles at a nearly constant radius. If this is done the bending and focusing magnets can be disposed in a circular or near circular beam path without covering the whole area of the ring. The focusing magnets are shaped so that their fields alternately increase and decrease radially across the beam (see Fig. 3), providing a feature called *alternating gradient focusing* (AG focusing). The overall result is that

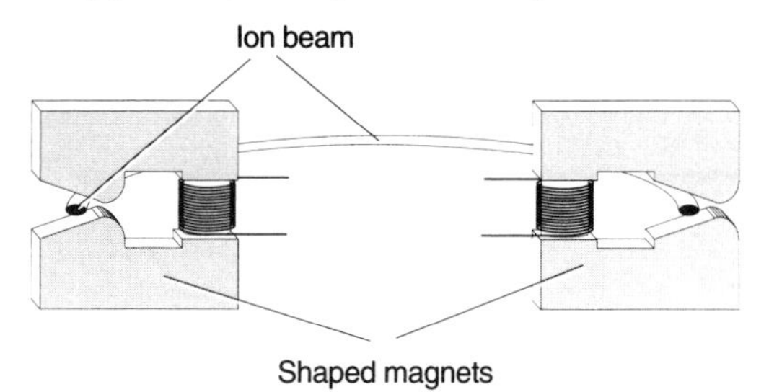

Fig. 3 **The focusing magnets of a synchrotron.**

continued on next page

particle accelerators (contd)

particles moving at larger or smaller radii because of their different momentum, tend to move back to the mean radius.

Giant synchrotrons using these features have been built with rings of two or more kilometres diameter to accelerate protons and to study their interactions with other protons and elementary particles.

Colliding beam accelerators

Because the energy of an accelerator, and with it its cost, must be increased a hundred-fold for a ten-fold increase in the rest mass of the particles, it has been necessary to design and build alternative ways of increasing the effective energy of the colliding particles. This has been achieved by arranging for two beams of particles, e.g. protons and antiprotons, to collide at a shallow angle to each other at a number of intersections round a ring. In order to increase the number of particles available such intersecting rings are connected to two storage rings where bunches of particles can be accumulated for periods of about a day before injection. Such a machine was used to detect for the first time the W and Z particles, which mediate the weak nuclear interaction.

Linear accelerators

One further type of accelerator should be mentioned: *linear accelerators* or linacs in which the particles from a suitable source are accelerated up a tube which may be several kilometres long. The essential components (Fig. 4) are a series of hollow electrodes, called *drift tubes* which are connected sequentially to opposite poles of an alternating voltage source. The particle will *drift* for half the period of the alternating current in each hollow electrode before being accelerated over into the next electrode. Because the particles are now moving faster the next tube must be longer to allow for the required drift time.

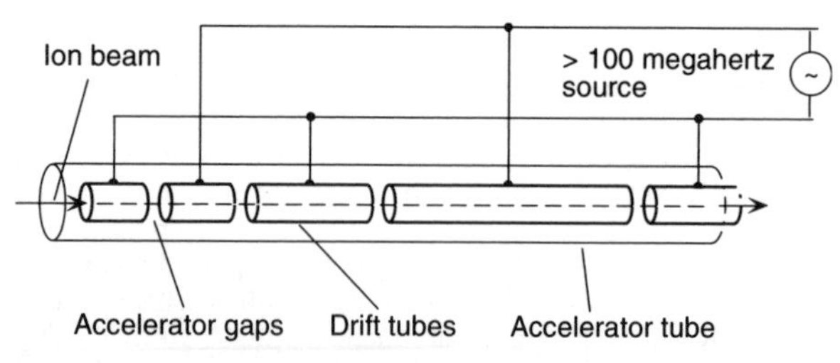

Fig. 4 **Diagram of part of a linear accelerator**.

peak dose Maximum absorbed radiation dose at any point in an irradiated body, usually at a small depth below the surface, due to secondary radiation effects.

peak value The maximum positive or negative value of an alternating quantity. Also *amplitude*, *crest value*, *maximum value*.

pebble-bed reactor A reactor with a cylindrical core into which spherical fuel pellets are introduced at the top and extracted at the base. » p. 35.

pellet Nuclear fusion fuel contained in concentric spheres of glass, plastic and other materials. They are hit by a burst of laser energy to produce fusion. » p. 42.

pellicle A strippable photographic emulsion used to form a *stack* in nuclear research emulsion techniques.

penetrating shower Cosmic-ray shower containing mesons and/or other penetrating particles. See **cascade shower**.

penetration factor Probability of incident particle passing through a nuclear potential barrier.

penetrometer A device for the measurement of the penetrating power of radiation by comparing its transmission through various absorbers.

pent-, penta- Prefixes from Gk. *pente*, five. Used in the construction of compound terms, e.g. *pentactinal*, 5-rayed.

peri- Prefix from Gk. *peri*, around.

period (1) Time taken for one complete cycle of an alternating quantity. Reciprocal of frequency. (2) In a *reactor*, the time in which the neutron flux changes by a factor of *e*.

periodicity See **frequency**.

period meter Instrument for measurement of reactor period.

period of decay See **half-life**.

period range See **start-up procedure**.

peripheral Situated or produced around the edge.

periscope An optical instrument comprising an arrangement of reflecting surfaces whose purpose is to enable an observer to view along an axis deflected or displaced with respect to the axis of the observer's eye; useful for viewing radioactive sources.

permanent implant An implant into biological tissue of radioactive material of short half-life, e.g. radon, arranged so that the prescribed dose is delivered by the time that the radioactive material has decayed, thus rendering removal of the source unnecessary.

permeability The rate of diffusion of gas or liquid under a pressure gradient through a porous material. Expressed, for thin material, as the rate per unit area and for thicker material, per unit area of unit thickness.

permeation The flow of a fluid through a porous material. If J_x is the flux as defined by *Fick's law of diffusion*, then

$$J_x = \frac{P_n (p_1 - p_2)^{1/2}}{l}$$

(Richardson's law) where P_n is the permeability, l is the thickness of material and $(p_1 - p_2)$ is the pressure difference across the material.

permissible dose See **maximum permissible dose**.

personal dosimeter Sensitive tubular electroscope using a metallized quartz fibre unit which is viewed against a calibrated scale. It is charged by a generator and discharged by ionizing radiation. Used by workers in places of potential radiation hazard. » p. 78.

personnel monitoring Monitoring for radioactive contamination of any part of an individual, his breath or excretions, or any part of his clothing.

phantom material Material which produces radiation absorption and back scatter very similar to human tissue. It is used in models to study appropriate doses, radiation scattering etc. A *phantom* is a reproduction of (part of) the body in this material. Also *tissue equivalent material*.

phase space A multidimensional space in which co-ordinates represent the variables required to specify the state of a system; particularly the six-dimensional space with three position and three momentum co-ordinates. Used for the study of particle systems.

phase velocity Velocity of propagation of any one phase state, such as a point of zero instantaneous field, in a steady train of sinusoidal waves. It may differ from the velocity of propagation of the disturbance, or *group velocity*, and, in transmission through ionized air, may exceed that of light. Also *wave velocity*.

-phore Suffix from Gk. *pherein*, to carry.

phosphorescence Luminescence which persists for more than 0.1 nanoseconds after excitation. See **fluorescence**.

phot CGS unit of illumination. 1 phot = 1 lumen $cm^{-2} = 10^4$ lumens m^{-2}.

photochemical effect Chemical effects of radiation, e.g. a light-catalysed reaction.

photodisintegration The ejection of a neutron, proton or other particles from an atomic nucleus following the absorption of a *photon*. A γ-ray photon with energy 2.23 MeV can cause a deuteron nucleus to emit both a neutron and a proton.

photoelasticity Phenomenon whereby strain in certain materials causes the material to become **birefringent**. Coloured fringes are observed when the transmitted light is viewed through crossed Nicol prisms.

photoelectric absorption That part of the absorption of a beam of radiation associated with the emission of photoelectrons. In general, it increases with increasing atomic number but decreases with increasing quantum energy when *Compton absorption* becomes more significant.

photoelectric constant The ratio of Planck's constant to the electronic charge. This quantity is readily measured by experiments on photoelectric emission and forms one of the principal methods by which the value of Planck's constant may be determined.

photoelectric photometer A photometer in which the light from the lamp under test is measured by the current from a photoelectric cell.

photoelectron spectroscopy The study of the energy of *photoelectrons* emitted from a material during irradiation by visible or ultraviolet light or X-rays, in order to analyse the properties of surfaces, interfaces and bulk materials. It can also be used to deduce binding energies for deep core levels in atoms with a high degree of precision.

Also *electron spectroscopy*, *photoionization spectroscopy*. See **ionization potential**.

photofission Nuclear fission induced by gamma rays.

photographic-emulsion technique Study of the tracks of ionizing particles as recorded by their passage through a nuclear emulsion.

photographic photometry Measurement of intensity of radiation by comparing a photographic image of the source with that of a standard source.

photoionization spectroscopy See **photoelectron spectroscopy**.

photomeson Meson resulting from interaction of photon with nucleus, usually a *pion*.

photometry The measurement of the luminous intensities of light sources and of luminous flux and illumination. See **photometer**.

photomultiplier Photocell with series of dynodes used to amplify emission current by electron multiplication, used in e.g. scintillation detectors. » p. 76.

photon Quantum of light of electromagnetic radiation of energy $E = h\nu$ where h is Planck's constant and ν is the frequency. The photon has zero rest mass, but carries momentum $h\nu/c$, where c is the velocity of light. The introduction of this 'particle' is necessary to explain the photoelectron effect, the Compton effect, atomic line spectra, and other properties of electromagnetic radiation. See **gauge bosons**. » p. 80.

photonegative Material of which electrical conductivity decreases with increasing illumination. See **photopositive**.

photoneutron Neutron resulting from interaction of photon with nucleus.

photopeak The energy of the predominant photons released during the decay of a radionuclide.

photopositive The opposite of **photonegative**.

photopic luminosity curve Curve giving the relative brightness of the radiations in an equal-energy spectrum when seen under ordinary intensity levels. See **scotopic luminosity curve**.

photoproton Proton resulting from interaction of photon with nucleus.

photosensitive Property of being sensitive to action of visible or invisible radiation.

physical optics That branch of the study of light dealing with phenomena such as diffraction and interference, which are best considered from the standpoint of the wave theory of light.

physics Study of electrical, luminescent, mechanical, magnetic, radioactive and thermal phenomena with respect to changes in energy states without change of chemical composition.

physio- Prefix from Gk. *physis*, nature.

physiological Relating to the functions of plant or animal as a living organism.

pick-up reaction Nuclear reaction in which incident particle collects nucleon from a target atom and proceeds with it.

pico- SI prefix for 1-million-millionth, or 10^{-12}. Formerly *micromicro-*. Abbrev. *p*.

picosecond The million-millionth part of a second, 10^{-12} s. Abbrev. *ps*.

pieze Unit of pressure in the metre-tonne-second system, equivalent to 10^3 N m^{-2} or 1 kN m^{-2}.

piezo- Prefix from Gk. *piezein*, to press.

piezoelectric effect, piezoelectricity Electric polarization arising in some anisotropic (i.e. not possessing a centre of symmetry) crystals (quartz, Rochelle salt, barium titanate) when subject to mechanical strain. An applied electric field will likewise produce mechanical deformation.

pile Original name for a reactor made from the pile of graphite blocks which formed the moderator of the original nuclear reactor which first went critical on the 2 December 1942 in Chicago, Illinois, USA.

pilotherm A thermostat in which the temperature control is brought about by the deflection of a bimetallic strip, thereby switching on and off the electric heating current.

pi meson See **meson**. Also *π-meson*.

pimpling Small swellings on the surface of fuel during burnup.

pin *Fuel pin*. Very slender fuel cans used e.g. in fast reactors or, in a group, to form certain types of fuel element, e.g. in water cooled reactors. See **fuel element**. » p. 24.

pinch effect In a plasma carrying a large current, the constriction arising from the interaction of the current with its own magnetic field, just as two wires each carrying a current in the same direction experience an attractive force. The principle is used in fusion machines to confine the plasma.

pion See **meson**.

pitch Distance between centres of adjacent fuel channels in a reactor.

pitchblende The massive variety of uraninite. Radium was first discovered in this mineral. This and helium are due to the disintegration of uranium.

PIXE Abbrev. for **Particle-Induced X-ray Emission**. See **trace element analysis** p. 251.

Planckian colour The colour or wavelength-intensity distribution of the light emitted by a black body at a given temperature. » p. 86.

Planck's constant The fundamental constant which is the basis of **Planck's law**. It has the dimensions of energy × time, i.e. action. The present accepted value is 6.626×10^{-34} J s.

Planck's law Basis of quantum theory, that the energy of electromagnetic waves is confined in indivisible packets or quanta, each of which has to be radiated or absorbed as a whole, the magnitude being proportional to frequency. If E is the value of the quantum expressed in energy units and ν is the frequency of the radiation, then $E = h\nu$, where h is known as **Planck's constant.** See **photon**.

Planck's radiation law An expression for the distribution of energy in the spectrum of a black-body radiator:

$$E_\nu \, d\nu = \frac{8 \pi h \nu^3}{c^3 (e^{h\nu/kT} - 1)} \, d\nu ,$$

where E_ν is the energy density radiated at a temperature T within the narrow frequency range from ν to $\nu + d\nu$; h is Planck's constant, c the velocity of light, e the base of natural logarithms and k Boltzmann's constant. » p. 86.

plane of polarization The plane containing the incident and reflected light rays and the normal to the reflecting surface. The magnetic vector of plane-polarized light lies in this plane. The electric vector lies in the *plane of vibration* which is that containing the plane-polarized reflected ray and the normal to the plane of polarization. The description of plane-polarized light in terms of the plane of vibration is to be preferred as this specifies the plane of the electric vector.

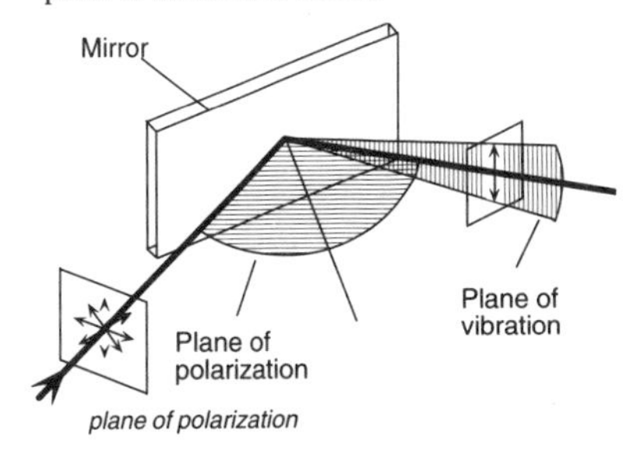

plane of polarization

planetary electron See **Bohr theory**.

plane wave A wave for which equiphase surfaces are planes.

planigraphy See **tomography, emission**.

plasma- Prefix from Gk. *plasma*, anything moulded. Also *plasmo-* and suffix *-plasm*.

plasma Ionized gaseous discharge in which there is no resultant charge, the number of positive and negative ions being equal, in addition to unionized molecules or atoms. The state of high-temperature gases in a fusion machine. » p. 38.

plasma heating In fusion research, plasmas may be heated by ohmic heating, compression by magnetic fields, injection of high-energy neutral atoms, and by **cyclotron resonance heating**. » p. 41.

plasma temperature Temperature expressed in degrees K (thermodynamic temperature) or electron-volts (kinetic temperature). 1 KeV = 10 000 K.

plasma torch A torch in which solids, liquids or gases are forced through an arc within a water-cooled tube, with consequent ionization; de-ionization on impact results in very high temperatures. Used for cutting and depositing carbides.

plasmoid Any individual section of a plasma with a characteristic shape.

plastics A generic name for certain organic substances, mostly synthetic or semisynthetic (casein and cellulose derivatives) condensation or polymerization products, also for certain natural substances (shellac, bitumen, but excluding natural rubber), which under heat and pressure become plastic, and can then be shaped or cast in moulds, extruded as rod, tube etc, or used in the formation of laminated products, paints, lacquers, glues etc. Plastics are **thermoplastic** or **thermosetting**. Adaptability, uniformity of composition, lightness and good electrical properties make plastic substances of wide application, though relatively low resistance to heat, strain and weather are, in general, limiting factors of consequence.

plate (1) Each of the two extended conducting electrodes which, with a a dielectric between, constitutes a capacitor. (2) US *anode* in valves.

plateau See **Geiger region**.

plateau length The voltage range which corresponds to the plateau of a Geiger counter tube.

plateau slope The ratio of the percentage change in count rate for a constant source, to the change of operating voltage. Measured for a median voltage corresponding to the centre of the Geiger plateau. Often expressed as percentage change in count rate for 100 volt change in potential. » p. 75.

plate out The coating (or plating) of exposed surfaces in a reactor with material held in suspension in the coolant. See **tramp uranium**.

plug Piece of absorbing material used to close the aperture of a channel through a reactor core or other source of ionizing radiation.

plutonium An element. Symbol Pu, at. no. 94, product of radioactive decay of *neptunium*. It has seven isotopes:

A	Abundance %	half-life	decay mode
237	All	45.3 d	ε
238	radioactive	87.74 y	α
239		24 100 y	α
240		6570 y	α
241		14.4 y	β^-
242		0.376 My	α
243		4.96 h	β^-

The fissile isotope plutonium-239, produced from uranium-238 by neutron absorption in a reactor, is the most important for the production of nuclear power and in weapons. » pp. 35, 65.

plutonium reactor A reactor in which plutonium-239 is used as the fuel.

Po Symbol for **polonium**.

pocket chamber A small ionization chamber used by individuals to monitor their exposure to radiation. It depends upon the loss of a given initial electric charge being a measure of the radiation received. Also *personal dosimeter*.

poise CGS unit of viscosity. The viscosity of a fluid is 1 poise when a tangential force of 1 dyne per unit area maintains unit velocity gradient between two parallel planes. Equal to 1 dyne cm^{-2} s, or in SI units 10^{-1} N m^{-2} s. Named after the physicist Poiseuille. Abbrev. *P*. See **coefficient of viscosity, viscosity**.

Poiseuille's formula An expression for the volume of liquid per second Q, which flows through a capillary tube of length L and radius R, under a pressure P, the viscosity of the liquid being η:

$$Q = \frac{\pi P R^4}{8 L \eta}.$$

poison Any contaminating material which, because of high-absorption cross-section, degrades intended performance, e.g. fission in a nuclear reactor, radiation from a phosphor. » p. 19.

Poisson's ratio One of the elastic constants of a material. It is defined as the ratio of the lateral contraction per unit breadth to the longitudinal extension per unit length, when a piece of the material is stretched. For most substances its value lies between 0.2 and 0.4. The relationship between Poisson's ratio ν, Young's modulus E, and the modulus of rigidity, G, is given by:

$$\nu = \frac{E}{2G} - 1.$$

polarity (1) Distinction between positive and negative electric charges (Franklin). (2) General term for difference between two points in a system which differ in one respect, e.g. potentials of terminals of a cell or electrolytic capacitor, windings of a transformer, video signal, legs of a balanced circuit, phase of an alternating current.

polarization (1) Non-random orientation of electric and magnetic fields of electromagnetic wave. (2) Change in a dielectric as a result of sustaining a steady electric field, with a similar vector character; measured by density of dipole moment induced.

polarized beam (1) In **electromagnetic radiation**, a beam in which the vibrations are partially or completely suppressed in certain directions. (2) In **corpuscular radiation**, a beam in which the individual particles have non-zero spin and in which the distribution of the values of the spin component varies with the direction in which they are measured.

polarizer Prism of double refracting material, or Polaroid plate, which passes only plane-polarized light, or produces it through reflection.

polar molecule A polar molecule with unbalanced electric charges, usually valency electrons, resulting in a dipole moment and orientation.

pole The part of a magnet, usually near the end, towards which the lines of magnetic flux apparently converge or from which they diverge, the former being called a *south pole* and the latter a *north pole*.

pole strength The force exerted by a particular magnetic pole upon a *unit pole* situated at unit distance from it. See **Coulomb's law for magnetism**.

poloidal field The magnetic field generated by an electric current flowing in a ring. Cf. **toroidal field, tokamak**. » p. 40.

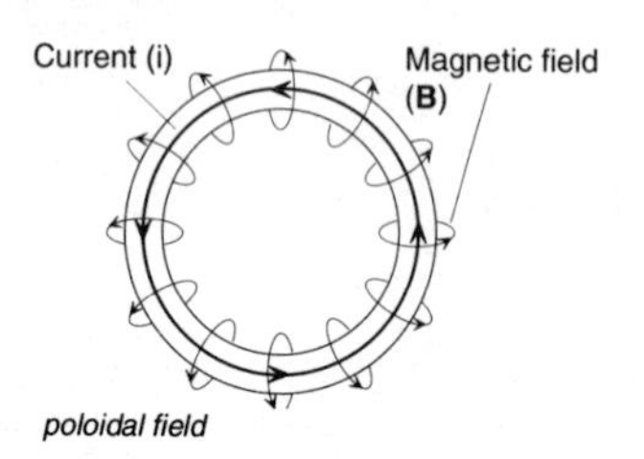

poloidal field

polonium A radioactive element. Symbol Po, at.no. 84. Important as an α-ray source relatively free from γ-emission. It has six isotopes:

A	Abundance %	half-life	decay mode
206	All radioactive	8.8 d	ε
207		5.8 h	ε
208		2.90 y	α
209		102 y	α
210		138.4 d	α
211		0.52 s	α

poly- Prefix from Gk, *polys*, many.

polynucleotide See **nucleic acid**.

pool reactor See **swimming-pool reactor**.

pool-type reactor A dual-cycle reactor in which all the components of the primary cooling circuit (heat exchangers, pumps etc) are contained within the primary pressure vessel, the situation in a pressurized water reactor. Cf. **loop-type reactor**.

population inversion According to the *Boltzmann principle* the number of particles with a higher energy will be less than the number with a lower energy for a system in equilibrium at a given temperature. In a non-equilibrium situation it is possible to invert this so that the number of particles in the higher energy level is greater than the number in the lower level. This is called *population inversion*. It is an essential process in the operation of *masers* and *lasers*. See **optical pumping**.

port Place of access to a system, used for introduction or removal of energy or material, e.g. *glove-box port*.

positive Said of point in circuit which is higher in electric potential than earth.

positive electricity Phenomenon in a body when it gives rise to effects associated with deficiency of electrons, e.g. positive electricity appears on glass rubbed with silk.

positive electron See **positron**.

positive ion An atom (or a group of atoms which are molecularly bound) which has lost one or more electrons, e.g. α-particle is a helium atom less its 2 electrons. In an electrolyte the positive ions (*cations*), produced by dissolving ionic solids in a polar liquid like water, have an independent existence and are attracted to the anode. Negative ions likewise are those which have gained one or more electrons.

positron *Positive electron*, of the same mass as and opposite charge to the (negative) electron. Produced in the decay of radioisotopes and *pair production* by X-rays of energy much greater than 1 MeV. The antiparticle to the **electron**.

positronium Combination of a positron and an electron forming a hydrogen atom-like system, the positron taking the place of the proton. The system is very short-lived (less than 10^{-7} s) and disappears with the emission of γ-ray photons.

post- Prefix from L. *post*, after.

posting The transfer of highly radioactive material around a plant in such a way as to never expose personnel to dangerous radiation.

potential Scalar magnitude, negative integration of the electric (or magnetic) field intensity over a distance. Hence all potentials are relative, there being no absolute *zero potential*, other than a convention, e.g. earth, or at infinite distance from a charge.

potential barrier Maximum in the curve covering two regions of potential energy, e.g. at the surface of a metal, where there are no external nuclei to balance the effect of those just inside the surface. Passage of a charged particle across the boundary should be prevented unless it has energy greater than that corresponding to the barrier. Wave-mechanical considerations, however, indicate that there is a definite probability for a particle with less energy to pass through the barrier, a process known as *tunnelling*. Also *potential hill*.

potential coefficients Parts of total potential of a conductor produced by charges on other conductors, treated individually.

potential difference (1) The difference in potential between two points in a circuit when maintained by an e.m.f. or by a current flowing through a resistance. In an electrical field it is the work done or received per unit positive charge moving between the two points. Abbrev. *pd*. SI unit is the **volt**. (2) The line integral of magnetic field intensity between two points by any path.

potential drop Difference of potential along a circuit because of current flow through the finite resistance of the circuit.

potential energy Universal concept of *energy* stored by virtue of position in a field, without any observable change, e.g. after a mass has been raised against the pull of gravity. A body of mass m at a height h above the ground possesses potential energy mgh, since this is the amount of work it would do in falling to the ground. In electricity, potential energy is stored in an electric charge when it is taken to a place of higher potential through any route. A body in a state of tension or compression (e.g. a coiled spring) also possesses potential energy.

potential temperature The temperature which a specimen of gas would have if it were brought to standard pressure adiabatically. The potential temperature is given by the expression:

$$\theta = T\left(\frac{P_0}{P}\right)^{\frac{\gamma-1}{\gamma}},$$

where T is the absolute temperature, P the pressure of the gas, P_0 is the standard pressure, and γ is the ratio of the specific heat capacities (= 1.40 for air).

potential trough Region of an energy diagram between two neighbouring hills, e.g. arising from the inner electron shells of an atom.

potential well Localized region in which the potential energy of a particle is appreciably lower than that outside. Such a well forms a trap for an incident particle which then becomes *bound*. Quantum mechanics shows that the energy of such a particle is quantized. Applied particularly to the variation of potential energy of a nucleon with distance from the nucleus.

Potter-Bucky grid Type of lead grid designed to avoid exposure of film to scattered X-radiation in diagnostic radiography. Mechanical oscillation of the grid eliminates reproduction of the grid pattern on the radiograph.

pound The unit of mass in the old UK system of units established by the Weights and Measures Act, 1856, and until 1963 defined as the mass of the Imperial Standard Pound, a platinum cylinder kept at the Board of Trade. In 1963 it was redefined as 0.453 592 37 kg. The US pound is defined as 0.453 592 427 7 kg. See **fundamental dynamical units** p. 161.

poundal Unit of force in the foot-pound-second system. The force that produces an acceleration of 1 ft s^{-2} on a mass of 1 pound. Abbrev. *pdl.* 32.2 pdl = 1 lbf (lb wt); 1 pdl = 0.138255 N. See **fundamental dynamical units** p. 161.

powder technology The technology covering the production and handling of powders, particle-size analysis and the properties of powder aggregates.

power Rate of doing work. Measured in joules per second and expressed in *watts* (1 W = 1 J s^{-1}). The foot-pound-second unit of power is the *horsepower*, which is a rate of working equal to 550 ft lbf per second. 1 horsepower is equivalent to 745.7 watts. 1 watt is equal to 10^7 CGS units, i.e. 10^7 erg s^{-1}.

power breeder A nuclear reactor which is designed to produce both useful power and fuel.

power coefficient The change of reactivity of a reactor with increase in power. In a heterogeneous reactor, due to temperature differences of the fuel, moderator and coolant, it differs from the isothermal temperature coefficient of reactivity which assumes temperatures throughout the core change by the same amount.

power control rod A device used for control of the power level of a nuclear reactor. Usually a neutron-absorbing rod containing cadmium or boron steel, but may be a fuel rod or part of the moderator. » p. 26.

power density Energy released per second per unit volume of a reactor core.

power ramping Fairly rapid increase in power of a reactor after a prolonged period at some lower level.

power range See **start-up procedure**.

power supply (1) Arrangement for delivering available power from a source, e.g. public mains, in a form suitable for valves or transistors etc, generally involving a transformer, rectifier, smoothing filter, circuit-breaker or other protection and frequently incorporating electronic regulation. In a full-wave supply, use is made of a full-wave rectifier and filter. (2) US for *mains*.

Pr Symbol for **promethium**.

practical units Obsolete system of electrical units, whereby the ohm, ampere and volt were defined by physical magnitudes. Replaced now by the SI system, in which these units are defined in terms of arbitrarily fixed units of length, mass, time and electric current. See **SI units** p. 237.

pre- Prefix from L. *prae-*, in front of, before.

precursor A nuclide which precedes another in a radioactive decay chain.

pressure See **atmospheric pressure**.

pressure in bubbles A spherical bubble of radius r, formed in a liquid for which the surface tension is T, contains air (or some other gas or vapour) at a pressure which exceeds that in the liquid in its immediate vicinity by $2T/r$. The excess pressure within a soap bubble in air is $4T/r$ since the soap film has two surfaces.

pressure of atmosphere See **atmospheric pressure**.

pressure suppression A method of reducing the pressure of a coolant liquid after an accident by passing the steam generated into cold water ponds and so condensing it.

pressure-tube reactor Reactor in which the fuel elements are contained in a large number of separate tubes through which the coolant water flows, rather than in a single pressure vessel, e.g. the Canadian **CANDU** and British **SGHWR** reactors. » p. 29.

pressure vessel Reactor containment vessel, usually made of thick steel or prestressed concrete, capable of standing high pressure and used in gas-cooled and light-water reactors. See **reactor vessel**. » p. 31.

pressurized-water reactor A reactor using water cooling at a pressure such that its boiling point is above the highest temperature reached. Abbrev. *PWR*. » p. 24.

primary circuit The circulation of the **primary coolant** through a reactor core.

primary coolant The fluid circulated through the reactor to remove heat. In *direct-cycle* reactors such as the boiling-water reactor this drives the turbines directly but in other types the heat is passed via a heat exchanger to a **secondary coolant**. » p. 24. Also *reactor coolant*.

primary energy Energy such as crude oil, natural gas and coal, which is used in its natural form to provide *secondary energy* like electricity.

primary ionization (1) In collision theory, the ionization produced by the primary particles, in contrast to total ionization, which includes the *secondary ionization* produced by delta rays. (2) In counter tubes, the total ionization produced by incident radiation without gas amplification.

primary radiation Radiation which is incident on the absorber, or which continues unaltered in photon energy and direction after passing through the absorber. Also *direct radiation*.

primary separation plant That part of a fuel reprocessing plant coming after the **head end**, where the uranium and plutonium are separated from each other and from other fission products but before final purification.

primary standard A standard agreed upon as representing some unit (e.g. length, mass, e.m.f.) and carefully preserved at a national laboratory. Cf. **secondary standard**.

primordial A term used to describe radionuclides thought to have been present near the time of the origin of the universe. See **cosmogonic**. » p. 105.

principal quantum number When **quantum mechanics** is applied to any particle moving in a central potential, e.g. the electron in a hydrogen atom, the angular momentum and the z-component of the angular momentum are quantized with quantum numbers l and m_l. For a particular l and m_l the solutions of the Schrödinger equation are well-behaved only if the energy is also quantized. These solutions are denoted by n, the *principal quantum number*. In many-electron atoms each electron is described by an orbital specified by n, l, and m_l and the spin quantum number m_s ; the assignment of the orbitals depends on the Pauli exclusion principle. In general the energy depends on both n and l. The value of n denotes the *electron shell* and that of l, the subshell that the electron occupies. » p. 121.

principle of equivalence A statement of the theory of relativity: an observer has no means of distinguishing whether his laboratory is in a uniform gravitational field or is in an accelerated frame of reference. See **relativity.**

principle of least action Principle stating that the actual motion of a conservative dynamical system between two points takes place in such a way that the action has a minimum value with reference to all other paths between the points which correspond to the same energy.

principle of relativity A universal law of nature which states that the laws of mechanics are not affected by a uniform rectilinear motion of the system of co-ordinates to which they are referred. Einstein's relativity theory is based on this principle, and on the postulate that the observed value of the velocity of light is constant and is independent of the motion of the observer. See **relativity**.

principle of superposition The resultant disturbance at a given place and time caused by a number of waves traversing the same space is the vector sum of the disturbances which would have been produced by the individual waves separately. The principle is the basis for the explanation of interference and diffraction effects.

prismatic Prism-shaped; composed of prisms. Applies to the fuel elements of certain types of gas-cooled reactors. » p. 35.

pro- Prefix from Gk. and L. *pro*, before in time or place, used in the sense of 'earlier', 'more primitive' or 'placed before'.

Probabilistic Risk Assessment See **risk assessment** p. 229.

probability density Quantum mechanics suggests that electrons must not be regarded as being located at a defined point in space, but as forming a cloud of charge surrounding the nucleus, the cloud density being a measure of the probability of the electron being located at this point. See **uncertainty principle**.

probe Portable radiation-detector unit, cable-connected to counting or monitory equipment.

process factor See **separation factor**.

production reactor A reactor designed for large-scale production of transmutation elements such as plutonium. They are characterized by the short residence time of the fuel pins which are therefore reprocessed before there has been time for a large buildup of fission products. They are generally military reactors.

proliferation In the nuclear policy context, the spread of nuclear weapons capability to countries not at present possessing such weapons.

promethium A radioactive element of the rare earth series. Symbol Pm, at.no. 61. It has nine isotopes:

A	Abundance %	half-life	decay mode
142	All	405 s	ε
143	radioactive	265 d	ε
144		349 d	ε
145		5.5 y	ε
146		17.7 y	ε
147		2.62 y	β^-
148		5.37 d	β^-
149		53.1 h	β^-
150		2.68 h	β^-

promoter A DNA region in front of the coding sequence of a gene which binds RNA polymerase and therefore signals the start of a gene. » p. 92.

prompt critical The condition in which a reactor has become critical solely with **prompt neutrons**. In this situation the reactor is difficult to control because the reactor **period** is much reduced. » p. 22.

prompt gamma The γ-radiation emitted at the time of fission of a nucleus.

prompt neutrons Nucleons released in a primary nuclear fission process with practically zero delay. » p. 18.

propagation Transmission of energy in the form of waves in the direction normal to a wavefront, which is generally spherical, or part of a sphere, or plane. Applies to acoustic, electromagnetic, water etc waves.

propagation constant Measure of diminution in magnitude and retardation in phase experienced by current of specified frequency in passing along unit length of a transmission line or waveguide, or through one section of a periodic lattice structure. Given by the natural logarithm of the ratio of output to input current or of acoustic particle velocity.

propagation loss The transmission loss for radiated energy traversing a given path. Equal to the sum of the *spreading loss* (due to increase of the area of the wavefront) and the *attenuation loss* (due to absorption and scattering).

propagation of light The transverse electromagnetic waves propagated through free space with a velocity of 2.9979×10^8 m s^{-1} and wavelengths about 400 to 800 nm. The ratio (velocity of light in free space/velocity in medium) is the refractive index of the medium. According to the special theory of relativity, the velocity of light is absolute and no body can move at a greater speed. » p. 82.

proportional counter A counter which uses the *proportional region* in a tube characteristic, where the gas amplification in the tube exceeds unity but the output pulse remains proportional to initial ionization.

proportional region The range of operating voltage for a counter tube in which the gas amplification is greater than unity and the output pulse is proportional to the energy released by the initial event.

protein Protein molecules consist of one, or a small number, of *polypeptide* chains each of which is a linear polymer of several hundred amino acids linked through their amino and carboxylate groups by *peptide bonds*. The amino acid side chains can have a positive or negative charge, a short aliphatic chain or an aromatic residue. Because the 20 amino acids can be arranged in nearly any sequence, the potential diversity of structure and function is enormous. The properties of each polypeptide depends on the amino acid sequence (its *primary structure*) which itself determines the correct folding of the chain in three dimensions in its *native conformation* and thus give its specific biological activity. Because the three dimensional structure is largely dependent on weak forces, it is usually readily disrupted by extremes of pH or heat with a resulting loss of biological activity. When present, covalent cross-links, particularly disulphide bonds, provide a more stable component of higher orders of structure but, in most proteins, are not themselves sufficient to stabilize the native conformation. » p. 91.

proter-, protero-. Prefixes from Gk. *proteros*, before, former.

protium Lightest isotope of hydrogen, of mass unity (^{1}H), most prevalent naturally. The other isotopes are deuterium (^{2}H) and tritium (^{3}H).

proto- Prefix from Gk. *protos*, first.

proton The nucleus of the hydrogen atom; of positive charge and atomic mass number of unity. With neutrons, protons form the nucleus of all atoms, the number of protons being equal to the atomic number. It is the lightest *baryon*, 1.007 276 amu, and the most stable. Beams of high-energy protons produced in particle accelerators are used to study elementary particles. » p. 10.

proton-proton chain A series of thermonuclear reactions which are initiated by a reaction between two protons. It is thought that the proton-proton cycle is more important than the carbon cycle in the cooler stars.

proton resonance A special case of nuclear magnetic resonance. Since the nuclear magnetic moment of protons is now well known, that of other nuclei is found by comparing their resonant frequency with that of the proton.

proton synchrotron A **synchrotron** in which the accelerated particles are protons. It is capable of producing particle energies of more than 200 GeV. See **particle accelerators** p. 206.

proto-oncogene A gene which is required for normal function of the organism, but which when altered can become an **oncogene**.

ps English transliteration of Gk. letter ψ. Pronounced *S*.

pseud-, pseudo- Prefixes from Gk. *pseudos*, false.

P-shell The *electron shell* in an atom corresponding to a principal quantum number of six. In naturally occurring elements it is never completely filled and is the outermost shell for most stable heavy elements. See **atomic structure** p. 121.

Ψ/J particle See **J/Ψ particle**.

Pu Symbol for **plutonium**.

pulsed inertial device Fusion system based on **inertial confinement**.

pulse ionization chamber Ionization chamber for the detection of individual particles by their primary ionization. It has to be followed by a very high gain stable amplifier.

pumping speed Rate at which a pump removes gas in creating a near-vacuum; measured in dm^3 per minute or cu. ft per minute, stp, against a specified pressure.

Purex TN for tri-*n*-butyl phosphate, diluted in kerosene which is used for extracting uranium and plutonium from other fission products in spent fuel. Cf. **butex**. » p. 60.

Purex process The process now almost universally used in the reprocessing of uranium and plutonium, having superseded the older *Butex process*.

PWR See **pressurized-water reactor**.

pycn-, pycno-, pykn-, pykno- Prefixes from Gk. *pyknos*, compact, dense.

pyrometer Instrument for measuring high temperatures.

Q

Q (1) Symbol for **charge**. (2) Symbol of merit for an energy-storing device, resonant system or tuned circuit. Parameter of a tuned circuit such that $Q = \omega L/R$, or $1/\omega CR$, where L = inductance, C = capacitance and R = resistance, considered to be concentrated in either inductor or capacitor. Q is the ratio of shunt voltage to injected emf at the resonant frequency $\omega/2\pi$. $Q = f_r/(f_1-f_2)$ where f_r is the resonant frequency and (f_1-f_2) is the bandwidth at the half-power points. For a single component forming part of a resonant system it equals 2π times the ratio of the peak energy to the energy dissipated per cycle. For a dielectric it is given by the ratio of displacement to conduction current. Also *magnification factor, Q-factor, quality factor, storage factor*.

QCD Abbrev. for **Quantum ChromoDynamics**.

QED Abbrev. for **Quantum ElectroDynamics**.

Q-shell The *electron shell* in an atom corresponding to a principal quantum number of seven. It is the outermost shell for heavy radioactive elements. See **atomic structure** p. 121.

***Q*-switching** A means of producing high instantaneous power from a laser. The cavity resonator has its reflectivity or '*Q*' controlled externally. Q is made small while the population inversion is built up to its peak value. The reflectivity is then increased and the resultant high Q produces an intense burst of energy which almost completely empties the high energy states in a time of about 10^{-8} s. Switching is by a *Kerr cell* shutter or by rotating one of the mirrors.

quadr-, quadri- Prefixes from L. *quattuor*, four.

quality In radiography, it indicates approximate penetrating power. Higher voltages produce higher quality X-rays of shorter wavelength and greater penetration. (This term dates from a period before the nature of X-rays was completely understood).

quality factor See **relative biological effectiveness.** » p. 103.

quantity of light Product of luminous flux and time during which it is maintained; usually stated in lumen-hours.

quantity of radiation Product of intensity and time of X-ray radiation. Not measured by energy, but by energy density and a coefficient depending on ability to cause ionization.

quantization In quantum theory, the division of energy of a system into discrete units (*quanta*), so that continuous infinitesimal changes are excluded.

quantum (1) General term for the indivisible unit of any form of physical energy; in particular the *photon*, the discrete amount of electromagnetic radiation energy, its magnitude being $h\nu$ where ν is the frequency and h is Planck's constant. See **graviton.** » p. 83. (2) An interval on a measuring scale, fractions of which are considered insignificant.

quantum chromodynamics Theory of strong interactions between elementary particles including the interaction that binds protons and neutrons to form a nucleus. It assumes that strongly interacting particles are made of *quarks* and that *gluons* bind the quarks together. Abbrev. *QCD*.

quantum efficiency Number of electrons released in a photocell per photon of incident radiation of specified wavelength.

quantum electrodynamics A relativistic quantum theory of electromagnetic interactions. It provides a description of the interaction of electrons, muons and photons and hence the underlying theory of all electromagnetic phenomena. Abbrev. *QED*.

quantum electronics Electronics concerned with the amplification or generation of microwave power in solid crystals, governed by quantum mechanical laws.

quantum field theory The overall theory of fundamental particles and their interactions. Each type of particle is represented by appropriate *operators* which obey certain commutation laws. Particles are the quanta of fields in the same way as photons are the quanta of the electromagnetic field. So *gluon* fields and *intermediate vector boson* fields can be related to *strong* and *weak* interactions. Quantum field theory accounts for the **Lamb shift**.

quantum gravity The theory that would unify gravitational physics with modern **quantum field theory**.

quantum mechanics A generally accepted theory replacing classical mechanics for microscopic phenomena. Quantum mechanics also gives results consistent with classical mechanics for macroscopic phenomena. Two equivalent formalisms have been developed: *matrix mechanics* (Heisenberg) and *wave mechanics* (Schrödinger). The theory accounts for a very wide range of physical phenomena. See **correspondence principle**, **statistical mechanics**.

quantum number One of a set, describing possible quantum states of a system, e.g.

nuclear spin. See **principal quantum number, spin**. See **atomic structure** p. 121.

quantum statistics Statistics of the distribution of particles of a specified type in relation to their energies, the latter being quantized. See **Bose-Einstein statistics, Fermi-Dirac statistics**.

quantum theory The theory developed from *Planck's law* to account for *black-body radiation*, the *photoelectric effect* and the *Compton effect* and to form the *Bohr theory* of the atom and its modification by Sommerfeld. Now superseded by **quantum mechanics**. » p. 83.

quantum voltage Voltage through which an electron must be accelerated to acquire the energy corresponding to a particular quantum.

quantum yield Ratio of the number of photon-induced reactions occurring, to total number of incident photons.

quark A type of fundamental particle that forms the constituents of *hadrons*. There are currently believed to be six types of quarks (and their antiquarks). In the simple quark theory, the baryon is composed of three quarks, an antibaryon is composed of three antiquarks, and a meson is composed of a quark and an antiquark. No quark has been observed in isolation. See **colour, flavour**.

quartz lamp A lamp which contains a mercury arc under pressure, a powerful source of ultraviolet radiation.

quasi-optical waves Electromagnetic waves of such short wavelength that their laws of propagation are similar to those of visible light.

quencher That which is introduced into a luminescent material to reduce the duration of phosphorescence.

quenching Process of inhibiting continuous discharge, by choice of gas and/or external valve circuit, so that discharge can occur again on the incidence of a further photon or particle in a counting tube. Essential in a **Geiger-Müller counter**. » p. 75.

quench time The time required to quench the discharge of a Geiger tube: **dead time** for internal quenching, **paralysis time** for electronic quenching, although dead time is often used synonymously for the other two terms.

quintal Unit of mass in the metric system, equal to 100 kg. Abbrev. *q*.

Q-value (1) Quantity of energy released in a given nuclear reaction. Normally expressed in MeV, but occasionally in atomic mass units. (2) Ratio of thermonuclear power output to power needed to maintain the plasma.

R

R Symbol (1) to indicate **Rankine scale**; (2) for **röntgen** unit in X-ray dosage.

R Symbol for (1) the *gas constant*; (2) the **Rydberg constant**.

Ra Symbol for **radium**.

rabbit See **shuttle**.

race-track Discharge tube or ion-beam chamber where particles are constrained to an oval path.

rad Former unit of radiation dose which is absorbed, equal to 0.01 J/kg of the absorbing (often tissue) medium. See **gray**.

radar Abbrev. for *RAdio Detection And Ranging*. In general, a system using pulsed radio waves, in which reflected *(primary radar)* or regenerated (*secondary radar*) pulses lead to measurement of distance and direction of target.

radiance Of a surface, the luminous flux radiated per unit area.

radiant flux The time rate of flow of radiant electromagnetic energy.

radiant-flux density A measure of the radiant power per unit area that flows across a surface. Also *irradiance*.

radiant heat Heat transmitted through space by *infrared radiation*.

radiant intensity The energy emitted per second per unit solid angle about a given direction.

radiating surface The effective area of a radiator available for the transmission of heat by radiation.

radiation The dissemination of energy from a source. The energy falls off as the inverse square of the distance from the source in the absence of absorption. The term is applied to electromagnetic waves (radio waves, infrared, light, X-rays, γ-rays etc) and to acoustic waves. It is also applied to emitted particles (α, β, protons, neutrons etc). See **Planck's radiation law**, **Stefan-Boltzmann law**, **Wien's laws for radiation from a black body**. » Chapter 9.

radiation area Area to which access is controlled because of a local radiation hazard.

radiation counter A counter used to detect individual particles or photons. » Chapter 8.

radiation danger zone A zone within which the **maximum permissible dose rate** or **maximum permissible concentration** is exceeded.

radiation diagram See **radiation pattern**.

radiation dose The amount of radiation absorbed by a substance.

radiation field See **field**.

radiation flux density Rate of flow of radiated energy through unit area of surface normal to the beam (for particles this is frequently expressed in number rather than energy). Also *radiation intensity*.

radiation hazard The danger to health arising from exposure to ionizing radiation, either due to external irradiation or to radiation from radioactive materials within the body.

radiation intensity See **radiation flux density**.

radiation length The path length in which relativistic charged particles lose e^{-1} of their energy by radiative collisions. See **bremsstrahlung**.

radiation pattern Polar or cartesian representation of distribution of radiation in space from any source and, in reverse, effectiveness of reception. Also *radiation diagram*.

radiation potential See **ionization potential**.

radiation pressure Minute pressure exerted on a surface normal to the direction of propagation of a wave. It is due to the rate of transfer of momentum by the wave. For electromagnetic waves incident on a perfect reflector this pressure is equal to the energy density in the medium. In quantum physics the radiation consists of *photons* and the radiation pressure is due to the transfer of the momentum of the photons as they strike the surface. Radiation pressures are very small, 10^{-5} N m^{-2} for sunlight at the Earth's surface. For sound waves in a fluid the pressure gives rise to 'streaming', i.e. a flow of the fluid medium.

radiation pyrometer A device for ascertaining the temperature of a distant source of heat, such as a furnace, by allowing radiation from the source to face, or be focused on, a thermojunction connected to a sensitive galvanometer, the deflection of the latter giving, after suitable calibration, the required temperature. For temperatures c. 500°C–1500°C. » p. 72.

radiation source Any device producing radiation of any kind, e.g. a lamp, an X-ray machine, a star, a nuclear reactor. See **sealed source**.

radiation trap (1) Beam trap for absorbing intense radiation beam with a minimum of scatter. (2) Maze or labyrinth formed by entry corridor with several right-angle bends, used for approach to multicurie radiation sources on some accelerating machines.

radiative collision A particle collision in which kinetic energy is converted into electromagnetic radiation. See **bremsstrahlung**.

radiator In radioactivity, the origin of α-, β- and/or γ-rays. Also *source*.

radioactivation analysis If a material undergoes bombardment by neutrons in a nuclear reactor, then frequently radioactive nuclei are produced. This artificial radioactivity can be studied to give information about the isotopes present in the material.

radioactive atom An atom which decays into another species by emission of an α- or β-ray (or by electron capture). Activity may be natural or induced.

radioactive chain See **radioactive series**.

radioactive decay See **disintegration constant, half-life, radioactive atom**.

radioactive decay by heavy ion emission Radioactive decay by the emission of nuclei heavier than the alpha particle. The probability of this occurring is very small but carbon-14 rather than α-decay has been observed from radium-223.

radioactive equilibrium Eventual stability of products of radioactivity if contained, i.e. rate of formation (quantitative) equals rate of decay. Particularly important between radium and radon.

radioactive isotope Naturally occurring or artificially produced isotope exhibiting radioactivity; used as a source for medical or industrial purposes. Also *radioisotope*. » p. 10.

radioactive series Most naturally occurring radioactive isotopes belong to one of three series that show how they are related through radiation and decay. Each series involves the emission of an α-particle, which decreases the *mass number* by 4, and β- and γ-decay which do not change the mass number. The natural series have members having mass number: (a) $4n$ (thorium series); (b) $4n + 2$ (*uranium-radium series*); (c) $4n + 3$ (*actinium series*). Members of the $4n + 1$ (*plutonium series*) can be produced artificially. Also *radioactive chain*.

radioactive standard A radiation source for calibrating radiation measurement equipment. The source has usually a long half-life and during its decay the number and type of radioactive atoms at a given reference time is known.

radioactive tracer Small quantity of radioactive preparation added to corresponding non-active material to *label* or *tag* it so that its movements can be followed by tracing the activity. (The chemical behaviour of radioactive elements and their non-active isotopes is identical.)

radioactivity Spontaneous disintegration of certain natural heavy elements (e.g. radium, actinium, uranium, thorium) accompanied by the emission of α-rays, which are positively charged helium nuclei; β-rays, which are fast electrons; and γ-rays, which are short X-rays. The ultimate end product of radioactive disintegration is an isotope of lead. See **artificial radioactivity, induced radioactivity**.

radiocarbon dating See **radiometric dating** p. 222.

radiocolloids Radioactive atoms in colloidal aggregates.

radioelement An element exhibiting natural radioactivity.

radio-frequency spectrometer Type of **mass spectrometer** used in the study of ions in plasmas.

radio-frequency spectroscopy See **electron spin resonance, nuclear magnetic resonance**.

radiogenic Said of stable or radioactive products arising from radioactive disintegration.

radiography Process of image production using X-rays.

radio heating See **high-frequency heating**.

radioisotope See **radioactive isotope**.

radioisotope thermoelectric generator Thermoelectric generator powered by heat from a radioactive source and suitable for long periods of maintenance-free operation on remote sites, e.g. for lighting marine navigational buoys or powering spacecraft. Also *Ripple* (UK) or *SNAP* (US) generators. See **space reactors** p. 239.

radiology The science and application of X-rays, gamma rays and other penetrating ionizing or non-ionizing radiations.

radioluminescence Luminous radiation arising from rays from radioactive elements, particularly in mineral form.

radiolysis Chemical decomposition of materials induced by ionizing radiation.

radiometer Instrument devised for the detection and measurement of electromagnetic radiant energy and acoustic energy, e.g. thermopile, bolometer, microradiometer. See **Crookes radiometer**. » p. 72.

radiometric dating See panel on p. 222.

radiometric surveying See **aerial radiometric surveying** p. 117.

radionuclide Any nuclide (isotope of an element) which is unstable and undergoes natural radioactive decay.

radionuclide imaging The use of radionuclide substances to image the normal or abnormal physiology or anatomy of the body. Technetium-99 is an important radionuclide used for diagnostic imaging in medicine.

radiometric dating

It is fairly easy to measure the amount of an unstable isotope relative to the amount of a parent or daughter product or of a stable isotope of the same element. Because we know the time needed for half the isotope to disappear or decay (the half-life) this ratio can be used to calculate when a fossil or artefact, like pottery, or a geological formation was formed, provided certain assumptions are made. The best known of these assumptions is that used in radiocarbon dating, namely that the ratio of radioactive carbon-14 to stable carbon-12 has remained constant in the atmosphere over a long period and that once carbon-14 has been incorporated into a fossil or artefact it can no longer exchange with the carbon-14 in the atmosphere. Carbon-14 is formed in the atmosphere by the bombardment of nitrogen-14 by cosmic-ray neutrons.

Radiocarbon dating

The half-life of carbon-14 is 5730 years and, if the assumptions about the stability of the Earth's atmosphere and of carbon-14 are correct, then a measurement of the amount of carbon-14 still present should give the age of the fossil or artefact. The problem is that only one atom of carbon-14 is present for every 10^{12} atoms of carbon-12 and so the methods of measurement must be very sensitive. It also means that times much longer than 50 000 years (10 half-lives) are difficult to study.

The original method of measuring carbon-14 was to count the beta particles which result from its decay. One gram of natural carbon from a present-day sample, such as a newly dead bone, will contain about 10^{10} atoms of carbon-14 which will produce 15 disintegrations per minute. If the bone had been 10 000 years old (two half-lives) this number would fall to four and it would take two days counting to give a result significant at the 1% level. This method is therefore time-consuming and not very accurate, particularly for older and smaller samples. More recently methods have been developed using **particle accelerators** (see p. 206), chiefly tandem Van de Graaff accelerators and cyclotrons which measure the ratio of carbon-14 to carbon-12 atoms directly. This is much more sensitive, extending the method to 100 000 years ago and milligram quantities. The main difficulties are in removing contaminating ions of the same mass number such as nitrogen-14 and CH_2, and quite complicated procedures are necessary to achieve acceptable degrees of purity.

continued on next page

radiopaque Opaque to radiation (esp. X–rays).

radiophotoluminescence Luminescence revealed by exposing to light a material which has been previously irradiated.

radioresistant Able to withstand considerable radiation doses without injury, as certain bacteria, e.g. *Micrococcus radiodurans*.

radiosensitive Quickly injured or changed by irradiation. The gonads, the blood-forming organs, and the cornea of the eye are most radiosensitive organs in Man.

radiotherapy Theory and practice of medical treatment of disease, particularly any of the forms of cancer, with large doses of X–rays or other ionizing radiations.

radiothermoluminescence Luminescence released upon heating a substance previously exposed to radiation. Now used for personal dosimetry.

radiotoxicity The amount of damage done to an organ or tissue by the ingestion of radioactive substance.

radium A radioactive metallic element, one of the alkaline earth metals. Symbol Ra, at.no. 88, r.a.m. 226, mp 700°C. Because of its chemical similarity to calcium, a large fraction of any radium ingested is deposited in bone, but large-scale surveys have shown that for the normal population only about 850 millibecquerels is to be found in the skeleton with a smaller quantity in the soft tissues. Radium is still used in some forms of cancer therapy in a procedure called *radiometric dating (contd)*

Radiocarbon dating covers the comparatively recent past; other methods are more useful in determining much longer times such as when the measured atoms were incorporated into a mineral, which may have happened as the Earth cooled. Their problem is that, while we know the present-day ratios, we do not know how many daughter nuclei were present to start with. This is overcome by using a second stable isotope of the same element as the daughter product and assuming that, originally when the mineral crystallized, the ratio of the two isotopes was the same. This method has been very fully investigated by comparing the two ratios: (1) the amount of the daughter, strontium-87 to the parent, rubidium-87 and (2) the ratio of strontium-87 to the stable isotope, strontium-86. For different minerals over a range of compositions, plotting one ratio against the other gives a straight line, indicating an age of 4.5×10^9 years.

There are a number of similar methods which may be more suitable for other minerals and situations. Some of these are summarized in the following table:

Procedure	Isotope	Half-life	Production	Principal use
Beryllium	^{10}Be	1.6 My	Cosmic rays	Age of deep marine deposits
Carbon	^{14}C	5730 y	Cosmic rays	Age of fossils, human artefacts
Chlorine	^{36}Cl	0.3 My	Cosmic rays	Water migration in glaciers etc
Fission track	^{238}U	4500 My	Primordial	Age since firing of glassy substances
Potassium/ Argon	^{40}K	1.3×10^9 y	Primordial	Age since minerals formed
Rubidium/ Strontium	^{87}Rb	4.8×10^{10} y	Primordial	Age since minerals formed; age of Earth
Thermal luminescence	Any			Time since last heated

Production by cosmic rays occurs in the atmosphere and is assumed to have stayed constant over long periods. Primordial production occurred at the time of the formation of the galaxy although the entrapment of the atoms concerned would have occurred when the minerals cooled. See **fission track dating.**

brachytherapy, where the small container (or radium needle) can be placed close to the tumour. There is a considerable dose to other organs and danger to medical staff who have to position the needle unless the needle can be remotely loaded. Radium has six isotopes:

A	Abundance %	half-life	decay mode
222	All	38 s	α
223	radioactive	11.4 d	α
224		3.66 d	α
225		14.8 d	β^-
226		1602 y	α
227		42 m	β^-

The metal is white and resembles barium in its chemical properties. It occurs in bröggerite, cleveite, carnotite, pitchblende and in certain mineral springs. Pitchblende and carnotites are the chief sources of supply.

radium cell A sealed container in the shape of a thin-walled tube (usually made of metal) normally loaded into larger containers, e.g. **radium needle**.

radium emanation See **radon**.

radium needle A container in form of a needle, usually platinum-iridium or gold alloy, designed primarily for insertion into tissue. Little used now.

radium therapy Radiotherapy by the use of the radiations from radium, now almost entirely superceded by X-rays.

radon A zero-valent, radioactive element, the heaviest of the noble gases. Symbol Rn, at.no. 86, r.a.m. 222, mp −150°C, bp −65°C. It is formed by the disintegration of radium and over half the natural ionizing radiation received by the UK population comes from radon, which can accumulate in buildings in areas where radium occurs in rocks. See **ionizing radiation**. It has seven isotopes:

A	Abundance %	half-life	decay mode
207	All	9.3 m	ε
210	radioactive	2.4 h	α
211		14.6 h	ε
212		24 m	α
218		35 ms	α
222		3.83 d	α
224		107 m	β^-

Also *radium emanation* (obsolete).

radon seeds Short lengths of gold capillary tubing containing radon used in treatment of malignant and non-malignant neoplasms.

raffinate Liquid layer in solvent extraction system from which the required solute has been extracted, e.g. in the chemical extraction of uranium from its ores, it is the liquid left after the uranium has been extracted by contact with an immiscible solvent.

Ramsauer effect Sharp decrease to zero of scattering cross-section of atoms of inert gases, for electrons of energy below a certain critical value.

random coincidence Simultaneous operation of two or more coincidence counters as a result of their discharge by separate incident particles arriving together (instead of common discharge by a single particle as is normally assumed in interpreting the readings).

range (1) Length of track along which ionization is produced in a nuclear particle. (2) Distance of effective operation of nuclear forces.

Rankine scale Absolute scale of temperature, based on degrees Fahrenheit. See **Fahrenheit scale**.

rare earth elements A group of metallic elements possessing closely similar chemical properties. They are mainly trivalent, but otherwise similar to the alkaline earth elements. The group consists of the lanthanide elements 57 to 71, plus scandium (21) and yttrium (39). Extracted from monazite, and separated by repeated fractional crystallization, liquid extraction, or ion exchange.

Rasmussen Report The influential US report on reactor safety issued in the mid-1970s. Also *Reactor Safety Study*. See **risk assessment** p. 229.

ratchetting Intermittent movement of fuel elements arising from thermal cycling and differential expansion effects.

ratemeter See **count ratemeter**.

rating Specified limit to operating conditions, e.g. current rating etc.

rationalized units Systems of electrical units for which the factor 4π is introduced in Coulomb's laws so that it will be absent from more widely used relationships. In **Heaviside-Lorentz units** (rationalized Gaussian units) this is done directly, thus modifying values for the unit charge and unit pole. In **MKSA** units it is done indirectly by modifying the values of the permittivity and permeability of free space.

ray General term for the geometrical path of the radiation of wave energy, always in a direction normal to the wavefront, but with possible reflection, refraction, diffraction, divergence, convergence and diffusion. By extension, also the geometrical path of a beam of particles in an evacuated chamber. This may be curved in an electric or magnetic field. See **particle**.

Rayleigh criterion Criterion for the resolution of interference fringes, spectral lines and images. The limit of resolution occurs when the maximum of intensity of one fringe or line falls over the first minimum of an adjacent fringe or line. For a telescope with a circular aperture of diameter D this criterion gives the smallest angular separation of the two images of point objects as $1.22\ \lambda/D$, where λ is the wavelength of the light.

Rayleigh-Jeans law An expression for the distribution of energy in the spectrum of a black-body radiator:

$$E_\nu\, d\nu = \frac{8\pi \nu^2 kT}{c^3}\, d\nu ,$$

where E_ν is the energy density radiated at a temperature T within a narrow range of frequencies from ν to $\nu + d\nu$, k is Boltzmann's constant and c the velocity of light. The formula holds only for low frequencies. See **Planck's radiation law**.

Rayleigh limit One-quarter of a wavelength, the maximum difference in optical paths between rays from an object point to the corresponding image point for perfect definition in a lens system.

Rayleigh scattering See **scattering**.

Rb Symbol for **rubidium**.

RBMK reactor Graphite-moderated, boiling-water-cooled, pressure-tube reactor unique to the USSR (the type involved in the Chernobyl accident). » p. 49.

reaction The equal and opposite force arising when a force is applied to a material system; in particular the force exerted by the supports or bearings on a loaded mechanical system.

reactor types

Reactors are classified according to their *fuel, moderator* and *coolant* (or less frequently according to their *size, power output* and *function*). Several hundred types of reactor have been tested or suggested based on the following alternatives: *fuel*: uranium-235; plutonium-239; uranium-233; *moderator*: light water; heavy water; graphite; beryllium; organic liquid (or none in fast reactors); *coolant*: gas; light water; heavy water; organic liquid; liquid metal. Fuels are classified as *natural, slightly enriched*, or *highly enriched* according to the extent to which the proportion of fissile material has been increased beyond its normal isotopic abundance.

The figure illustrates some of these alternatives with lines representing the connections between the main groupings. » Chapter 3.

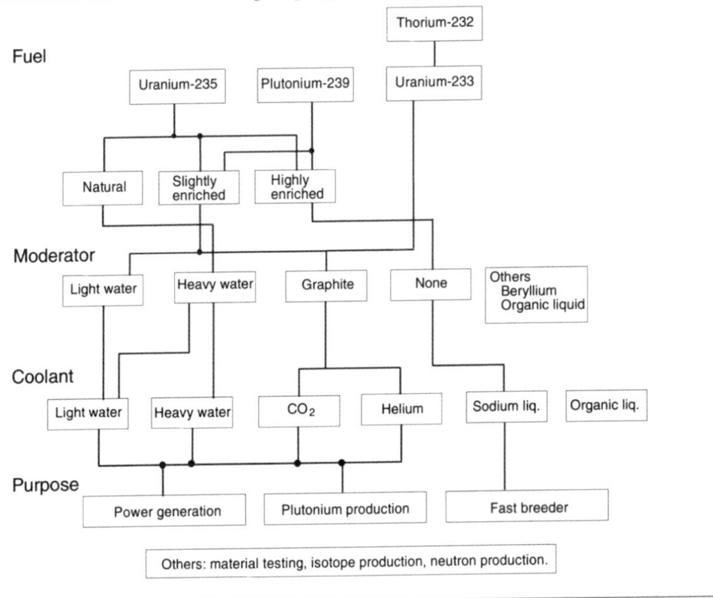

reaction rate The rate of fission in a nuclear reactor.

reactivity The departure of the multiplication constant of a reactor from unity, measured in different ways, i.e. **cent**, **dollar**, **nile**, **inhour**, or simply *per cent reactivity*. » p. 20.

reactivity worth Change of reactivity of a reactor caused by the addition or removal of a material or piece of equipment. The control rods of a reactor could be *worth* 10% of the total reactivity. Also *worth*. » p. 20.

reactor noise Random statistical variations of neutron flux in reactor.

reactor oscillator Device which produces periodic variations of reactivity by mechanical oscillation of neutron absorbing sample in reactor core. Used to measure reactor properties, or nuclear cross-section of sample.

Reactor Safety Report See **Rasmussen Report, risk assessment** p. 229.

reactor simulator Digital computer which simulates variations in reactor neutron flux produced by changes in any operating parameter. Useful for training and for investigating reactor effects.

reactors in space See **space reactors** p. 239.

reactor trip Rapid reduction of reactor power to zero by emergency insertion of control mechanisms and (in some cases) removal of liquid moderator. Also *scram*.

reactor types See panel on p. 225.

reactor vessel The vessel in which the core, moderator, coolant and control rods are situated. See **pressure vessel**. » p. 24.

Réaumur scale A temperature scale ranging from 0°R to 80°R (freezing point and boiling point of pure water at normal pressure).

reciprocity principle Principle that the interchange of radiation source and detector will not change the level of radiation at the latter, whatever the shielding arrangement between them.

recoil atom An atom which experiences a sudden change of direction or reversal, after the emission from it of a particle or radiation. Also *recoil nucleus*.

recoil nucleus See **recoil atom**.

recoil particles Particles arising through collision or ejection, e.g. Compton recoil electrons.

recombination Neutralization of ions in gas, by combination of charges or transfer of electrons. Important for ions arising from the passage of high-energy particles.

recovery rate The rate at which recovery occurs after radiation injury. It may proceed at different rates for different tissues.

recovery time (1) For a Geiger tube, the period between the end of the dead time and the restoration of full normal sensitivity. (2) For a counting system, the minimum time interval between two events recorded separately.

recti- Prefix from L. *rectus*, straight.

recycle Repetition of fixed series of operations, e.g. biodegradation of organic material followed by regrowth; applied also to isotope separation.

reduced mass The quantity

$$\frac{mM}{(M+m)},$$

used in the study of the relative motion of two particles, masses m and M, about their common centre of gravity. This is used in place of the smaller mass; movement of the larger is then ignored.

reflectance See **reflection factor**.

reflected wave, reflected ray A wave turned back from a discontinuity in a continuous medium.

reflection Return of neutrons to reactor core after a change of direction experienced in the shield surrounding the core.

reflection factor The ratio which the luminous flux reflected from a surface bears to that falling upon it. Also *coefficient of reflection, reflectance*.

reflection laws For wave propagation: (1) incident beam, reflected beam and normal to surface are coplanar; (2) the beams make equal angles with the normal.

reflectivity Proportion of incident energy returned by a surface of discontinuity.

reflector A layer of material (e.g. graphite) designed to scatter or reflect neutrons back into a reactor without absorbing many.

refraction Phenomenon which occurs when a wave crosses a boundary between two media in which its phase velocity differs. This leads to a change in the direction of propagation of the wavefront in accordance with *Snell's law*.

refractive index The absolute refractive index of a transparent medium is the ratio of the phase velocity of electromagnetic waves in free space to that in the medium. It is given by the square root of the product of the complex relative permittivity and complex relative permeability. Symbol n. See **refraction.**

refractory metals Term applied to those transition group elements in the periodic table which have high melting points. They include chromium, titanium, platinum, tantalum, tungsten and zirconium.

regeneration Reprocessing of nuclear fuel by removal of fission products.

Regge trajectory A graph relating spin angular momentum at energy for a nuclear particle. Possible quantized values of spin correspond to large discrete energy increments on the graph. This enables recurrences of nuclear particles to be predicted: the extra energy corresponding to the greater rest mass expected to be associated with such particles. A *recurrence* is a particle identical in all respects, except energy (or mass) and spin momentum, with a known particle, and is regarded as being a higher energy equivalent of the normal particle.

region of limited proportionality Range of operating voltages for a counter tube in which the gas amplification depends on the number of ions produced in the initial ionizing events as well as on the voltage. For larger initial events the counter saturates. » p. 75.

regulating rod Fine **control rod** of reactor.

regulatory body Organizations set up by governments to oversee the proper use of method or technology. Particularly important in the nuclear industry, where such bodies have power to license use, construction and disposal.

relative abundance See **abundance**.

relative atomic mass Mass of atoms of an element formerly in *atomic weight units* but now more correctly given on the *unified scale* where 1u is 1.660×10^{-27} kg. Abbrevs. *RAM, r.a.m.* See **atomic weight**.

relative biological effectiveness The inverse ratio of the absorbed dose of ionizing radiation to the absorbed dose of 200 kV X-rays which would produce an equivalent biological damage. Definition dates from early radiological measurements. *Quality factors* are rounded off values used in the definition of the sievert. Abbrev. *rbe*. » p. 103.

relative density The ratio of the mass of a given volume of a substance to the mass of an equal volume of water at a temperature of 4°C. Originally *specific gravity*.

relative plateau slope See **plateau slope**.

relative stopping power See **stopping power**.

relativistic Said of any deviation from classical physics and mechanics based on relativity theory.

relativistic mass equation When a particle is accelerated up to a velocity (v) which is more than a small fraction of the phase velocity of propagation of light in a vacuum (c), it is said to be *relativistic* with mass increased according to the formula

$$m = \frac{m_0}{\sqrt{1 - \frac{v^2}{c^2}}},$$

where m_0 = rest mass (at low velocities). Required to be considered in cyclotron, betatron and linear accelerator design.

relativistic particle A particle with a velocity comparable with that of light.

relativity Theory based on the equivalence of observation of the same event from different frames of reference having different velocities and accelerations. Introduced in 1905 and generalized a decade later, Einstein's *special relativity* was verified by observations and the precession of the perihelion of the planet Mercury. Important results of this restricted theory include the **relativistic mass equation**, the **mass-energy equation** and **time dilation**.

relaxation time Most processes which exhibit decay are assumed to follow an exponential law. For example, the drift velocity $< v_x >$ of electrons in a conductor will decay to zero after the field is removed according to

$$< v_x > = < v_x >_0 \, e^{-t/\tau},$$

where $< v_x >_0$ is the velocity at time $t = 0$, and τ is the *relaxation time* for the decay.

rem Abbrev. for **röntgen equivalent man**.

remote handling equipment Apparatus developed to enable an operator to manipulate highly radioactive materials from behind a suitable shield, or from a safe distance.

reprocessing See **fuel reprocessing**.

reproducibility Precision with which a measured value can be repeated in a process or component.

reserves Block of ore proved by investigation to warrant extraction. More generally the similarly proved availability of e.g. uranium ore worldwide with presently available technology and at current prices. Cf. **resources**.

reservoir Any volume in an isotope separation plant which is for the purpose of storing material or to ensure smooth operation.

residence time The term given by dividing the volume of any reservoir or pool or processing unit by the rate of flow through it. The expression gives the average time spent in the pool by the substance or organism under consideration.

residual activity In a nuclear reactor the remaining activity after the reactor is shut down following a period of operation. » p. 60.

resistance In electrical and acoustical fields, the real part of the impedance characterized by the dissipation of energy as opposed to its storage. Electrical resistance may vary with temperature, polarity, field illumination, purity of materials etc. SI unit is the **ohm**. Reciprocal of *conductance*.

resistivity Intrinsic property of a conductor, which gives the resistance in terms of its dimensions. If R is the resistance in ohms, of a wire l m long, of uniform cross-section a m^2, then $R = \rho\, l/a$, where the resistivity ρ is in ohm metres (*not* ohm m^{-3}). Also, erroneously, *specific resistance*.

resolution time The minimum time interval needed for two events to be recorded separately by a detector of ionizing radiation. It is also the maximum time between two events recorded as *coinciding*.

resolution-time correction A correction applied to the observed counting rate for random events such as those measured by a detector of ionizing radiation. It allows for those events which are not recorded because of the detector's finite resolution time.

resolving time See **resolution time**.

resonance If a vibrating system, mechanical or acoustical, is set into forced vibrations by a periodic driving force and the applied frequency is at or near the natural frequency of the system, then vibrations of maximum velocity amplitude result. This is called resonance and it corresponds to minimum mechanical or acoustical impedance. The *sharpness of resonance* is measured by the ratio of the dissipation to the inertia of the system which also determines the rate of decay of the vibrating system when it is impulsed. In nuclear physics it refers to the peaks of absorption which occur when a subatomic particle collides with a nucleus and is a function of the thermal energy of the nucleus and the mass energy of the impinging particle. See **quality factor**, **decay factor**. » p. 14.

resonance escape probability In a reactor, the probability of a fission neutron slowing down to thermal energy without experiencing resonance. » p. 20.

resonance heating See **magnetic pumping**.

resonance level An excited level of the compound system which is capable of being formed in a collision between two systems, such as between a nucleon and a nucleus.

resonance radiation Emission of radiation from gas or vapour when excited by photons of higher frequency.

resonances During nuclear reactions very unstable mesons or hyperons are frequently created. These decay through the strong interaction, with a half-life of the order of 10^{-23} s. Consequently such particles are undetectable and their formation as an intermediate step in the reaction can only be inferred from indirect measurements. These temporary states are known as *resonances* to distinguish them from metastable particles with half-lives of the order of 10^{-1} s, which are detectable.

resonance scattering See **scattering**.

resources Ores estimated to be present worldwide from geological date and which are reasonably assured of being recoverable with present technology and at economic prices. Cf. **reserves**.

rest mass Mass of a particle measured by an observer at rest relative to it. In nuclear reactions, the total *rest mass* of the particle involved need not be conserved. Cf. **mass**. See **relativistic mass equation**.

rest-mass energy Rest-mass energy is c^2 times the *rest mass* of the particle, where c is the speed of light.

retro- Prefix from L. *retro*, backwards, behind.

reversed field pinch A toroidal magnetic trap in which the toroidal field changes sign in the outer region of a plasma discharge. Found in one type of experimental fusion reactor. Abbrev. *RFP*.

reversible Said of a process whose effects can be reversed so as to bring a system to its original thermodynamic state.

RFP Abbrev. for **Reversed Field Pinch**.

rheo- Prefix from Gk. *rheos*, current, flow.

rheology The science of the flow of matter. The critical study of elasticity, viscosity and plasticity.

rig An experimental setup designed to investigate a particular property, e.g. the effect of neutrons on cladding. Often arranged, in the nuclear industry, in association with a reactor or other nuclear plant.

ripple Small wave on the surface of a liquid for which the controlling force is not gravity, as for large waves, but surface tension. The velocity of ripples diminishes with increasing wavelength, to a minimum value which for water is 23 cm s^{-1} for a wavelength of 1.7 cm.

Ripple generator See **radioisotope thermoelectric generator**.

risk assessment See panel on p. 229.

r.m.m. Abbrev. for *relative molecular mass*.

Rn Symbol for **radon**.

rod A rod-shaped reactor fuel element, control absorber or sample which is intended for irradiation in a reactor.

röntgen The older unit of X-ray or gamma dose, for which the resulting ionization liberates a charge of each sign of 2.58×10^{-4} coulombs per kilogram of air. Symbol R. » p. 103.

röntgen equivalent man Former unit of biological dose given by the product of the absorbed dose in R and the relative biological efficiency of the radiation. Abbrev. *rem*. Now replaced by **effective dose equivalent**, unit the sievert (Sv). » p. 103.

röntgenology US for **radiology**.

röntgen rays See **X-rays**.

rotameter An indicating and measuring device for the rate of flow of gases and liquids.

Ru Symbol for **ruthenium**.

rubidium A metallic element in the first group of the periodic system, one of the alkali metals. Symbol Rb, at.no. 37, r.a.m. 85.47, mp 38.5°C, bp 690°C, rel.d. 1.532. The element is widely distributed in nature, but occurs only in small amounts; the chief source is carnallite. The metal is slightly radioactive and it has nine isotopes.

A	Abundance %	half-life	decay mode
82		1.25 m	ε
83		86.2 d	ε
84		32.9 d	ε
85	72.17		
86		18.8 d	β^-
87	27.83		
88		17.8 m	β^-
89		15.2 m	β^-
90		153 s	β^-

rubidium-strontium dating A method of

risk assessment

A description of the safety of a plant in terms of the frequency and consequence of any possible accident. The safety of the first nuclear power plants was assessed by identifying design-based accidents and determining whether the engineered safeguards could prevent them. The consequences of the maximum credible accident were also calculated. This *deterministic* approach has failed on a number of occasions because essentially trivial malfunctions and operator error have resulted in quite unforeseen accidents.

Designers and nuclear safety agencies have therefore increasingly turned to a much more complete analysis of all aspects of a plant's operation called *Probabilistic Risk Assessment* (PRA), which can be split naturally into four stages. (1) The identification of accidents which might occur. This will include everything from a weld failure in the main coolant circuit to a fuse blowing in the control board. (2) Estimating the frequency with which a possible accident occurs, i.e. its probability per unit time. (3) Determining the quantified consequences of each accident to the plant itself, its operators, the surrounding population and the environment. (4) Using the risk information, i.e. the frequencies and consequences, as a guide in making decisions about e.g. whether the plant is acceptably safe, how it could most cost-effectively be made safer, or the nature of the weak points of the design.

These simple concepts are applied to nuclear reactors at three levels of analysis:

Damage to the reactor core. This involves identifying the initiating events and then, assuming the success or failure of each emergency system, constructing an *event tree* (see Fig. 1) which will have different endpoints depending on which system failed or succeeded. The probability of each event is estimated from the history of previous incidents and the events may be very detailed, down to the level of e.g. individual pumps or particular operator errors.

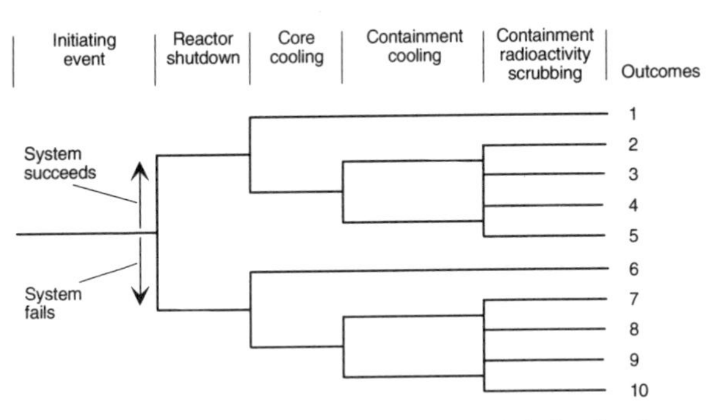

Fig. 1 **Illustrative event tree**. The conventions include indicating the success of a system by an upwards line at the start of the system's position.

Effect on the containment barrier. Again an event tree has to be made for the possible accident sequences and will involve e.g. fuel cladding, the reactor vessel and coolant, as well as the containment building. It will result in endpoints which can be divided into different categories depending on the proportions and amounts of the radioactivity released. This analysis will involve knowledge of the physics and chemistry of the fission products and of their movements. It should give information about improving the design of the containment barriers.

Dispersion of radioactivity and its effects. This requires the modelling of how a cloud of radioactivity will move under different conditions, its fallout onto

continued on next page

land and sea, and its subsequent ingestion by humans or incorporation into foodchains. It will require the analysis of risk to society in an area perhaps extending over hundreds of miles. From the results, the number of people who may be affected can be calculated, and predictions made about the area of land to be evacuated and the cost of decontamination.

Probabilistic risk assessment is a very ambitious endeavour because it seeks to include every possible accident sequence. There are inevitably areas where there is insufficient information, which has lead to criticism of the whole concept of this kind of analysis. This has, in turn, prompted a more detailed investigation of the gaps in knowledge that have been revealed and to attempts to quantify the degree of uncertainty which may, in places, be very large.

There are a number of areas of high uncertainty including the modelling of complex physics and the effects of external disasters like fire and earthquake. The latter are called *hazards* in PRA. Two others will be discussed here: dependent failures and human error. Dependent failures are those in which any two events are connected in some way and they therefore invalidate the usual rule that probabilities can be multiplied together. They include two emergency systems relying on a common power supply, the common mode failure of two valves of identical design and the siting of two key safety systems in the same room liable to flooding. Because there are necessarily many duplicate and backup components in reactor safety systems, dependent failures of this kind are important but they are very difficult to quantify because they can arise in many different ways and there is little historical information on these kinds of faults.

Human errors such as pressing the wrong switch are easy to quantify but human actions following a misapprehension about the true state of a reactor or from simple malice are much more difficult to predict. There is cause for concern in this area because some major accidents have resulted from misapprehension at the least.

History and Applications

The first attempt at complete identification of accident sequences and frequencies was made in the UK in the late 1960s in order to work out a policy for the siting of the Advanced Gas-cooled Reactors. In the 1970s full PRAs were carried out on two Pressurized-Water Reactors and two Boiling-Water Reactors and provoked an intense discussion (The Reactor Safety Study or Rasmussen Report) which was further stimulated when, after the Three Mile Island accident (» p. 49), it was found that the fundamental system weakness in the auxiliary feedwater system had been identified, although the precise accident sequence had not been foreseen.

It had been hoped at the time of the Reactor Safety Report that the safety of a reactor would be determined primarily by its conceptual design but further work has shown that, from the safety point of view, it is the detailed systems which are most important. Many reactors have detailed system weaknesses which can be discovered and corrected by a risk-based analysis. It is at this level of detailed improvement that PRA is now of prime importance.

In the UK the analysis was used as part of the risk assessment for the later Advanced Gas-cooled Reactors but it has been most important in the Sizewell 'B' project, the first UK pressurized-water reactor, where the analyses have been part of the safety case for all stages from the conceptual design to those for its construction and operation.

Probabilistic Risk Assessment offers a new approach to characterizing nuclear power plant safety because of its complete and quantified nature. However, it does have major uncertainties which means that its role in the overall safety case remains that of providing an important input into the safety decision rather than making the decision itself.

determining the age in years of geological material, based on the known decay rate of rubidium-87 to strontium-87. See **radiometric dating** p. 222.

Russell-Saunders coupling Extreme form of coupling between orbital electrons of atoms. The angular and spin momenta of the electrons combine and the combined momenta then interact. Also *l-s coupling*.

ruthenium A metallic element. Symbol Ru, at.no. 44, r.a.m. 101.07, mp 2400°C, rel.d. 12.26. It has 14 isotopes:

A	Abundance %	half-life	decay mode
94		52 m	ε
95		1.65 h	ε
96	5.5		
97		2.88 d	ε
98	1.86		
99	12.7		
100	12.6		
101	17.0		
102	31.6		
103		39.4 d	β^-
104	18.7		
105		4.44 h	β^-
106		372 d	β^-
107		3.8 m	β^-

The metal is silvery-white, hard and brittle. It occurs with the platinum metals in osmiridium, and is used in certain platinum alloys.

Rutherford atom Earliest modern concept of atomic structure, in which all the positive charge and nearly all the mass of the atom is in the nucleus. Electrons, equal in number to the atomic number, occupy the rest of the volume and make the atom electrically neutral.

Rutherford scattering See **scattering**.

Rydberg constant The frequency of atomic spectrum lines in a given series are related by the Rydberg formula. The Rydberg constant, *R*, involved was first deduced from spectroscopic data but has since been shown to be a universal constant.

$$R = \frac{2\pi^2 e^4}{c\,h^3} M_r ,$$

where M_r is the reduced mass of the electrons, *e* is the electronic charge, *c* is the velocity of light and *h* is Planck's constant.

Rydberg formula A formula, similar to that of Balmer, for expressing the wave numbers (ν) of the lines in a spectral series:

$$\nu = R\left[\frac{1}{(n+a)^2} - \frac{1}{(m+b)^2}\right],$$

where *n* and *m* are integers and *m, n, a* and *b* are constants for a particular series, and *R* is the *Rydberg constant*.

S

s Symbol for **second** (time).

s Symbol for (1) distance along a path; (2) solubility; (3) specific entropy.

σ Symbol for (1) **conductivity**; (2) normal stress; (3) nuclear **cross-section**; (4) Stefan-Boltzmann constant; (5) surface charge density; (6) **surface tension**; (7) **wave number**.

S Symbol for (1) **sulphur**; (2) siemens.

S Symbol for (1) area; (2) **entropy**; (3) Poynting vector.

Σ Symbol for *sum of.*

saddle point Point on plot of potential energy against distortion for nucleus at which fission will occur, instead of return to equilibrium.

safety factor Of a fusion system, the **aspect ratio** multiplied by the ratio of toroidal to poloidal field. This requires to be greater than unity for magneto-hydrodynamic stability. Symbol *q*. » p. 40.

safety rods Rods of neutron-absorbing material capable of rapid insertion into a reactor core to shut it down, in case of emergency.

salting-out The removal of an organic compound from an aqueous solution by the addition of salt.

samarium A hard and brittle metallic element. Symbol Sm, at.no. 62, r.a.m. 150.35, mp 1350°C, bp 1600°C, rel.d. 7.7. It has 14 isotopes:

A	Abundance %	half-life	decay mode
142		72.5 m	ε
143		8.83 m	ε
144	3.1		
145		340 d	ε
146		103 My	α
147	15.1		
148	11.3		
149	13.9		
150	7.4		
151		90 y	β^-
152	26.6		
153		46.8 h	β^-
154	22.6		
155		22.4 m	β^-

It is feebly, naturally radioactive, but the stable isotope samarium-149, produced by the decay of promethium-149 in fission reactors, has a very high neutron capture cross-section of 5.3×10^4 barns and is therefore a reactor poison, but one which does not have a large effect on reactor kinetics.

sampling Selection of an irregular signal over stated fractions of time or amplitude (pulse height).

sandwich Photographic nuclear research emulsion forming a series of thin layers with intervening layers in which some event is to be studied.

sandwich irradiation The irradiation of tissues from opposite sides.

Sargent diagram Log-log plot of radioactive decay constant against maximum β-ray energy, for various β-emitters. Most of the points relating to natural heavy radioisotopes lie on one or other of two straight lines.

saturated vapour A vapour which is sufficiently concentrated to exist in equilibrium with the liquid form of the same substance.

saturation activity The maximum level of artificial radioactivity, induced in a given sample, by a specific level of irradiation when the rate of formation equals the rate of decay.

scattering General term for irregular reflection or dispersal of waves or particles, e.g. in acoustic waves in an enclosure leading to diffuse reverberant sound, **Compton effect** on electrons, light in passing through material, electrons, protons and neutrons in solids, radio waves by ionization. See **forward scatter**, **back scatter**. Particle scattering is termed *elastic* when no energy is surrendered during scattering process, otherwise it is *inelastic*. If due to electrostatic forces it is termed **coulomb scattering**; if short-range nuclear forces are involved it becomes *anomalous*. Long-wave electromagnetic wave scattering is *classical* or **Thomson scattering**, while for higher frequencies *resonance* or *potential* scattering occurs according to whether the incident photon does or does not penetrate the scattering nucleus. Coulomb scattering of α-particles is *Rutherford scattering*. For light, scattering by fine dust or suspensions of particles is *Rayleigh scattering*, while that in which the photon energy has changed slightly, due to interaction with vibrational or rotational energy of the molecules, is *Raman scattering*. *Shadow scattering* results from interference between scattered and incident waves of the same frequency. See **atomic scattering**.

scattering amplitude Ratio of amplitude of scattered wave at unit distance from scattering nucleus to that of incident wave.

scattering cross-section Effective impenetrable cross-section of scattering nucleus for incident particles of low energy. The radius of this cross-section is the *scattering length*.

scattering mean free path The average distance travelled by a particle between successive scattering interactions. It depends upon the medium traversed and the type and energy of the particle.

schizo- Prefix from Gk. *schizein*, to cleave.

science The ordered arrangement of ascertained knowledge, including the methods by which such knowledge is extended and the criteria by which its truth is tested. The older term *natural philosophy* implied the contemplation of natural processes *per se*, but modern science includes such study and control of nature as is, or might be, useful to mankind. *Speculative science* is that branch of science which suggests hypotheses and theories, and deduces critical tests whereby unco-ordinated observations and properly ascertained facts may be brought into the body of science proper.

Schrödinger equation Fundamental equation of *wave mechanics*. Solutions of this equation are *wave functions* for which the square of the amplitude expresses the probability density for a particle or a set of particles. If the system is isolated then a *time-independent* form of the equation is applicable. The solution for this version for bound particles shows that the energy for the system must be quantized.

scintillation Minute light flash caused when α-, β- or γ-rays strike certain phosphors, known as *scintillators*. The latter are classed as liquid, inorganic, organic or plastic according to their chemical composition.

scintillation camera An imaging device which may have either a single sodium iodide crystal or multiple crystals, which is capable of detecting and recording the spatial distribution of an internally administered radionuclide. Also *gamma camera*.

scintillation counter Counter consisting of a *phosphor* or *scintillator*, e.g. NaI(Tl), which, when radiation falls on it, emits light which is detected and amplified by a photomultiplier, the height of the pulses from which are proportional to the energy of the event. These pulses are further amplified and passed to a single-channel or multichannel pulse height analyser, to measure the energy and intensity of the radiation. Also *scintillation spectrometer*. » p. 76.

scintillation spectrometer See **scintillation counter**.

scoop The device used for removing gas from a centrifuge in an enrichment plant. » p. 56.

scotopic luminosity curve The curve giving relative brightness of the radiations in an equal-energy spectrum when seen at a very low intensity level. See **photopic luminosity curve**.

scram General term for emergency shutdown of a plant, esp. of a reactor when the safety rods are automatically and rapidly inserted to stop the fission process.

scram rod An emergency safety rod used in a reactor.

screening See **fluoroscopy**.

screening constant A number which when subtracted from the atomic number (Z) of an atom, gives the *effective* atomic number so that X-ray spectra may be described by 'hydrogen-like' formulae. It arises from the nuclear charge ($+Ze$) being screened by the inner electron shells.

sealed face production line A method for handling and fabricating radioactive material in which the remotely controlled equipment is sited behind a continuous impervious face with suitable inspection and other ports, as in the figure below. Materials are brought to the site in sealed containers and automatically transferred to the production line and, after fabrication, moved to sealed containers for any further operations. Alternative to separate **glove boxes**. See figure below.

sealed source A radioactive source for e.g. medical or calibration purposes, in a radiation leak-proof container. It is opened by remote

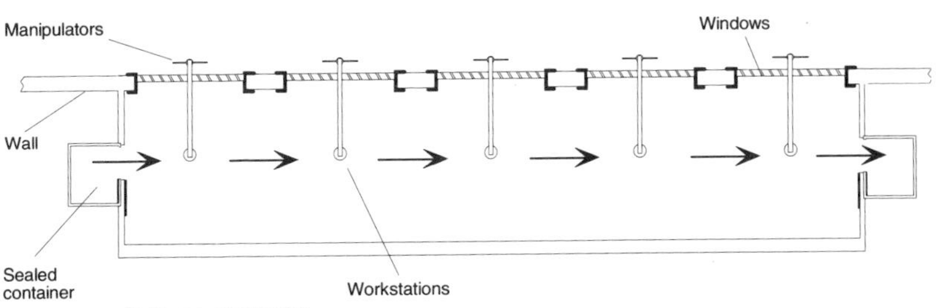

sealed face production line

control only in its shielded site of operation.

seal plug A removable plug at the end of the coolant tubes in a CANDU-type reactor which allows access for the refuelling machine. » p. 29.

second (1) 1/60 of a minute of time, or 1/86 400 of the mean solar day. Since 1965 defined, in terms of the resonance vibration of the caesium-133 atom, as the interval occupied by 9 192 631 770 cycles. This was adopted in 1967 as the SI unit of time-interval. Abbrev. *s*. (2) Unit of angular measure, equal to 1/60 of a minute of arc; indicated by the symbol ″. (3) In duodecimal notation 1/12 of an inch; indicated by ″. (4) Unit for expressing flow times in capillary viscometers (Redwood, Saybolt Universal, or Engler), e.g. an Engler second is that viscosity which allows 200 cm^3 of fluid through an Engler viscometer in 1 second.

secondary coolant In a reactor, a separate stream of coolant which is converted to steam by the **primary coolant** in a heat exchanger (steam generator) to power the turbines. » p. 24.

secondary radiation The radiation produced by the interaction of a primary radiation and an absorbing medium.

secondary shutdown system In the unlikely event of failure of control rods to enter and shut down reactor, a system such as, in a gas-cooled reactor, the insertion of nitrogen which absorbs neutrons much more strongly than does carbon dioxide. See **emergency shutdown system**.

secondary standard A copy of a **primary standard** for general use in a standardizing laboratory.

sector disk Rotating disk with angular sector removed, interposed in path of beam of radiation. Used to produce known attenuation, or to chop or modulate intensity of transmitted beam.

secular equilibrium Radioactive equilibrium where parent element has such long life that activities remain effectively constant for long periods.

Segrè chart A chart on which all known nuclides are represented by plotting the number of protons vertically against the number of neutrons horizontally. Stable nuclides lie close to a line which rises from the origin at 45° and gradually flattens at high atomic masses. Nuclides below this line tend to be β-emitters whilst those above tend to decay by positron emission or electron capture. Data for half-life, cross-section, disintegration energy etc are frequently added.

selection rules Restrictions on the transitions between quantum states of atoms, molecules or nuclei. The rules are derived theoretically by quantum mechanics. See **nuclear selection rules**.

selective absorption Absorption of light, limited to certain definite wavelengths, which produces so-called absorption lines or bands in the spectrum of an incandescent source, seen through the absorbing medium. See **atomic absorption spectroscopy**, **Kirchhoff 's law**.

self-absorption See **self-shielding**.

self-consistent field The energy levels in many-electron atoms cannot be calculated exactly. An approximation method is used that starts by assuming the electrons occupy levels similar to that of hydrogen, postulates the electrostatic field in which the electrons exist, and then calculates a new set of energy levels from which a new field is calculated. The process is repeated until the system is *self-consistent*.

self-quenching Said of counter tubes which do not depend on an external circuit for quenching, the residual gas providing sufficient resetting for the next operation of detecting a further photon or particle.

self-rectifying Said of an X-ray tube when an alternating voltage is applied directly between target and cathode.

self-regulating Said of a system when departures from the required operating level tend to be self-correcting, and in particular of a nuclear reactor where changes of power level produce a compensating change of reactivity, e.g. through negative temperature coefficient of reactivity. Also *load following*.

self-scattering The scattering of radioactive radiation by the body of the material which is emitting the radiation.

self-shielding In large radioactive sources, the absorption in one part of the radiation arising in another part. Also *self-absorption*.

semi- Prefix from L. *semi*, half.

semiconductor detector A reverse-biassed diode detector for radiation. » p. 76.

semiconductor diode laser A laser in which the lasing medium is *p*- or *n*-type semiconductor diode, e.g. gallium arsenide. Capable of continuous output of a few milliwatts at wavelengths in the range 700–900 nm.

sensitive time Period for which conditions of supersaturation in a cloud chamber or bubble chamber are suitable for formation tracks.

sensitive volume The portion of an ionization chamber or counter tube across which the electric field is sufficiently intense for incident radiation to be detected. The portion of living cells believed to be susceptible to ionization damage. See **target theory**.

sensitivity General term for ratio of response (in time and/or magnitude) to a driving force or stimulus, e.g. galvanometer response to a current, minimum signal required by a ratio of output level to illumination in a camera tube or photocell.

separation energy The energy required to separate 1 nucleon from a complete nucleus.

separation factor The ratio of the abundance of isotope at the end of a separation system or unit to that at the start of the process. It is usually only slightly greater than unity. » p. 56.

separation potential A dimensionless function used in the definition of the **separative work** of an uranium enrichment plant. It is given by

$$V(N) = (2N - 1)\ln\left[\frac{N}{1-N}\cdot\frac{1-N_0}{N_0}\right] + \frac{(1-2N_0)(N-N_0)}{N_0(1-N_0)},$$

where N and N_0 are the concentrations of the product and initial materials respectively.

separative efficiency In a single stage, the ratio of actual concentration to the change in the theoretical value.

separative element One unit of a cascade forming a complete isotope separation plant.

separative power The quantity of material a **separative element** is capable of enriching. It is given by the expression

$$\frac{\theta}{1-\theta}\cdot L\frac{(\alpha-1)}{2} \text{ moles per second,}$$

where α is the **separation factor** and θ, the **cut**. L is the number of moles/second of material.

separative work Measures the amount of separation an enrichment plant can achieve and defined as

$$PV(N_P) + WV(N_W) - FV(N_F),$$

where P, W, F are the masses of product, waste and feed (i.e. initial) materials, $V(N_P)$, $V(N_W)$ and $V(N_F)$ are the *value functions* of isotope concentrations N_P, N_W, N_F respectively. Measured in *separative work units* (kg SWU or tonnes SWU). It is not the weight of enriched material drawn from the plant.

serial radiography A technique for making a number of radiographs of the same subject in succession.

SGHWR Abbrev for *Steam Generating Heavy-Water Reactor*.

sharpness of resonance The rapidity with which resonance phenomena are shown and then disappear as the frequency of excitation of a constant driving force is varied through the resonant frequency.

sharp series Series of optical spectrum lines observed in the spectra of alkali metals. Has led to energy levels for which the orbital quantum number is zero being designated *s-levels*.

shearing See **chopping**. » p. 61.

sheath The can protecting a nuclear fuel element.

shed Minute unit of nuclear cross-section, 10^{-52} m^2 or 10^{-24} barn.

shell See **atomic structure** p. 121.

shell model A model of the nucleus of the atom with a nucleon moving independently in the common field representing the effect of the other nucleons. This leads to the nucleons being arranged in shells as for the electron shells in atomic structure.

shield Screen used to protect persons or equipment from electric or magnetic fields, X-rays, heat, neutrons etc. In a nuclear reactor the shield surrounds it to prevent the escape of neutrons and radiation into a protected area.

shielded box **Glove box** protected by lead walls and lead-glass windows, and with facilities for manipulation of contents by remote handling equipment.

shielded nuclide A nuclide, found among fission fragments, but not known to be formed as a result of beta decay. It is therefore assumed to have been a direct fission product .

shielding Protective use of low atomic number materials to thermalize strong neutron beams, or concrete, lead or other heavy materials to shield against gamma radiation when it might be harmful to operator or measuring system. Concrete is commonly used for large areas and lead for smaller.

shielding windows Dense glass blocks or liquid-filled tanks used as windows for inspecting the interior of shielded boxes.

shield pond, shielding pond Deep tank of water used to shield operators from highly radioactive materials stored and manipulated at the bottom.

shim rod Coarse control rod of reactor. It is usually positioned so that the reactor will be just critical when the rod is near the centre of its travel path. It is designed to move slowly unless it is also used as a **safety rod**, when a magnetic clutch allows it to drop rapidly into the core.

shiva Powerful laser capable of producing pulses of up to 15 kJ of energy, for use in nuclear fusion experiments. » p. 59.

shock heating Heating, esp. of a plasma, by the passage of a shock wave.

shock wave Wave of high amplitude, in which the group velocity is higher than the phase velocity, leading to a steep wavefront. This happens, for example, when an explosive is detonated. The speed at which the chemical reaction travels through the material is higher than the speed of sound in the material, hence a shock wave occurs. Strong shocks cause luminosity in gases, and so are useful for spectroscopic work. Also *blast wave*.

short-range forces Non-coulomb forces which act between nucleons and are responsible for the stability of the nucleus.

shower Result of impact of a high-energy cosmic ray and photons so that a very large number of ionizing particles and photons are produced, directed downwards in a narrow cone.

shower unit The mean path length for reduction of 50% of the energy of cosmic rays as they pass through matter.

shroud tube A tube which lies between the position of the control rod in the core and the access for the control-rod actuator, guiding and restraining the control rod when it is partly or fully withdrawn from the core. » p. 34.

shunt Addition of a component to divert current in a known way, e.g. from a galvanometer, to reduce temporarily its effective sensitivity.

shutdown Reduction of power level in nuclear reactor to lowest possible value by maintaining core in subcritical condition.

shutdown amplifier See **trip amplifier**.

shutdown heating, shutdown power Heat coming from the continued decay of the fission products formed in a reactor. See **decay heat**. » p. 46.

shuttle Container for samples to be inserted in, and withdrawn from, nuclear reactors, where they are made radioactive by irradiation with neutrons. Also *rabbit*.

SI Abbrev. for *Système International* (d'Unités). See **SI Units** p. 237.

sievert The SI unit for *dose equivalent*, measured in **grays** times a *quality factor* for the type of radiation and a *weighting factor* for the tissue irradiated. Numerically equivalent to gray for electrons and X-rays irradiating the whole body. Symbol Sv. See **effective dose equivalent**. » p. 103.

sigma particle Hyperon triplet, rest mass equivalent to 1190 MeV, hypercharge 0, isotopic spin 1.

sigmoid curve An S-shaped curve which is often obtained in dose-effect curves in radiobiological studies.

simple harmonic motion The motion of a particle (or system) for which the force on the particle is proportional to its distance from a fixed point and is directed towards the fixed point. The particle executes an oscillatory motion about the point. The motion satisfies the equation

$$(d^2x/dt^2) = -\omega^2\xi,$$

where x is the displacement of the particle and ω is a constant for the motion. The majority of small amplitude oscillatory motions are simple harmonic, e.g. the oscillations of a mass suspended by a spring, the swing of a pendulum, the vibrations of a violin string, the oscillations of atoms or molecules in a solid, the oscillations of air as a sound wave passes. When such a motion takes place in a resistive medium, e.g. air, the oscillations die away with time; the motion is then said to be *damped*. Abbrev. *s.h.m.*

simultaneity Basic consequence of *relativity*. Two events that are simultaneous according to one observer if a clock with him records the same time, may occur at different times according to another observer in another reference frame moving relative to the first.

sintering The fritting together of small particles to form larger particles, cakes or masses; in case of ores and concentrates, it is accomplished by fusion of certain constituents. As used in powder metallurgy, sintering consists in mixing metal powders having different melting points, and then heating the mixture to a temperature approximating the lowest mp of any metal included. In sintered carbides, powdered cobalt, having the lowest mp, acts as the binder holding together the unmelted particles of the hard carbides.

sinusoidal wave A wave whose profile is a pure sine curve. Also *harmonic wave*.

site licence Permission to operate a nuclear site given by the relevant authority and without which work cannot proceed.

SI units See panel on p. 237.

skiagram, skiagraph Obsolete term for a *radiograph*, the film produced after exposure to X-rays.

skin dose Absorbed or exposure radiation dose received by or at the skin of a person exposed to sources of ionization. Cf. **tissue dose**.

SLAC Abbrev for *Stanford Linear Acceleration Centre*, USA. A *linear accelerator*

SI units

The *international system of units*, the SI (Système International d'Unités), is a coherent system of metric units, with seven *base* and two *supplementary* units from which other units are derived.

	physical quantity	SI unit	Symbol	Definition of SI unit
Base units	length	metre	m	
	mass	kilogram	kg	
	time	second	s	
	electric current	ampere	A	
	thermodynamic temperature	kelvin	K	
	amount of substance	mole	mol	
	luminous intensity	candela	cd	
Supplementary units	plane angle	radian	rad	
	solid angle	steradian	sr	
Some derived units	energy	joule	J	$m^2 kg s^{-2}$
	force	newton	N	$mkg s^{-2}$
	pressure	pascal	Pa	$m^{-1} kg s^{-2}$
	power	watt	W	$m^2 kg s^{-3}$
	electric charge	coulomb	C	s A
	electric potential difference	volt	V	$m^2 kg s^{-3}A^{-1}$
	electric resistance	ohm	Ω	$m^2 kg s^{-3}A^{-2}$
	electric capacitance	farad	F	$m^{-2} kg^{-1}s^4A^2$
	magnetic flux	weber	Wb	$m^2 kg s^{-2}A^{-1}$
	magnetic flux density	tesla	T	$kg s^{-2}A^{-1}$
	frequency	hertz	Hz	s^{-1}
Other units with special names	length	ångström	Å	10^{-10} m
	length	micrometre	μm	10^{-6} m
	acceleration of free fall	gal	Gal	$10^{-2} m s^{-2}$
	pressure	bar	bar	10^5 Pa
	magnetic flux density	gamma	γ	10^{-9} T

The special SI units for ionizing radiations, such as becquerel, gray and sievert are defined on p. 103. The *micron* is sometimes used for micrometre and *milligal* (mGal, $10^{-5}m s^{-2}$) is also commonly used. Decimal multiples can be used with the following prefixes and symbols.

multiple	prefix	symbol	multiple	prefix	symbol
10^{-1}	deci	d	10	deca	da
10^{-2}	centi	c	10^2	hecto	h
10^{-3}	milli	m	10^3	kilo	k
10^{-6}	micro	μ	10^6	mega	M
10^{-9}	nano	n	10^9	giga	G
10^{-12}	pico	p	10^{12}	tera	T
10^{-15}	femto	f			

The SI system is an *MKSA* (metre-kilogram-second-ampere) system, for which earlier equivalents in other systems are shown in **fundamental dynamical units** p. 161.

which at the centre gives electron beams of energy about 50 GeV. See **particle accelerators** p. 206.

s-levels See **sharp series**.

slice The cross-sectional portion of the body which is scanned for the production of images in e.g. **computer aided tomography** and **magnetic resonance imaging**.

slowing-down area In reactor physics calculation, one sixth of the mean square distance travelled by neutrons from their source to reach thermal energy. See **Fermi age**.

slowing-down density In reactor theory, the rate at which neutrons slow down past a given energy, per unit volume.

slowing-down length The square root of the slowing-down area.

slowing-down power Increase in **lethargy of neutrons** per unit distance travelled in medium.

slow neutron See **neutron**. » p. 18.

slug (1) Unit of mass in the *gravitational system* of units. A force of 1 lbf (pound-force) acting on a mass of 1 slug gives it an acceleration of 1 ft s^{-2}. See **fundamental dynamical units** p. 161. (2) Unit of fuel in nuclear reactor, either rod or slab of fissile material encased in a hermetic can of Al, Be, Magnox, Zr or stainless steel. Also *cartridge*. See **fuel rod**.

slumping The movement of molten fuel; not necessarily as the result of an accident but most dramatically seen after the Chernobyl accident in the Ukraine.

slurry reactor A reactor in which fuel or **blanket** material exists as a slurry carried by the coolant fluid.

Sm Symbol for **samarium**.

smear test A method of estimating the loose, i.e. easily removed, radioactive contamination upon a surface. Made by wiping the surface and monitoring the swab.

smoke control area An area, statutorily defined, in which emission of smoke from chimneys is prohibited.

societal risk The chance of radiation damage to large numbers of people, the result of bomb testing or nuclear accidents.

sodium A metallic element, one of the alkali metals. Symbol Na, at.no. 11, r.a.m. 22.9898, mp 97.5°C, bp 883°C, rel.d. 0.978, valency 1. Sodium does not occur in nature in the free state and is a soft silvery-white metal, which reacts violently with water. Sodium is the seventh commonest element in the Earth's crust, with an abundance of 2.27%, and 1.06% in seawater. Its principal mode of occurrence is in complex silicates, especially in igneous and metamorphic rocks. It is also present as various salts in evaporite deposits which are commercial sources of sodium. It has eight isotopes:

A	Abundance %	half-life	decay mode
20		0.11 s	ε
21		22.5 s	ε
22		2.60 y	ε
23	100		
24		15.0 h	β^-
25		60 s	β^-
26		1.1 s	β^-
27		0.30 s	β^-

In nuclear engineering it is mainly used as a coolant in fast-breeder reactors, where its high thermal conductivity, low neutron capture cross-section and high boiling point make it almost the only coolant capable of removing the heat from the tightly packed cores of these reactors. » p. 36.

sodium-cooled reactor A reactor in which liquid sodium is used as the primary coolant, as in the **fast reactor**. » p. 36.

sodium iodide scintillation crystal A high-density photon absorber that converts the energy of a photon from radioactivity to a light photon.

soft radiation General term used to describe radiation whose penetrating power is very limited, e.g. low-energy X-rays.

solar constant The total electromagnetic energy radiated by the Sun at all wavelengths per unit time through a given area, normal to the solar beam, at the mean distance of the Earth and after correction for loss by absorption in the Earth's atmosphere. Its current value is 1.37 kW m^{-2}. It is not, in fact, truly constant, and variations of the order of 0.1% are detectable.

solid solution Usually a *primary solid solution*, but the term may also be applied to the case when an intermediate constituent dissolves one of its components.

solid-state detector A detector of ionizing radiation which uses an energy-sensitive solid-state device. See **lithium-drifted silicon detector**. » p. 76.

solid-state physics Branch of physics which covers all properties of solid materials, including electrical conduction in crystals of semiconductors and metals, superconductivity and photoconductivity.

solvent degradation The solvents used in fuel reprocessing plants have limited lifetimes because of radiation from the spent fuel. » p. 60.

somatic effects of radiation The biolog-

space reactors

The first nuclear reactor-powered space vehicle was launched by the USA in 1961 as a navigational satellite. It did not use a conventional nuclear reactor but *radioisotope thermoelectric generators* in which the radiation from highly enriched plutonium heated the electrical junction between two dissimilar metals having a cold junction at space temperature, i.e. a *thermocouple*. Each reactor held less than 20000 curies of plutonium-238 and the many thermocouples, forming the *thermopile*, converted about 4% of the heat directly into electricity. Much larger versions are now used in so-called *SNAP* (USA) and *Ripple* (UK) generators..

The USSR has also launched a large number of *Radar Oceanic Reconnaissance Satellites*, RORSATs, with very small conventional reactors which produce about 2 kW of electrical power. Solar panels could not be used because of the drag which they produce at the low orbits needed. These reactors use heat from fission instead of from radioactive decay to heat the thermocouple junctions and there has been concern about their eventual return to Earth. Although RORSATs are designed so that their reactors are boosted into high orbit at the end of their lives, two out of 31 have re-entered the atmosphere accidentally and one of these broke up and scattered radioactive debris over Canada's North West Territories in 1978. After that incident a backup fuel ejection system was included in the design so that the fuel would break up into small pieces. This was not needed in the second incident in 1988 when a backup booster was able to lift the reactor into a high orbit.

The USSR has also developed a newer design of space reactor, called TOPAZ, which produces 10 kW of power by *thermionic energy conversion* in which the reactor heats a cathode from which the electrons are expelled and pass to a cold anode and so generate a current.

All these reactors and radiation sources are essentially unshielded and concern has been expressed not only about the effects of collision either with the ground or with space debris but also about the gamma radiation emitted which generates signals powerful enough to disrupt the communications of other satellites and the observations of ground stations.

ical effects which do not affect the gonads and are therefore not transmitted to the offspring. Somatic effects are much more common than the genetic effects of radiation.

Sommerfeld atom Atomic model developed from Bohr atom, but allowing for elliptic orbits with radial, azimuthal, magnetic and spin quantum numbers. Modern theories modify this by regarding the electrons as forming a cloud, the density of which is described in terms of their wave function.

Soret effect See **thermal diffusion**.

sound velocity See **velocity of sound**.

source See **radiator**.

source range See **start-up procedure**.

sources of neutrons Fast neutrons are obtained by (1) nuclear transformations and (2) fission in nuclear reactors. To obtain slow neutrons, a moderator must be used. For laboratory sources see **neutron source**.

source strength Activity of radioactive source expressed in disintegrations per second.

space reactors See panel on p. 239.

space-time Normal 3-dimensional space plus dimension of time, modified by gravity in relativity theory.

spallation Any nuclear reaction when several particles result from a collision, e.g. cosmic rays with atoms of the atmosphere; chain-reaction in a nuclear reactor or weapon.

spallation neutron source A powerful pulsed neutron source for research. Protons, accelerated in a synchrotron, are focused on a target of uranium-238; 25 to 30 neutrons per proton are released, moderated and collimated into beams.

spark chamber Radiation detector for rendering visible the tracks of ionizing particles by the sparks formed following the ionization. Consists of a stack of parallel metal plates with the electric field between them raised nearly to the breakdown point.

spark spectra The most important way of exciting spectra is by means of an electric spark. The high temperature reached will generate the spectrum lines of multiply ionized atoms as well as those that are uncharged and singly ionized. (As distinct from *arc spectrum*.) Also the evaporation of metal from the electrodes leads to additional lines not associated with the gas through which the discharge takes place.

special relativity See **relativity**.

special unitary groups A scheme which predicts that as far as *strong interactions* are concerned, elementary particles can be grouped into multiplets, the particles in each multiplet being considered as different states of the same particle. This unitary group SU(3) has been successful in correlating the range of particles and in predicting the existence of hitherto undiscovered particles, notably the particle Ω^-. The *isospin* unitary group SU(2) is a subgroup of SU(3). Other groups are being explored in connection with the explanation of strong interactions in terms of *quarks*.

specific Term used generally to indicate that the property described relates to unit mass of the substance involved, e.g. specific entropy is entropy per kilogram.

specific activity See **activity**.

specific charge Charge/mass ratio of elementary particle, e.g. the ratio e/m_e of the electronic charge to the rest mass of the electron = 1.759×10^{11} coulomb per kilogram.

specific gravity See **relative density**.

specific heat capacities of gases Gases have two values of specific heat capacity: c_p, the s.h.c. when the gas is heated and allowed to expand against a constant pressure, and c_v, the s.h.c. when the gas is heated while enclosed within a constant volume.

specific heat capacity The quantity of heat which unit mass of a substance requires to raise its temperature by one degree. This definition is true for any system of units, including SI, but whereas in all earlier systems a unit of heat was defined by putting the s.h.c. of water equal to unity, SI employs a single unit, the joule, for all forms of energy including heat, which makes the s.h.c. of water 4.1868 kJ kg^{-1} K^{-1}. See **mechanical equivalent of heat**.

specific ionization Number of ion pairs formed by ionizing particle per cm of path. Also the *total specific ionization* to avoid confusion with the *primary specific ionization* which is defined as the number of ion clusters produced for unit length of track.

specific latent heat See **latent heat**.

specific power US for **fuel rating**.

specific volume The volume of unit mass; the reciprocal of density.

spectral series Group of related spectrum lines produced by electron transitions from different initial energy levels to the same final one. The recognition and measurement of series has been of great importance in atomic and quantum theories.

spectral-shift-controlled reactor A reactor in which the loss of reactivity which would occur on burnup is compensated by *softening* the neutron spectrum, e.g. by varying the heavy-water/light-water ratio in the reactor coolant, or by change of coolant temperature. Abbrev. *SSCR*.

spectrometer Instrument used for measurements of wavelength or energy distribution in a heterogeneous beam of radiation.

spectroradiometer A spectrometer for measurements in the infrared.

spectroscope General term for instrument (spectrograph, spectrometer etc) used in spectroscopy. The basic features are a slit and collimator for producing a parallel beam of radiation, a prism or grating for 'dispersing' different wavelengths through differing angles of deviation, and a telescope, camera or counter tube for observing the dispersed radiation. » p. 80.

spectroscopy The practical side of the study of spectra, including the excitation of the spectrum, its visual or photographic observation, and the precise determination of wavelengths.

spectrum Arrangement of components of a complex colour or sound in order of frequency or energy, thereby showing distribution of energy or stimulus among the components. A *mass spectrum* is one showing the distribution in mass, or in mass-to-charge ratio of ionized atoms or molecules. The mass spectrum of an element will show the relative abundances of the isotopes of the element.

spectrum colours The continuous range of merging colours which is produced when white light is split into its component wavelengths: red, orange, yellow, green, blue, indigo, violet.

spectrum line Isolated component of a spectrum formed by radiation of almost uniform frequency. Due to photons of fixed energy radiated as the result of a definite electron transition in an atom of a particular element.

speed (1) The rate of change of distance with time of a body moving in a straight line or in a continuous curve (cf. **velocity**, a vector expressing both magnitude and direction). Units of speed are metres per second ($m\ s^{-1}$), feet per second ($ft\ s^{-1}$), miles per hour (mph), kilometres per hour ($km\ h^{-1}$), knots etc. (2) Angular velocity, expressed in revolutions per minute, radians per second etc.

speed of light The constancy and universality of the speed of light *in vacuo* is recognized by *defining* it (1983) to be exactly $2.997\,924\,58 \times 10^8\ m\ s^{-1}$. This enables the SI fundamental unit of *length*, the metre, to be defined in terms of this value.

speed of sound See **velocity of sound**.

spent fuel Reactor fuel element which must be replaced due to (1) burnup or depletion, (2) poisoning by fission fragments, (3) swelling and/or bursting. The fissile material is not exhausted and so-called spent fuel is normally subsequently reprocessed.

spin The intrinsic angular momentum of an electron, nucleus or elementary particle. Spin is quantized in integral multiples of half the *Dirac unit*. Those particles with a spin of odd multiples are *fermions* (e.g. electrons, proton, neutron) and those with even multiples are *bosons* (e.g. photon, phonon). It is the quantized electron spin angular momentum combined with the orbital angular momentum that gives rise to the fine structure in atomic line spectra.

spin polarization When applied to a beam of particles, polarization denotes the preferential orientation of the particle spin.

spin quantum number Contribution to the total angular momentum of the electron of that due to the rotation of the electron about its own axis.

spin wave The spin of the magnetic ions in ferro- and ferrimagnetic and in antiferromagnetic materials are orientated along preferred directions. Coherent deviations of the spins from these directions are propagated in space and time like waves.

split-flow reactor A reactor in which the coolant enters at the central section and flows outwards at both ends (or vice versa).

splitting ratio See **cut**.

spontaneous emission Process involving the emission of energy in an atomic system without external stimulation. Spontaneous emission is a strictly quantum effect. Cf. **laser**, **stimulated emission**.

spontaneous fission Nuclear fission occurring without absorption of energy. The probability of this increases with increasing values of the fission parameter Z^2/A (Z = atomic number, A = relative atomic mass) for the fissile nucleus. *Induced* fission is that caused by the impact of nuclear particles. » p. 13.

spread factor See **distribution coefficients**.

SPS Abbrev. for *Super Proton Synchrotron*. A particle accelerator at CERN, Geneva, Switzerland. Output beam of 500 GeV protons for *fixed-target* experiments. Modified to perform 270 GeV antiproton colliding-beam experiments. Used to discover, in 1983, W^+, W^- and Z *gauge bosons* that mediate weak interactions.

spurious coincidences Apparent coincidences recorded by a coincidence counting system, when a single particle has not passed through both or all the counters in the system. They usually result from the almost simultaneous discharge of two counters by different particles.

spurious counts Counts arising in counter tubes from voltage leakages in the counter and defects in external quenching circuits.

spurious pulse A pulse arising from the self-discharge of a particle counter leading to erroneous signals.

square law The law of inverse squares expressing the relation between the amount of radiation falling upon unit area of a surface and the distance of the surface from the source.

SQUID Abbrev. for *Superconducting Quantum Interference Device*. A family of devices capable of measuring extremely small currents, voltages and magnetic fields. Based on two quantum effects in superconductors: (1) *flux quantization* and (2) the *Josephson effect*. SQUIDs can detect changes in magnetic flux densities of ≈ 1 nT (10^{-9} T). Examples are the study of fields generated by the action of the human brain and in magnetobiological research.

Sr Symbol for **strontium**.

stable Used to indicate the incapability of following a stated mode of spontaneous change, e.g. *beta stable* means incapable of ordinary beta disintegration but capable of isomeric transition or alpha disintegration etc.

stable equilibrium The state of equilibrium of a body when any slight displacement increases its potential energy. A body in stable equilibrium will return to its original position after a slight displacement.

stack Pile of photographic plates exposed to radiation together, and used to study tracks of ionizing particles.

stage Unit of cascade in isotope separation plant, consisting of single separative element, or group of these elements, operating in parallel on material of same concentration. » p. 56.

stage separation factor See **separation**

factor.

stainless steel A widely used material in nuclear reactors because of its good anti-corrosion properties. Despite its high thermal neutron cross-section, an alloy containing iron (54.5%), chromium (20%), nickel (25%) with 0.5% niobium is the fuel cladding in gas-cooled reactors because it can operate at high temperatures (> 800°C). This alloy has the following properties:

σ_{cap} mb	mp °C	Therm. conduct. $W\,m^{-1}K^{-1}$	density $g\,cm^{-3}$
320	1690	33	8.0

standard Establishment unit of measurement, or reference instrument or component, suitable for use in calibration of other instruments. Basic standards are those possessed or laid down by national laboratories or institutes (*NPL, BSI* etc).

standard atmosphere Unit of pressure, defined as 101 325 N/m^2, equivalent to that exerted by a column of mercury 760 mm high at 0°C. Abbrev. *atm.*

standard chamber Ionization chamber used for calibration of radioactive sources, or of absolute values of exposure doses.

standard temperature and pressure See **stp, standard atmosphere.**

standing-off dose The absorbed dose after which occupationally exposed radiation workers must be temporarily or permanently transferred to duties not involving further exposure. Doses are normally averaged over 13-week periods and standing-off would then continue for the remainder of the corresponding period.

stand pipe The connection between the charge face and the interior of a reactor vessel, giving access to the fuel channels, e.g. for refuelling.

Stark effect Splitting of atomic energy levels, and of corresponding emission spectrum lines, by placing source in region of strong electric field. Cf. **Zeeman effect**.

Stark-Einstein equation The energy absorbed per mole for a photochemical reaction is

$$E = N h \nu ,$$

where N is Avogadro's number, h is Planck's constant and ν is the frequency of the absorbed light.

start-up, starting-up time The time required by an instrument or system (e.g. nuclear reactor, chemical plant etc) to reach equilibrium operating conditions.

start-up procedure The procedure followed when bringing a nuclear reactor into operation. It involves four successive stages: (*a*) *source range*, where a neutron source is introduced to generate the required neutron flux (this may not be necessary in a reactor which has already been operating); (*b*) *counter range*, where reactor is just critical but counters are required to monitor neutron-flux changes; (*c*) *period range*, where changes in reactivity are monitored on period meter; (*d*) *power range*, where reactor is operating within its designed power ratings.

state The energy level of a particle as specified by the appropriate quantum numbers.

state function A quantity in thermodynamics which has a unique value for each state of a system. Internal energy, entropy and enthalpy are examples. The value associated with a given state is independent of the process used to bring about that state.

statistical error Radiation detectors measure the average count for random events. As a result of statistical fluctuations, the average count N has a statistical error of $\sqrt{N}$.

statistical mechanics Theoretical predictions of the behaviour of a macroscopic system by applying statistical laws to the behaviour of component particles. *Quantum mechanics* is an extension of classical statistical mechanics introducing the concepts of the quantum theory, esp. the Pauli exclusion principle. *Wave mechanics* is a further extension based on the Schrödinger equation, and the concept of particle waves.

steady state A state which is in dynamic equilibrium, with entropy at its maximum.

steam Water in the vapour state; formed when specific latent heat of vaporization is supplied to water at boiling point. The specific latent heat varies with the pressure of formation, being approximately 2257 $kJ\,kg^{-1}$ at atmospheric pressure.

steam driers The last drops of water in steam can be removed by making the steam change direction abruptly, when the momentum of the water carries it on to a surface which leads it back into the boiling water. Used in boiling-water reactors. » p. 28.

Stefan-Boltzmann law Law stating that the total radiated energy from a black body per unit area per unit time is proportional to the fourth power of its absolute temperature, i.e. $E = \sigma T^4$ where σ (Stefan-Boltzmann constant) is equal to $5.6696 \times 10^{-8}\,W\,m^{-2}\,K^{-4}$.

stellarator A toroidal fusion device in which the magnetic fields are generated entirely by conductors placed around the torus. See **tokamak**. » p. 38.

step function A function which makes an instantaneous change in value from one constant value to another.

Stern-Garlach experiment Atomic beam experiment which provided fundamental proof of quantum theory prediction that magnetic moment of atoms can only be orientated in certain fixed directions relative to an external magnetic field.

sthene Unit of force in the metre-tonne-second system, equivalent to 10^3 newtons.

sticking probability Probability of an incident particle, which reaches the surface of a nucleus, being absorbed and forming a compound nucleus.

stiction Abbrev. for *STatic frICTION*. See **friction**.

stilb Unit of luminance, equal to 1 cd cm^{-2} or 10^4 cd m^{-2} of a surface.

stimulated emission Process by which an incident photon of frequency ν stimulates an atom to make a transition from energy E_2 to energy E_1 where

$$\nu = (E_2 - E_1)/h,$$

h being Planck's constant. The atom is left in the lower energy state as *two* photons of the same frequency emerge, the incident one and the emitted one. An essential process in the operation of a **laser**.

stokes The CGS unit of kinematic viscosity (10^{-4} m^2 s^{-1}). Abbrev. *St*.

stopping equivalent Thickness of a standard substance which would produce the same energy loss as the absorber under consideration. The standard substance is usually air at stp but can be Al, Pb, H_2O etc.

stopping power Loss, resulting from a particle traversing a material. The *linear stoppage power* S_L is the energy loss per unit distance and is given by $S_L = -dE/dx$, where x is path distance and E is the kinetic energy of the particle. The *mass stopping power* S_M is the energy lost per unit surface density traversed and is given by $S_M = S_{L/r}$, where ρ is the density of the substance. If A is taken as the relative atomic mass of an element and n the number of atoms per unit volume, then the *atomic stopping power* S_A of the element is defined as the energy loss per atom per unit area normal to the motion of the particle, and is given by $S_A = S_L/n = S_M A/N$, where N is Avogadro's number. The *relative stopping power* is the ratio of the stopping power of a given substance to that of a standard substance, e.g. air or aluminium. » p. 100.

storage The keeping of radioactive waste material in a facility, either made specially or naturally occurring, with the intention of treating it further or of *disposing* of it, normally elsewhere. Storage implies further treatment, disposal does not.

storage factor See *Q*.

stp Abbrev. for *Standard Temperature and Pressure*; a temperature of 0°C and a pressure of 101 325 N m^{-2}. Also *STP*. See **standard atmosphere**.

straggling Variation of range or energy of particles in a beam passed through absorbing material, arising from random nature of interactions experienced. Additional straggling may arise from instrumental effects such as noise, source thickness and gain instability.

strangeness A property that characterizes *quarks* and so *hadrons*. The strangeness of leptons and gauge bosons is zero. Strangeness is conserved in strong and electromagnetic interactions between particles but not in weak interactions. K-mesons and hyperons which have non-zero strangeness are termed *strange* particles.

stray radiation Direct and secondary radiation from irradiated objects which is not serving a useful purpose.

streaming effect See **channelling effect**.

stress corrosion The development of cracks caused by both chemical action and tensile stress.

stringer Group of reactor fuel elements strung together for insertion into one channel of the core.

stripped atom Ionized atom from which at least one electron has been removed.

stripper The section of an isotope-separation plant which removes the selected isotope from the waste stream.

stripping A phenomenon observed in deuteron (or heavier nuclei) bombardment in which only a portion of the incident particle merges with the target nucleus, the remainder proceeding with most of its original momentum practically unchanged in direction.

stripping foil In a tandem Van de Graaf accelerator, ions can be made to change their sign by passing through a foil of an element whose outer electrons are loosely bound. If the foil is placed at a high voltage position they will be first attracted to the foil and then accelerated away from it. See **particle accelerators** p. 206.

strong interaction An interaction between particles involving *baryons* and *mesons* completed in a time of the order of 10^{-23} s. It is the strong interaction that binds protons and neutrons together in the nuclei of atoms. The underlying fundamental interaction to these processes is the strong interaction between the constituent *quarks*.

strong interaction between quarks A

fundamental interaction mediated by *gluons*. It is the interaction for the binding together of *quarks* (and *antiquarks*) in hadrons.

strontium A metallic element. Symbol Sr, at.no. 38, r.a.m. 87.62, mp 800°C, bp 1300°C, rel.d. 2.54. Silvery-white in colour, it is found naturally in *celestine* and in *strontianite*; it also occurs in mineral springs. It has 13 isotopes:

A	Abundance %	half-life	decay mode
81		22 m	ε
82		25.0 d	ε
83		32.4 d	ε
84	0.56		
85		64.8 d	ε
86	9.8		
87	7.0		
88	82.6		
89		50.5 d	β^-
90		28.8 y	β^-
91		9.5 h	β^-
92		2.7 h	β^-
93		7.4 m	β^-

It has similar chemical qualities to calcium, which considerably increases the radiobiological danger of the radioactive isotope, strontium-90, which is a major long-lived fission product from uranium and found in the fallout from nuclear explosions. The radiogenic isotope strontium-87 is produced by radioactive decay of rubidium-87, ^{87}Rb, and is used (as the ratio $^{87}Sr/^{86}Sr$) in **radiometric dating** p. 222.

strontium unit Unit used to measure the concentration of radioactive strontium-90 in calcium; 1 SU = 10^{-12} Ci g^{-1}.

sub- Prefix from L. *sub*, under, used in the following senses: (1) deviating slightly from, e.g. *subtypical*, not quite typical; (2) below, e.g. *subvertebral*, below the vertebral column; (3) somewhat, e.g. *subspatulate*, somewhat spatulate; (4) almost, e.g. *subthoracic*, almost thoracic in position.

subcritical Said of an assembly of fissile material which has a multiplication factor for neutrons of less than unity.

submicron A particle of diameter less than a **micron**. Visible by ultramicroscope.

sulphur A non-metallic element occurring in many allotropic forms. Symbol S, at.no. 16, r.a.m. 32.06, valencies 2, 4, 6. The rhombic (β-) sulphur is a lemon yellow powder. It has: mp 112.8°C, rel.d. 2.07. Monoclinic (β-) sulphur has a deeper colour than the rhombic form. It has: mp 119°C, bp 444.6°C, rel.d. 1.96. There are nine isotopes:

A	Abundance %	half-life	decay mode
30		1.2 s	ε
31		2.6 s	ε
32	95.02		
33	0.75		
34	4.21		
35		87.4 d	β^-
36	0.017		
37		5.0 m	β^-
38		170 m	β^-

Chemically, sulphur resembles oxygen, and can replace the latter in many compounds, organic and inorganic. It is abundantly and widely distributed in nature with an abundance in the Earth's crust of 340 ppm, and 900 ppm in seawater.

superconductivity Property of some pure metals and metallic alloys of having negligible resistance to the flow of an electric current at very low temperatures. Each material has its own critical temperature, T_c above which it is a normal conductor. When a current is established, it persists almost indefinitely. Magnetic fields can destroy the superconductivity, their strength depending on how far below the critical temperature the material is. Generally, T_c has been < 20 K, but in 1986–7 a class of materials with *perovskite* structures has been discovered which have $T_c \approx$ 90 K.

supercritical Said of an assembly of fissile material for which the neutron multiplication factor is greater than unity.

superficial radiation therapy X-ray therapy (usually by soft radiation produced at less than 140 kVp) of the skin or of any surface of the body made accessible.

super-, supra Prefixes from L. *super*, over, above.

superior Placed above something else; higher, upper (as the *superior* rectus muscle of the eyeball).

supersonic Faster than the speed of sound in that medium. Erroneously used for ultrasonic.

supersymmetry Theory which attempts to link all four fundamental forces, and postulates that each force emerged separately during the expansion of the very early universe.

supervoltage therapy Application of voltage, over a million volts, to X-ray tubes or accelerators in therapy. Also *megavoltage therapy*.

surface absorption coefficient See **absorption coefficient**.

surface irradiation Irradiation of a part of

the body by applying a mould or applicator loaded with radioactive material to the surface of the body.

surface sterilization Radiation with low-energy rays which penetrate thin surface layers only, e.g. with ultraviolet rays.

surface tension A property possessed by liquid surfaces whereby they appear to be covered by a thin elastic membrane in a state of tension, the surface tension being measured by the force acting normally across unit length in the surface. The phenomenon is due to unbalanced molecular cohesive forces near the surface. Units of measurement are dyne cm^{-1}, N m^{-1}. See **capillarity**, **liquid-drop model**, **pressure in bubbles**.

surge tank A suitably pressurized tank connected to e.g. the secondary steam-generating circuit of a reactor. It is able to compensate for changes in flow rate through the pumps by accepting or supplying condensate to the system.

survival curve A curve showing the percentage of organisms surviving at different times after they have been subjected to large radiation dose. Less often, a curve showing percentage of survivals at a given time against the size of dose.

Sv Abbrev. for **Sievert**.

swelling Change of volume of fuel rods which may occur during irradiation.

swimming-pool reactor A reactor in which the fuel elements are immersed in a deep pool of water which acts as coolant, moderator and shield. Also *pool reactor*.

swirl vane separator A **steam drier** in which the central steam flow rotates propellor-like blades, causing the impinging water droplets to be flung on to an outer surface and lead down to the boiling water again.

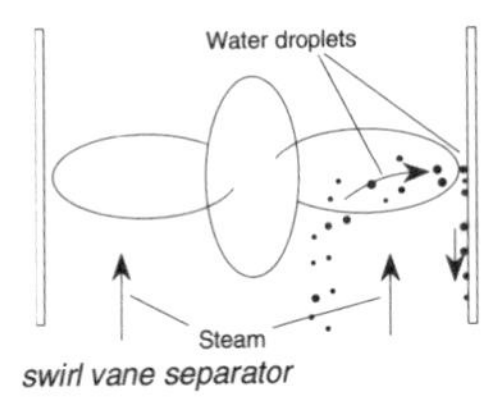

swirl vane separator

syn-, sym- Prefixes from Gk. *syn*, with, generally signifying fusion or combination.

synchrocyclotron A *cyclotron* in which the frequency of the accelerating voltage is varied to ensure that, despite the relativistic increase of mass of the particle with speed, the particles still arrive in synchronism with the accelerating voltages. Energies up to 700 MeV for protons can be achieved. The output is not continuous but is emitted in bursts of particles lasting about 100 µs. See **particle accelerators** p. 206.

synchrotron Machine for accelerating charged particles to very high energies. The particles move in an orbit of constant radius guided by a magnetic field. The acceleration is provided at one point in their orbit by a high-frequency electric field whose frequency increases to insure that particles of increasing velocity arrive at the correct instant to be further accelerated. Proton synchrotrons can produce energies greater than 200 GeV. Electron synchrotrons give energies up to 12 GeV. See **particle accelerators** p. 206.

synchrotron radiation (1) Electrons accelerated in a *synchrotron* produce a very intense, highly collimated, polarized beam of electromagnetic radiation, whose wavelength ranges continuously from 10^{-2} mm to 10^{-2} nm. Used with a monochromator, it is an important source for research purposes. (2) A theory of the origin of cosmic radio waves. It is suggested that the electrons moving in an orbit in a magnetic field are accelerated as in a synchrotron, but on a vastly larger scale.

synroc Nuclear waste products incorporated into a mixture of crystalline structures known to be stable over geological time.

system Generally, anything formed of parts placed together or adjusted into a regular or connected whole.

Système International d'Unités See **SI Units** p. 237.

systems analysis Complete analysis of all phases of activity of an organization, and development of a detailed procedure for all collection, manipulation and evaluation of data associated with the operation of all parts of it.

Szilard-Chalmers process A process in which a nuclear transformation occurs with no change of atomic number, but with breakdown of the chemical bond. This leads to formation of free active radicals from which material of high specific activity can be separated chemically.

T

t Symbol for tonne (metric **ton**).

T Symbol for (1) **tera-**, 10^{12}; (2) **tesla**.

T15 Large **tokamak** experiment, Moscow, USSR. » p. 38.

tagged atom See **labelled atom.** Also *radioactive tracer*.

tailings (1) Rejected portion of an ore; waste, *gangue*. (2) Portion washed away in water concentration. May be impounded in a *tailings dam* or pond, or stacked dry on a dump.

tails The depleted uranium produced at an enrichment plant, containing typically 0.25% uranium-238. » p. 56.

talbot Unit of luminous energy, such that 1 lumen is a flux of 1 talbot per second.

Tamman's temperature The temperature at which the mobility and reactivity of the molecules in a solid become appreciable. It is approximately half the melting point in kelvins.

tandem mirror See **magnetic mirror**.

tank reactor Covered type of **swimming-pool reactor**.

tare The weight of a vessel, wrapping, or container, which subtracted from the gross weight gives the net weight.

target Material irradiated by beam from accelerator.

target theory Proposed explanation of radiobiological effects, in which only a small sensitive region of each cell is susceptible to ionization damage.

tauon A lepton (τ^-) and its antilepton (τ^+) of exceptionally high mass, 1.78 GeV. Discovered at the Stanford Linear Accelerator Centre (SLAC) in electron-positron colliding beam experiments. Their existence necessitated the postulation of the *top* and *bottom* quarks to preserve the *lepton-quark symmetry*.

TBP Abbrev. for **Tri-*n*-Butyl Phosphate**.

technology The practice, description and terminology of any or all of the applied sciences which have practical value and/or industrial use.

tele- Prefix from Gk. *tele*, afar, at a distance.

teleradiography A technique to minimize distortion in taking X-ray photographs by placing X-ray tube some distance from the body.

teletherapy Treatment by X-rays from a powerful source at a distance, i.e. by high-voltage X-ray tubes or radioactive sources, such as cobalt-60, up to 80 000 GBq.

telo- Prefix from Gk. *telos*, end.

temperature A measure of whether two systems are relatively hot or cold with respect to one another. Two systems brought into contact will, after sufficient time, be in thermal equilibrium and will have the same *temperature*. A *thermometer* using a temperature scale established with respect to an arbitrary zero (e.g. **Celsius scale**) or to absolute zero (**Kelvin thermodynamic scale of temperature**) is required to establish the relative temperatures of two systems. See Zeroth law of **thermodynamics**.

temperature coefficient The fractional change in any particular physical quantity per degree rise of temperature.

temperature cycle Method of processing thick photographic nuclear research emulsions to ensure uniform development. They must be soaked in solutions at refrigerated temperatures and then warmed for the required processing period.

tensor force A non-central nuclear force which depends on the spin orientation of the nucleons.

tenth-value thickness Thickness of absorbing sheet which attenuates intensity of beam of radiation by a factor of 10.

tera- Prefix denoting 10^{12} times, e.g. a *terawatt-hour* is 10^{12} watt-hours. Symbol T.

term diagram The energy-level diagram for an isolated atom in which levels are usually represented by corresponding quantum numbers.

tesla SI unit of magnetic flux density or magnetic induction equal to 1 weber m^{-2}. Equivalent definition: the magnetic induction for which the maximum force it produces on a current of unit strength is 1 newton. Symbol T.

tetra- Prefix from Gk. *tetra*, four, e.g. *tetracyte*, four cells.

texture The mode of union or disposition, in regard to each other, of the elementary constituent parts in the structure of any body or material.

TFTR Abbrev. for *Tokamak Fusion Test Reactor.* Large **tokamak** experiment at Princeton, New Jersey, USA. » p. 38.

Th Symbol for **thorium**.

thawing A term used to describe the beginning of the fusion process of a solid. The corresponding temperature is the *thaw point* of the solid.

theories of light Interference and diffraction phenomena are explained by the *wave theory*, but when light interacts with matter, the energy of the light appears to be concentrated in *quanta*, called *photons*. The *quan-*

tum and *wave theories* are supplementary to each other. » p. 80.

therapeutic ratio The ratio between tumour lethal dose and tissue tolerance. In radio-resistant tumours, the tumour lethal dose equals, or is greater than, the dose required to destroy normal tissues.

therm-, thermo- Prefixes from Gk. *therme*, heat.

thermal capacity The amount of heat required to raise the temperature of a system through 1 degree. SI unit is J K^{-1}. See **molal specific heat capacity**, **specific heat capacity**.

thermal column Column or block of moderator in reactor which guides large thermal neutron flux to given experimental region.

thermal conductivity A measure of the rate of flow of thermal energy through a material in the presence of a temperature gradient. If (dQ/dt) is the rate at which heat is transmitted in a direction normal to a cross-sectional area A when a temperature gradient (dT/dx) is applied, then the thermal conductivity is

$$k = -\frac{(dQ/dt)}{A\,(dT/dx)}\ .$$

The SI unit is W m^{-1} K^{-1}. Materials with high electrical conductivities tend to have high thermal conductivities.

thermal cross-section Effective nuclear cross-section for neutrons of thermal energy. » p. 12.

thermal cycle An operating cycle by which heat is transferred from one part to another. In reactors, separate heat transfer and power circuits are usual to prevent the fluid flowing through the former, which becomes radioactive, from contaminating the power circuit. » p. 24.

thermal diffusion Process in which a temperature gradient in a mixture of fluids tends to establish a concentration gradient. Has been used for isotope separation. Also *Soret effect*.

thermal diffusivity Thermal conductivity divided by the product of specific heat capacity and density; more generally applicable than thermal conductivity in most heat transfer problems. Unit is metre squared per second.

thermal effusion The leaking of a gas through a small orifice, the gas being at a low pressure so that the mean free path of the molecules is large compared with the dimensions of the orifice.

thermal excitation Collision processes between particles by which atoms and molecules can acquire extra energy.

thermalization Process of slowing fast neutrons to thermal energies. In reactors normally the function of the moderator. » p. 19.

thermal leakage factor Ratio of number of thermal neutrons lost from reactor by leakage, to the number absorbed in reactor core. Also used is the *thermal non-leakage probability*, i.e. the fraction of thermal neutrons which do not leak out of the reactor. » p. 22.

thermal neutron See **neutron**.

thermal reactor A reactor for which the fission chain reaction is propagated mainly by thermal neutrons and therefore contains a moderator. Formerly sometimes *slow reactor*. » p. 18.

thermal resistance Resistance to the flow of heat. The unit of resistance is the *thermal ohm*, which requires a temperature difference of 1°C to drive heat at the rate of 1 watt. If the temperature difference is θ°C, the resistance S thermal ohms, and the rate of driving heat W watts, then $\theta = SW$.

thermal response Of a reactor, its rate of temperature rise if no heat is withdrawn by cooling. Its reciprocal is the *thermal inertia*.

thermal shield Inner shield of a reactor, used to protect biological shield from excess heating.

thermal shock The shock resulting when a body is subjected to sudden changes in temperature. The ability of the material to withstand its effects is called its *thermal shock resistance*.

thermal siphon The system causing flow round a vertical loop of fluid. When the bottom of one column is heated, the fluid rises, cools at the top, and falls down the other column. Used, e.g. for heating buildings, isotope separators.

thermal unit See **British-**, **calorie**, **joule**.

thermal utilization factor Probability of thermal neutron being absorbed by fissile material (whether causing fission or not) in infinite reactor core. Symbol f. » p. 22.

thermal vibration The motion of atoms, vibrating about their equilibrium positions in a crystalline solid. The motion increases as the temperature increases. Representing the vibrations by a set of *harmonic oscillators* which have zero point energy at 0 K, the oscillators increase their energy by discrete amounts, a *quantum* of energy, as the temperature increases. The specific heat of a solid can be explained in these terms. Detailed information about the thermal vibrations can be obtained from neutron diffraction, X-ray diffraction and other techniques.

thermion A positive or negative ion emitted

from incandescent material.

thermocouple If two wires of different metals are joined at their ends to form a loop, and a temperature difference between the two junctions unbalances the *contact emf* potentials, a current will flow round the loop. If the temperature of one junction is kept constant, that of the other is indicated by measuring the current. Conversely, a current driven round the loop will cause a temperature difference between the junctions.

thermoduric Resistant to heat.

thermodynamics The mathematical treatment of the relation of heat to mechanical and other forms of energy. Its chief applications are to heat engines (steam engines and IC engines; see **Carnot theorem**) and to chemical reactions. *Laws of thermodynamics*: *Zeroth law*: if two systems are each in thermal equilibrium with a third system then they are in thermal equilibrium with each other. This statement is tacitly assumed in every measurement of temperature. *First law*: the total energy of a thermodynamic system remains constant although it may be transformed from one form to another. This is a statement of the principle of the conservation of energy. *Second law*: heat can never pass spontaneously from a body at a lower temperature to one at a higher temperature (Clausius) *or* no process is possible whose only result is the abstraction of heat from a single heat reservoir and the performance of an equivalent amount of work (Kelvin-Planck). *Third law*: the entropy of a substance approaches zero as its temperature approaches absolute zero.

thermodynamic scale of temperature See **Kelvin thermodynamic scale of temperature**.

thermography The use of radiant heat emitted by the body to construct images of increased heat emission which can indicate tumours or inflammation. Reduced heat emission indicates reduced blood supply.

thermoluminescence Release of light by previously irradiated phosphors upon subsequent heating.

thermoluminescent dating Radiation emitted by the decay of an unstable isotope in crystalline material like quartz will cause electrons to be trapped in regions of imperfection in the crystal lattice. Subsequent heating will cause these electrons to be released as light (thermoluminescence) which can be measured. Knowledge of the amount of radioactivity present in the crystal will then allow the time to be determined since the crystals were last heated, e.g. as in the firing of pottery. See **radiometric dating** p. 222.

thermoluminescent dosimeter An instrument which registers integrated radiation dose, the read-out being obtained by heating the element and observing the thermoluminescent output with a photomultiplier. It has the advantage of showing very little fading if read-out is delayed for a considerable period, and forms an alternative to the conventional film badge for personnel monitoring.

thermometer An instrument for measuring temperature. A thermometer can be based on any property of a substance which varies predictably with change of temperature. For instance, the *constant volume gas thermometer* is based on the pressure change of a fixed mass of gas with temperature, while the *platinum resistance thermometer* is based on a change of electrical resistance. The commonest form relies on the expansion of mercury or other suitable fluid with increase in temperature.

thermometric scales See **Celsius scale**, **Centigrade scale**, **Fahrenheit scale**, **international practical temperature scale**, **Kelvin thermodynamic scale of temperature**, **Rankine scale**, **Réaumur scale**, **fixed points**.

thermometry The measurement of temperature.

thermonuclear bomb See **hydrogen bomb**. » p. 68.

thermonuclear energy Energy released by a *nuclear fusion* reaction that occurs because of the high thermal energy of the interacting particles. The rate of reaction increases rapidly with temperature. The energy of most stars is believed to be acquired from exothermic thermonuclear reactions. In the hydrogen bomb, a fission bomb is used to obtain the initial high temperature required to produce the fusion reactions. For laboratory experiments designed to release thermonuclear energy, see **JET**, **stellarator**, **tokamak** . » pp. 38, 41.

thermonuclear reaction A reaction involving the release of thermonuclear energy.

thermostat An apparatus which maintains a system at a constant temperature which may be preselected. Frequently incorporates a *bimetallic strip*.

theta pinch Cylindrical plasma constricted by an external current flowing in the θ direction to produce a solenoidal magnetic field.

thick source Radioactive source with appreciable self-absorption.

thick target A target which is not penetrated by primary or secondary radiation beam.

thick-wall chamber Ionization chamber in which build-up of ion current is produced by contribution of knock-on particles arising from wall material. » p. 74.

thimble ionization chamber A small cylindrical, spherical or thimble-shaped ionization chamber, volume less than 5 cm^3 with air-wall construction. Used in radiobiology. » p. 78.

thin source Radioactive source with negligible self-absorption.

thin target A target penetrated by primary radiation beam so that detecting instrument(s) may be used on opposite side of target to source.

thin-wall chamber Ionization chamber in which the number of knock-on particles arising from the wall material and absorption therein is small. » p. 74.

Thomson scattering The scattering of electromagnetic waves by free electrons. On a classical interpretation, the electron is set into oscillatory motion by the transverse electric field of the wave and radiates at the same frequency as the wave. The scattering cross-section of an electron is

$$\sigma = \frac{8}{3}\pi\,(e^2/4\pi\,\varepsilon_0\, m\, c^2)^2 = 0.66\times 10^{-28}\ \mathrm{m}^2.$$

where m and e are the mass and charge of the electron, c is the velocity of light and ε_0 the permittivity of free space.

thorides Naturally occurring radioactive isotopes in the radioactive series containing thorium.

thorium A metallic radioactive element, dark-grey in colour. Symbol Th, at.no. 90, r.a.m. 232.0381, mp 1845°C, rel.d. 11.2. Its abundance in the Earth's crust is 8.1 ppm and there are few independent thorium minerals. Its commercial sources are *monazite*, which occurs widely in beach sands where it is derived from acid igneous rocks and pegmatites, and *thorite*. It has six isotopes:

A	Abundance %	half-life	decay mode
228		1.91 y	α
229		7300 y	α
230		75 400 y	α
231		25.52 h	β^-
232	100		
233		22.3 m	β^-

Thorium-232 is fissile on capture of fast neutrons and is also fertile, uranium-235 (fissile with slow neutrons) being formed from thorium-232 by neutron capture and subsequent beta decay.

thorium reactor A breeder reactor in which fissile uranium-233 is bred in a blanket of fertile thorium-232.

thorium series The series of nuclides which result from the decay of thorium-232. The mass numbers of the members of the series are given by $4n$, n being an integer. The series ends in the stable isotope lead-208. See **radioactive series**.

thoron Thorium emanation, an isotope of radon (radon-220); half-life 54.5 s; a radioactive decay product of thorium. Symbol Tn. See **emanations**.

THORP Abbrev. for *THermal (reactor mixed) Oxide Reprocessing Plant*, at Sellafield, Cumbria, UK. » p. 61.

threshold dose The smallest dose of radiation that will produce a specified result.

threshold energy Minimum energy that can just initiate a given endoergic nuclear reaction. Exoergic reactions may also have threshold energies.

threshold limit value The maximum concentration of a named pollutant that a worker should be exposed to in a given period of time. Times vary for particular pollutants and are modified as standards change. In the UK standards are supervised by the Factory Inspectorate.

thymine One of the two pyrimidine bases in DNA in which it pairs with adenine. See **genetic code**. » p. 90.

time-of-flight spectrometer Way of measuring a neutron spectrum in which the energy or speed of neutrons is determined by the time taken by the neutrons to travel a known distance. A *chopper* admits neutrons in short bursts and the travel time is determined either by a second chopper which passes only neutrons of the correct velocity or by electronic delay measuring equipment. See **neutron velocity selector**.

time of operation Time between the occurrence of a primary ionizing event and the occurrence of the count in the detector system.

tissue dose Absorbed *depth dose* of radiation received by specified tissue. Cf. **skin dose**.

tissue equivalent material See **phantom material**.

Tn Symbol for **thoron**.

tokamak (1) In fusion, a toroidal apparatus for containing plasma by means of two magnetic fields: (a) a strong toroidal magnetic field created by coils surrounding the vacuum chamber and (b) a weaker poloidal field created by an intense electric current through the plasma. The resultant magnetic field is in the form of a helix surrounding the intense electric current through the plasma, as shown in the figure. Tokamak is an acronym of the Russian words meaning toroidal magnetic chamber. Cf. **poloidal field**, **toroidal field**. (2) General name for the fusion reactors which use this principle. » p. 38.

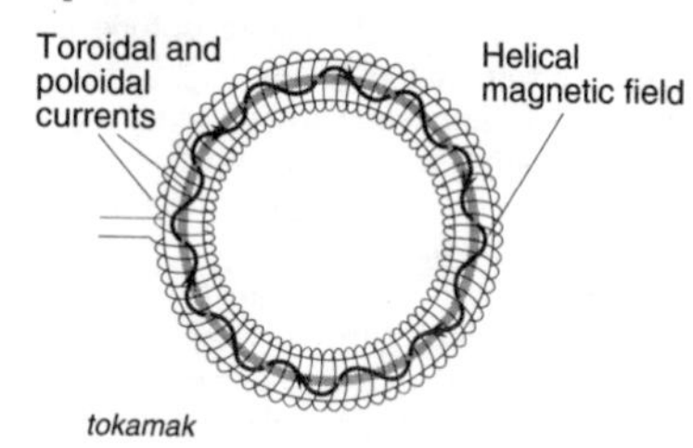

tokamak

tolerance dose Maximum dose which can be permitted to a specific tissue during radiotherapy involving irradiation of any other adjacent tissue.

tomography, emission Transverse section reconstruction of the radionuclide distribution within the body, obtained by acquiring images or **slices** of the head or body. This may be done by using **coincidence detection** (positrons) or single photon detection from gamma-ray emitters.

ton A unit of mass for large quantities. The *long ton*, once commonly used in the UK, is 2240 lb. The *short ton*, commonly used in the USA, is 2000 lb. The *metric ton* or *tonne* (1000 kilograms; symbol t) is 2204.6 lb. In the UK the *short ton* was used in metalliferous mining, the *long ton* in coal mining.

tonne Metric ton, 1000 kg. See **ton**.

topness A property that characterizes *quarks* and so hadrons. The topness of leptons and gauge bosons is zero. Topness is conserved in strong and electromagnetic interactions but not in weak interactions. Also known as *truth*.

toroidal field Magnetic field generated by a current flowing in a solenoid round a torus. Cf. **poloidal field, tokamak**. » p. 38.

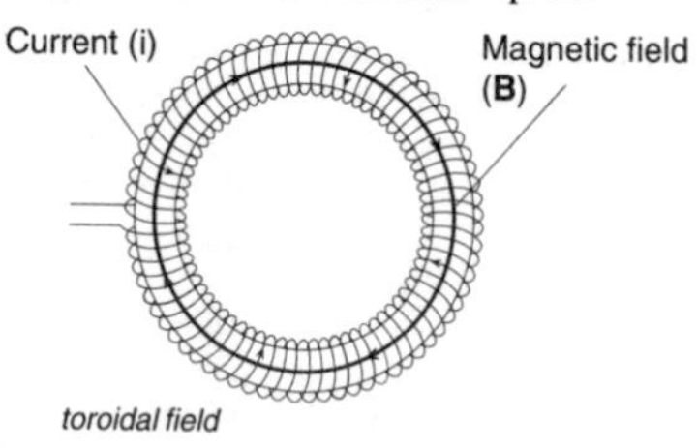

toroidal field

toroidal surface A lens surface in which the curvature in one plane differs from that in a plane at right angles.

torr Unit of low pressure equal to head of 1 mm of mercury or 133.3 N m^{-2}.

torus A surface of revolution shaped like the ring of an anchor or an American doughnut. It is generated by rotating a circle about a non-intersecting coplanar line as axis. The shape of the containment vessel in a tokamak fusion reactor. Adj. *toroidal*. Also *anchor ring*. » p. 38.

total absorption coefficient The coefficient which expresses energy losses due to both absorption and scattering. Relevant to narrow radiation beams. Preferred term *attenuation coefficient*.

total body burden (1) The summation of all radioactive materials contained in any person. (2) The maximum total amount of radioactive material any person may be permitted to contain.

total cross-section The sum of the separate cross-sections for all processes by which an incident particle can be removed from a beam. If all the atoms of an absorber have the same total cross-section, then it is identical with the **atomic absorption coefficient**. » p. 12.

total internal reflection Complete reflection of incident wave at boundary with medium in which it travels faster, under conditions where Snell's law of refraction cannot be satisfied. The angle of incidence at which this occurs (corresponding to an angle of refraction of 90°) is known as the *critical angle*.

toti- Prefix from L. *totus*, all, whole.

Townsend avalanche Multiplication process whereby a single charged particle, accelerated by a strong field, causes, through collision a considerable increase in ionized particles. » p. 75.

Townsend discharge A **Townsend avalanche** initiated by an external ionizing agent.

trace element analysis See panel on p. 251.

tracer atom Labelled atom introduced into a system to study structure or progress of a process. Traced by variation in isotopic mass or as a source of weak radioactivity which can be detected, there being no change in the chemistry. See **radioactive isotope**.

tracer compound A compound in which a small proportion of its molecules is labelled with a radioactive isotope.

tracer element One of the **radioelements**, e.g. radiophosphorus, used for experiments in which its radioactive properties enable its location to be determined and followed. Tracer technique may be applied to physiological, biological, pathological and technological experiments. For some purposes stable isotopes, e.g. carbon-13 and heavy hydrogen (see **deuterium**), are more conveniently used than radioisotopes.

track Path followed by particle esp. when rendered visible in photographic emulsion by cloud chamber, bubble chamber or spark

trace element analysis

One of the consequences of having neutrons and other atomic particles readily available has been the development of techniques in which very small samples are bombarded and their elemental composition measured from the energies of the gamma rays and X-rays emitted.

One method is called ***neutron activation analysis*** (abbrev. *NAA*) in which the sample is first irradiated with thermal neutrons to convert stable nuclei into radioactive ones by neutron capture and beta decay. After irradiation the sample is placed in a germanium crystal gamma detector and the energies and intensities of the gamma rays emitted carefully measured. The results can then be used to calculate how much of the original stable element was present.

The method is both very sensitive and essentially non-destructive. It is therefore very useful for very small samples of rare and valuable material such as paint from an old picture or glaze from porcelain, the results helping to determine the age and authenticity of the original. Sensitivity depends on the number of counts coming from the radioactive isotope and varies from about a millionth of a gram in the irradiated volume of the sample for zirconium to as little as a million times less for manganese. Only about 12 elements, including the eight lightest cannot be measured accurately in this way.

A second rather similar method is ***particle-induced X-ray emission*** (abbrev. *PIXE*) but here protons and sometimes alpha particles or heavy ions are made to bombard a similar thin sample. Bombardment causes the innermost electrons in the K- or L-shell to be removed and the atom ionized. This is followed by outer electrons falling into the vacancies with the emission of X-rays, whose energies and amounts are measured. The sensitivity of the method decreases as Z, the mass number, increases; it therefore complements neutron activation analysis. The overall precision of the method is about the same as neutron activation analysis although some combinations of elements in the sample may give confusing results because of the overlap in energies between the X-rays produced.

chamber.

tramp metal Stray metal pieces which are accidentally entrained in processed materials and must be removed before the material leaves the process.

tramp uranium Uranium dissolved from exposed fuel and plated out onto the structure in a liquid-cooled nuclear reactor.

transaxial tomography A process whereby serial radiographs are taken transverse to the vertical axis of the body. See **computer-aided tomography**.

transfer port The aperture through which items are inserted into or removed from a dry box, glove box, or shielded box (e.g. by sealing into plastic sac attached to the rim of the port).

transfer RNA An RNA molecule about 80 nucleotides long, with complementary sequences which result in several short *hairpin-like* structures. The loop at the end of one of these carries the anticodon triplet, which binds to the codon of the mRNA. The corresponding amino acid is bound to the 3′ end of the molecule. Abbrev. *tRNA*.

transformation See **atomic transmutation**.

transformation constant See **disintegration constant**.

transient equilibrium Radioactive equilibrium between daughter product(s) and parent element of which activity is decaying at an appreciable rate. Characterized by ratios of activity, but not magnitudes, being constant.

transient overpower Running a reactor accidentally at a power greater than the design level but without loss of coolant.

transition In atomic and nuclear physics, the terms used to describe the change from one quantum state to another. A transition from a higher to a lower energy state may be accompanied by the *emission* of a *photon*, while a transition from a lower to a higher state requires the *absorption* of a photon. Transitions are governed by *selection rules* which forbid some transitions.

transition element See **transition metal.**

transition metal One of the group of metals which have an incomplete inner electron shell. They are characterized by large atomic magnetic moments. Also *transition element*.

transition probability In atomic and nuclear physics, the probability per unit time that a system in quantum state k will undergo a transition to quantum state l. For an atom in an excited state there will be certain probability that the atom will spontaneously undergo a transition to a lower energy state. In the presence of photons of energy equal to the difference in energy between the states, there will be in addition a stimulated transition probability.

transmission See **transmittance**.

transmission coefficient Probability of penetration of a nucleus by a particle striking it.

transmission experiment An experiment in which radiation transmitted by a **thin target** is measured, to investigate the interaction which takes place. Such experiments are used in the measurement of total cross-sections for neutrons.

transmissivity See **transmission coefficient**.

transmittance Ratio of energy transmitted by a body to that incident on it. If scattered emergent energy is included in the ratio, it is termed *diffuse transmittance*, otherwise *specular transmittance*. Also *transmission*.

transmittivity Transmittance of unit thickness of non-scattering medium.

transmutation See **atomic transmutation**.

transparency Proportion of energy or number of incident photons or particles which pass through window of ionization chamber or Geiger counter.

transport Rate at which desired material is carried through any section of processing plant, e.g. isotopes in isotope separation.

transport cross-section Reciprocal of **transport mean free path**.

transport mean free path If **Fick's Law** of diffusion is applicable to the conditions in a nuclear reactor then the mean free path is three times the diffusion coefficient of neutron flux. In practice the theory usually has to be modified to take account of anisotropy of scattering and persistence of velocities.

transport theory Rigorous theoretical treatment of neutron migration which must be used under conditions where **Fick's Law** does not apply. See **diffusion theory**.

transuranic elements The artificial elements with atomic numbers 93 and upwards, which possess heavier and more complex nuclei than uranium, and which can be produced by the neutron bombardment of uranium. More than 12 of these have been produced, including neptunium, plutonium, curium, lawrencium etc.

trapped mode Propagation in which the radiated energy is substantially confined within a tropospheric duct.

trapping region Three-dimensional space in which particles from the Sun are guided into paths towards the magnetic poles, giving rise to **aurora**, and otherwise forming ionized shells high above the ionosphere. Also *magnetic tube*.

travelling wave A wave carrying energy continuously away from the source.

trega-. Prefix signifying 10^{12} times. It is replaced in the SI system of unit notation by the prefix **tera-**.

tri-. Prefix from L, *tres*, Gk. *tria*, three.

tri-*n*-butyl phosphate The solvent which, diluted with kerosene is now used for separating uranium and plutonium from spent fuel in the Purex process. Abbrev. *TBP*.

trip Automatic shutdown of reactor power initiated by signal from one of the safety circuits when one of the operational characteristics of the reactor deviates beyond a certain limit.

trip amplifier An amplifier operating the trip mechanism of a nuclear reactor. Also *shutdown amplifier*.

triple-axis neutron spectrometer An instrument used in neutron spectroscopy for determining the energies of neutrons scattered in a particular direction from a crystal. See **neutron elastic scattering**, **neutron inelastic scattering**.

triple point The temperature and pressure at which the three phases of a substance can coexist. The triple point of water is the equilibrium point (273.16 K and 610 N m^{-2}) between pure ice, air-free water and water vapour obtained in a sealed vacuum flask. It is one of the fundamental **fixed points** of the international practical scale of temperature. See **Kelvin thermodynamic scale of temperature**.

tritium The radioactive isotope of **hydrogen** of mass no. 3, symbol T, r.a.m. 3.0221, of half-life 12.3 years. It is very rare, the abundance in natural hydrogen being one atom in 10^{17} but tritium can be produced artificially by neutron absorption in lithium. Used in thermonuclear weapons and fusion reactors. It also can be used to label any compound containing hydrogen and it is consequently of great importance in biology.

tritium unit A unit corresponding to the presence of one atom of tritium in 10^{18} atoms of hydrogen. Pure tritium is therefore 10^{18} tritium units.

triton The tritium nucleus, consisting of one

proton combined with two neutrons.

tRNA See **transfer RNA**.

true coincidences Coincidences produced by a single particle discharging both or all counters in a nuclear counting device. Cf. **spurious coincidences**.

true (real) absorption coefficient The absorption coefficient applicable when scattered energy is not regarded as absorbed. Applicable to broad-beam conditions.

true watts Power dissipated in a.c. circuit.

truth See **topness**.

tube of force Space enclosed by all the lines of force passing through a closed contour. Of unit magnitude when it contains unit flux.

tunable dye laser A laser in which the excited material is a dye in an organic solvent; the dye fluoresces over a wide range of wavelengths, so the laser output can be tuned by altering the parameters of the system. » p. 42.

tunnelling See **potential barrier**.

turnover (1) In isotope separation, the total flow of material entering a given stage in a cascade. (2) Rate of renewal of a particular chemical substance in a given tissue.

twin One of a pair of two related entities similar in structure or function; often synonymous with *double*.

two-body force A type of interaction between 2 particles which is unmodified by the presence of other particles.

two-group theory Simplified treatment of neutron diffusion in which only two energy groups are considered. Only the high-energy fission and the fully thermalized neutrons are considered, with the partly thermalized neutrons neglected.

U – Z

u Symbol for unit of unified scale.

U Symbol for **uranium**.

U Symbol for (1) **potential difference**; (2) tension.

ultra- Prefix from L. *ultra*, beyond.

ultrasonography Use of reflected high-frequency sound waves to image organs of the body. Widely used in the diagnosis of disease of the abdomen and heart and in the management of pregnancy.

ultraviolet radiation Electromagnetic radiation in a wavelength range from 400 nm to 10 nm approximately, i.e. between the visible and X-ray regions of the spectrum. The *near* ultraviolet is from 400 to 300 nm, the *middle* from 300 to 200 nm and the *extreme* from 200 to 190 nm.

ultraviolet spectrometer An instrument similar to an optical spectrometer but employing non-visual detection and designed for use with ultraviolet radiation.

uncertainty principle The principle that there is a fundamental limit to the precision with which a position co-ordinate of a particle and its momentum in that direction can be simultaneously known. Also, there is a fundamental limit to the knowledge of the energy of a particle when it is measured for a finite time. In both statements, the product of the uncertainties in the measurements of the two quantities involved must be greater than the *Dirac constant*. The principle follows from the wave nature of particles. Also *Heisenberg principle*, *indeterminacy principle*.

unexcited Said of an atom in its ground state. See **excitation**.

unified field theory See **field theory**.

unified model of the nucleus A model incorporating many valuable features of both *collective-* and *independent-particle* models.

unit A dimension or quantity which is taken as a standard of measurement.

unit charge, unit quantity of electricity In SI units, 1 **coulomb**. In unrationalized MKSA units, the electric charge which experiences a repulsive force of 1 newton when placed 1 metre from a like charge *in vacuo*. Similarly, in the CGS electrostatic units, the force is 1 dyne when 1 centimetre apart. See **SI units**, p 237.

unsealed source A radioactive source of any kind from which the radiation can escape. It is therefore of very low activity such as a tracer being used during a medical investigation.

unstable equilibrium The state of equilibrium of a body when any slight displacement decreases its potential energy. The instability is shown by the fact that, having been slightly displaced, the body moves farther away from its position of equilibrium.

unstable oscillation Any oscillation, in a mechanical body, electrical circuit etc which increases in amplitude with time.

uptake In radiobiology, the quantity (or proportion) of administered substance subsequently to be found in a particular organ or tissue.

uracil *2,6-dioxypyrimidine*. One of the four bases in RNA and the only one which does not occur in DNA. Pairs with adenine. See **genetic code**.

uranium A hard grey metal. Symbol U, at. no. 92, r.a.m. 238.03, mp 1150°C, rel.d. 18.68. It has seven isotopes:

A	Abundance %	half-life	decay mode
233		0.159 My	α
234		0.245 My	α
235	0.720	700 My	
236		23.42 My	α
237		6.75 d	β^-
238	99.275	4500 My	
239		23.5 m	β^-

Because the half life of uranium-235 is very much less than that of uranium-238 the relative abundance of these two isotopes has varied over time with uranium-235 being about 3% 2×10^9 years ago. See **Oklo natural fission reactor** p. 203. » p. 55 and Chapter 2.

uranium enrichment Processes for increasing the content of uranium-235 in uranium for reactor use. The principal methods which are either in use or have been proposed are gaseous diffusion, gas centrifuge, jet nozzle, plasma centrifuge, chemical exchange and laser isotope separation. Diffusion and the gas centrifuge have been most generally employed although in the early days magnetic separation was used (*Calutron* method). See **isotope separation**. » p. 56.

uranium hexafluoride A volatile compound of uranium with fluorine, used in the gaseous diffusion process for separating the uranium isotopes. Very corrosive.

uranium-radium series The series of radioactive isotopes which result from the decay of uranium-238. The mass numbers of the members of the series are $4n+2$, where n is an integer. Series ends in the stable isotope, lead-206. See **radioactive series**.

v Symbol for (1) **velocity**; (2) specific volume of a gas.

V Symbol for **volt**.

V Symbol for (1) **potential**; (2) **potential difference**; (3) **electromotive force**; (4) **volume**.

vacancy Unoccupied site for ion or atom in crystal.

vacuum Literally, a space totally devoid of any matter. Does not exist, but is approached in interstellar regions. On Earth, the best vacuums produced have a pressure of about 10^{-8} N m^{-2}. Used loosely for any pressure lower than atmospheric, e.g. train braking systems, 'vacuum' cleaners etc.

vacuum evaporation Under normal conditions, there is an equilibrium of molecular exchange at the surface of a solid body – molecules leaving the surface and others captured by it. Under vacuum conditions and in space, these molecules are lost.

value function See **separation potential**.

van der Waals' equation An equation of state which takes into account the effect of intermolecular attraction at high densities and the reduction in effective volume due to the actual volume of the molecules: $(P + a/v^2)(v - b) = RT$, a and b being constant for a particular gas. See **gas laws**.

van der Waals' forces Weak attractive forces between molecules or crystals, represented by the co-efficient a in *van der Waals' equation*. They vary inversely as the sixth power of the interatomic distance, and are due to momentary dipoles caused by fluctuations in the electronic configuration of the molecules.

vapour A gas which is at a temperature below its critical temperature and can therefore be liquefied by a suitable increase in pressure.

vaporization See **entropy of fusion**.

vapour pressure The pressure exerted by a vapour, either by itself or in a mixture of gases. The term is often taken to mean saturated vapour pressure, which is the vapour pressure of a vapour in contact with its liquid form. The saturated vapour pressure increases with rise of temperature.

velocity (1) The rate of change of displacement of a moving body with time; a vector expressing both magnitude and direction. Cf. **speed** which is scalar. (2) For a wave, the distance travelled by a given phase divided by the time taken. Symbol v.

velocity of light See **speed of light**.

velocity of sound In dry air at stp 331.4 m s^{-1} (750 miles per hour). In fresh water, 1410 m s^{-1}, and in sea water, 1540 m s^{-1}. The above values are used for sonar ranging but do not apply to explosive shock waves. They must be corrected for variations of temperature, humidity etc.

vibration-rotation spectrum The infra-red end of the electromagnetic spectrum which arises from vibrational and rotational transitions within a molecule.

virgin neutrons Neutrons which have not yet experienced a collision and therefore retain their energy at birth.

virtual process As a consequence of the *uncertainty principle*, it is possible for the conservation of mass and energy to be violated for a time t by an amount E such that $Et \approx$ Dirac's constant (h-bar). A transition to a higher quantum state could take place provided this condition was satisfied, but the transition could not be observed and is called a *virtual process*. A particle created in such a process is called a *virtual particle*. This is an important mechanism of nuclear forces.

virtual quantum In higher order perturbation theory, a matrix element which connects an initial state with a final state involves intermediate states in which energy is not conserved. A photon or quantum in one of these states is designated a *virtual quantum*. This concept enables the coulomb energy between two electrons to be regarded as arising from the emission of virtual quanta by one of the electrons and their absorption by the other.

virus A particulate infective agent smaller than accepted bacterial forms, invisible by light microscopy, incapable of propagation in inanimate media and multiplying only in susceptible living cells, in which specific cytopathogenic changes frequently occur. Causative agent of many important diseases of Man, lower animals and plants, e.g. poliomyelitis, foot and mouth disease, tobacco mosaic. See **bacteriophage**.

viscoelastic A solid or liquid which when deformed exhibits both viscous and elastic behaviour through the simultaneous dissipation and storage of mechanical energy. Shown typically by polymers.

viscometer An instrument for measuring viscosity. Many types of viscometer employ **Poiseuille's formula** for the rate of flow of a viscous fluid through a capillary tube.

viscosity The resistance of a fluid to shear forces, and hence to flow. Such shear resistance is proportional to the relative velocity between the two surfaces on either side of a layer of fluid, the area in shear, the **coefficient of viscosity** of the fluid and the reciprocal of the thickness of the layer of fluid. For comparing the viscosities of liquids, various scales have been devised, e.g.

Redwood No. 1 seconds (UK), *Saybolt Universal seconds* (US), *Engler degrees* (Germany). See **kinematic viscosity**.

visible radiation Electromagnetic radiation which falls within the wavelength range of 780 to 380 nm, over which the normal eye is sensitive. » p. 80.

vitreous state A non-crystalline solid or rigid liquid, formed by supercooling the melt. Also *glassy state*.

vitrification The incorporation of radioactive waste products (particularly from nuclear fuel processing) into glass. Also *glassification*. Other techniques under study include ceramics, glass-ceramics, composite materials (e.g. glass beads in a metal matrix) and synthetic minerals.

volt The SI unit of *potential difference*, electrical potential, or emf, such that the potential difference across a conductor is 1 volt when 1 ampere of current in it dissipates 1 watt of power. Equivalent definition: if, in taking a charge of 1 coulomb between two points in an electric field, the work done on or by the charge is 1 joule, the potential difference between the points is 1 volt. Named after Count Alessandro Volta (1745-1827). Symbol V.

voltage The value of an emf or pd expressed in *volts*.

voltage drop (1) Diminution of potential along a conductor, or over an apparatus, through which a current is passing. (2) The possible diminution of voltage between two terminals when current is taken from them.

volt-amperes Product of actual voltage (in volts) and actual current (amperes), both r.m.s., in a circuit.

volume The amount of space occupied by a body; measured in cubic units. Symbol *V*.

volume ionization The mean ionization density in any given volume without reference to the specific ionization of the particles.

volumetric heat See **molal specific heat capacity**.

w Symbol for **work**.

W Symbol for **watt**.

W Symbol for (1) **weight**; (2) **work**.

wall effect (1) The contribution of electrons liberated in the walls of an ionization chamber to the recorded current. (2) The reduction in the count rate recorded with a Geiger tube due to ionizing particles not having the energy to penetrate the walls of the tube.

wall energy The energy per unit area stored in the domain wall bounding two oppositely magnetized regions of a ferromagnetic material.

wall-less counter A low-level proportional counter, the cathode of which consists of a cylindrical cage of thin wires, parallel to the cathode, which considerably reduces the background arising from electrons ejected by gamma rays. The counting volume may also be accurately defined by use of special *field tubes* at each end of the counter.

war gas Any gaseous chemical substance used in warfare (or in riot control) to produce poisonous or irritant effects upon the human body.

waste (1) Depleted material rejected by an isotope separation plant. (2) Unwanted radioactive material for disposal. » p. 60.

waste disposal See **disposal**.

waste storage See **storage**.

water A colourless, odourless, tasteless liquid, mp 0°C, bp 100°C. It is hydrogen oxide, H_2O, the liquid probably containing associated molecules, H_4O_2, H_6O_3 etc. On electrolysis it yields two volumes of hydrogen and one of oxygen. It forms a large proportion of the Earth's surface, occurs in all living organisms, and combines with many salts as water of crystallization. Water has its maximum density of 1000 kg m^{-3} at a temperature of 4°C. This fact has an important bearing on the freezing of ponds and lakes in winter, since the water at 4°C sinks to the bottom and ice at 0°C forms on the surface. Besides being essential for life, water has a unique combination of solvent power, thermal capacity, chemical stability, permittivity and abundance. See **heavy water**, **triple point**.

water equivalent The mass of water which would require the same amount of heat as a body to raise its temperature by one degree. It is its **thermal capacity** (the product of its mass and its *specific heat capacity*) divided by the s.h.c. of water (4.186 kJ $kg^{-1}K^{-1}$).

water monitor A monitor for measuring the level of radioactivity in a water supply, similar to, but much more sensitive than, an effluent monitor.

water reactor Nuclear reactor in which water (including heavy water) is the moderator and/or coolant. » p. 24.

watt SI unit of power equal to 1 joule per second. Thus, 1 horsepower (hp) equals 745.70 watts. Symbol W. See **SI units** p. 237.

watt-hour A unit of energy, being the work done by 1 watt acting for 1 hour, and thus equal to 3600 joules.

wave A time-varying quantity which is also a function of position. The characteristic of a wave is to transfer energy from one point to another without any particle of the medium being permanently displaced; particles merely oscillate about their equilibrium positions. In electromagnetic waves it is the changes in electric and magnetic fields which represent the wave disturbance. The progress of the wave is described by the passage of a *waveform* through the medium with a certain velocity, the *phase* or *wave velocity*. The energy is transferred at the *group velocity* of the waves making the waveform.

wave equation A differential equation which describes the passage of harmonic waves through a medium. The form of the equation depends on the nature of the medium and on the process by which the wave is transmitted. The solutions to the equation depend on the circumstances in which the wave is propagated. See **Schrödinger equation**.

waveform The shape, contour or profile of a wave; described by a phase relationship between successive particles in a medium. A waveform may be *periodic*, *transient* or *random*. Also *waveshape*.

wavefront Imaginary surface joining points of constant phase in a wave propagated through a medium. The propagation of waves may conveniently be considered in terms of the advancing wavefront, which is often of simple shape, such as a plane, sphere or cylinder.

wave function Mathematical equation representing the space and time variations in amplitude for a wave system. The term is used particularly in connection with the Schrödinger equation for particle waves.

waveguide Electromagnetic waves in the microwave region can be transmitted efficiently from a source to other parts of a circuit by means of hollow metal conductors called waveguides. The transmission can be described by the patterns of electric and magnetic fields produced inside the guide, different modes being characterized by different electric and magnetic field configurations. Dielectric guides operate similarly but generally have higher losses.

wave interference Relatively or completely stationary patterns of amplitude variation over a region in which waves from the same source (or two different coherent sources) arrive by different paths of propagation. *Constructive interference* arises when the two waves are in phase and their amplitudes add; *destructive interference* arises when they are out of phase and their amplitudes partly or totally neutralize each other.

wavelength (1) Distance, measured radially from the source, between two successive points in free space at which an electromagnetic or acoustic wave has the same phase; for an electromagnetic wave it is equal in metres to c/f where c is the velocity of light (in m s^{-1}) and f is the frequency (in Hz). Symbol λ. (2) Distance between two similar and successive points on a harmonic (sinusoidal) wave, e.g. between successive maxima or minima. (3) For electrons, neutrons and other particles in motion when considered as a *wave train*, $\lambda = h/p$, p is the momentum of the particle and h is Planck's constant. See **wave mechanics**, **de Broglie wavelength**. » p. 80.

wavelength of light The wavelength of visible light lies in the range from 400 to 700 nm approximately.

wave mechanics The modern form of the *quantum theory* in which events on an atomic or nuclear scale are explained in terms of the interactions between wave systems as expressed by the *Schrödinger equation*. For a bound particle, e.g. an electron in an atom, standing wave solutions are found for which only certain wavelengths are permitted, and consequently the energy is quantized. See **statistical mechanics**.

wave number In an electromagnetic wave, the reciprocal of the wavelength, i.e. the number of waves in unit distance.

wave-particle duality Light and other electromagnetic radiations behave like a wave motion when being propagated, and like particles when interacting with matter. Interference, diffraction and polarization effects can be described in terms of waves. The photoelectric effect and the Compton effect can be described in terms of *photons*, quanta of energy $E = h\nu$ where h is Planck's constant and ν is the frequency.

waveshape See **waveform**.

wave theory Macroscopic explanation of diffraction, interference and optical phenomena as an electromagnetic wave, predicted by Maxwell and verified by Hertz for radio waves. See **theories of light**.

wave velocity See **phase velocity**.

weak interaction A fundamental interaction between particles mediated by *intermediate vector bosons*. Weak interactions involve neutrinos or antineutrinos or both and are completed in about 10^{-10} s. This kind of interaction is responsible for radioactive β-decay.

weber The SI unit of magnetic flux. An emf of 1 volt is induced in a circuit through which the flux is changing at a rate of 1 weber per second. 1 weber equals 1 volt-second equals 1 joule per ampere. Equivalent definition: 1 weber is the magnetic flux through a surface over which the integral of the normal component of the magnetic induction is 1 tesla m^2. Symbol Wb. See **SI units** p. 237.

weight The gravitational force acting on a body at the Earth's surface. Units of measurement are the newton, dyne or pound-force.

Weight = mass × acceleration due to gravity,

and must therefore be distinguished from **mass**, which is determined by the quantity of material and measured in pounds, kilograms etc. Symbol *W*.

Weinberg and Salam's theory A unified theory of *weak* and *electromagnetic* interactions between particles. It predicted the behaviour of the W^+, W^- and Z^0 intermediate vector bosons as the agents for the weak interaction. These particles were discovered later, in 1983, using the CERN SPS (super proton synchrotron) modified to produce colliding-beam experiments with 270 GeV protons and 270 GeV antiprotons. Also *electroweak theory*.

well See **potential well**.

well counter A counter used for measuring radioactive fluids placed in a cylindrical container surrounded by the detecting element (hollow scintillation crystal or sensitive volume of special Geiger tube). » p. 74.

Westcott convention An approximation applied to the neutron flux in thermal reactor design. The flux is divided into a thermal component with Maxwellian distribution and a fast component with distribution proportional to *dE/E*, where *E* is the neutron energy. Sometimes a third component covering thermalization region may be included.

Westcott flux Theoretical neutron flux defined as equal to the reaction rate of a detector with cross-section which is unity for thermal electrons (velocity 2200 m s^{-1}) and varies inversely with the neutron velocity.

wet steam A steam-water mixture, such as results from partial condensation of dry saturated steam on cooling.

white light Light containing all wavelengths in the visible range at the same intensity. This is seen by the eye as white. The term is used, however, to cover a wide range of intensity distribution in the spectrum and is applied by extension to continuous spectra in other wavelength bands (e.g. *white noise*). Also *white radiation*.

white radiation See **white light**.

whole-body monitor Assembly of large scintillation detectors, heavily shielded against background radiation, used to identify and measure the gamma radiation emitted by the human body.

width The spread of uncertainty in a specified energy level, arising as a result of Heisenberg's **uncertainty principle**, proportional to the instability of the state concerned.

Wien's laws for radiation from a black body (1) *Displacement law*: $\lambda_m T$ = constant (0.0029 metre kelvin). (2) *Emissive power* ($E\lambda$): within the maximum intensity wavelength interval $d\lambda$,

$$E_\lambda = CT^5 d\lambda,$$

where $C = 1.288 \times 10^{-5}$ W m^{-2} K^{-5} and T is temperature in kelvins. (3) *Emissive power* (dE) in the interval $d\lambda$ is

$$dE = A\lambda^{-5} \exp(-B/\lambda T)\, d\lambda.$$

λ_μ is wavelength at E_{max}, A = 4.992 mJ , B = 0.0144 mK.

Wigner effect Changes in physical properties of graphite resulting from the displacement of lattice atoms by high-energy neutrons and other energetic particles in a reactor. It results in the building up of stored energy (Wigner energy) in the change of crystal lattice dimensions and hence in the change of overall bulk size.

Wigner energy Energy stored within a crystalline substance, due to the **Wigner effect**. » p. 47.

Wigner force Ordinary (non-exchange) short-range force between nucleons.

Wigner nuclides Those isobars of odd mass number in which the atomic number and neutron number differ by one.

Wilson chamber **Cloud chamber** of expansion type.

WIMS Abbrev. for **Winfrith Improved Multigroup Scheme**.

window Thin portion of wall or radiation counter through which low-energy particles can penetrate.

Winfrith Improved Multigroup Scheme Widely used set of computer codes for predicting the properties of thermal reactors.

work One manifestation of energy. The work done by a force is defined as the product of the force and the distance moved by its point of application along the line of action of the force. For example, a tensile force does work in increasing the length of a piece of wire; work is done by a gas when it expands against a hydrostatic pressure. As for all forms of energy, the SI unit of work is the *joule*, performed when a force of 1 newton moves its point of application through 1 metre along the line of action of the force. Alternatively, when 1 watt of power is expended for 1 second,

1 joule of work is done. Symbols w, W. See **joule**.

work function The minimum energy that must be supplied to remove an electron so that it can just exist outside a material under vacuum conditions. The energy can be supplied by heating the material (*thermionic* work function) or by illuminating it with radiation of sufficiently high energy (*photoelectric* work function). Also *electron affinity*.

work-hardening The increase in strength and hardness (i.e. resistance to deformation) produced by working metals. It is most pronounced in cold-working, and in the case of metals such as iron, copper, aluminium and nickel. Lead, tin and zinc are not appreciably hardened by cold-working, because they can recrystallize at room temperature.

worth See **reactivity worth**.

X Symbol for reactance.

Xe Symbol for **xenon**.

xeno- Prefix from Gk. *xenos*, strange, foreign.

xenon A zero-valent element, one of the noble gases, present in the atmosphere in the proportion of 1:170 000 by volume. Symbol Xe, at.no. 54, r.a.m. 131.30, mp −140°C, bp −106.9°C, critical temp. +16.6°C, density at stp 5.89 g dm^{-3}. It has 17 isotopes:

A	Abundance %	half-life	decay mode
121		40.1 m	ε
122		20.1 h	ε
123		2.08 h	ε
124	0.096		
125		17 h	ε
126	0.090		
127		36.4 d	ε
128	1.92		
129	26.4		
130	4.1		
131	21.2		
132	26.9		
133		5.25 d	β^-
134	10.4		
135		9.1 h	β^-
136	8.9		
137		3.82 m	β^-

The isotope xenon-135, is of considerable importance in the design of nuclear reactors because of its exceptionally high neutron capture cross-section of 2.7×10^6 barns, which causes **xenon poisoning**.

xenon override The provision of a means of compensating for the effect of xenon poisoning after power reduction in a reactor.

xenon poisoning In a fission reactor about 5% of the xenon-135 comes from direct fission and the remainder from the decay of iodine-135 with a half-life of 7.2 hours. In a reactor running at a steady state, production and depletion balance out but when power is *reduced* the long half-life of the iodine-135 results in a continued buildup of xenon-135 which reaches its peak after about 10 hours. Depending on the reactor this peak may be as much as 1.4 times the initial steady-state value. This results in increased capture of neutrons, a further loss of power and a potentially unstable state in the reactor, as occurred at Chernobyl. See **xenon override**. » p. 49.

xer-, xero- Prefixes from Gk. *xeros*, dry.

xeroradiography Radiography in which a xerographic, and not photographic, image is produced.

X-ray laser A laser with an output in the X-ray region of the spectrum. Using highly ionized selenium plasma as the lasing medium, laser output at 20.6 and 20.9 nm has been recently reported.

X-ray photon A quantum of X-radiation energy given by $h\nu$, where ν is the frequency and h is Planck's constant.

X-rays Electromagnetic waves of short wavelength (c. 10^{-3} to 10 nm) produced when high-speed electrons strike a solid target. Electrons passing near a nucleus in the target are accelerated and so emit a continuous spectrum of radiation (*bremsstrahlung*) ranging up from a *minimum wavelength*. In addition, the electrons may eject an electron from an inner shell of a target atom, and the resulting transition of an electron of a higher energy level to this level produces radiation of specific wavelengths. This is the *characteristic* X-ray spectrum of the target and is specific to the target element. X-rays may be detected photographically or by a counting device. They penetrate matter which is opaque to light; this makes X-rays a valuable tool for medical investigations. See **Compton effect**, **K-capture**, **L-capture**, **Moseley's law**, **synchrotron radiation** » p. 82.

X-ray therapy The use of X-rays for medical treatment.

Y Symbol for admittance.

yard Unit of length in the foot-pound-second

system formerly fixed by a line standard (Weights and Measures Act of 1878), redefined in 1963 as 0.9144 m.

yield (1) Ion pairs produced per quantum absorbed or per ionizing particle. (2) See **fission yield**.

yoke Part or parts of a magnetic circuit not embraced by a current-carrying coil, esp. in a generator or motor, or relay.

Yukawa potential A potential function of the form

$$V = \frac{V_\theta \exp(-kr)}{r},$$

r being distance. Characterizes the meson field surrounding a nucleon. The exponential tail of the Yukawa potential extends with appreciable strength to larger values of r than does that of the coulomb potential.

Z Symbol for impedance.

Zeeman effect The splitting of spectrum lines into a number of components by strong magnetic fields. The field splits the atomic energy levels into several components associated with different quantized orientations of the total magnetic moment with respect to the field. Cf. **Stark effect**.

zero-energy reactor See **zero-power reactor**.

zero-point energy Total energy at the absolute zero of temperature. The uncertainty principle does not permit a simple harmonic oscillator particle to be at rest exactly at the origin, and by the quantum theory, the ground state still has one half-quantum of energy, i.e. $h\nu/2$, and the corresponding kinetic energy.

zero-point entropy As follows from the third law of thermodynamics, the entropy of a system in equilibrium at the absolute zero must be zero.

zero-power reactor An experimental reactor for reactor physics studies with an extremely low neutron flux so that no forced cooling is required and there is insignificant build-up of fission products.

zircaloy TN for an alloy of zirconium with small amounts of tin, iron and chromium, used to clad fuel elements in water reactors. There are two kinds commonly used, zircaloy-2 in boiling-water reactors and zircaloy-4 in pressurized-water reactors. The latter has less nickel in the alloy and is significantly less brittle after irradiation. Each contains 98.2% zirconium and has neutron capture cross-sections of 193 millibarns. The fuel cladding is made by compressing and sintering the powdered alloy in the absence of air which would otherwise cause spontaneous combustion. zircaloy-2 has the following properties:

σ_{cap} mb	mp °C	Therm. conduct. W $m^{-1}K^{-1}$	density g cm^{-3}
193	2090	15	6.57

» p. 24.

zirconium A metallic element. Symbol Zr, at.no. 40, r.a.m. 91.22, mp 2130°C, rel.d. 4.15. It has 12 isotopes:

A	Abundance %	half-life	decay mode
87		1.6 h	ε
88		83.4 d	ε
89		78.4 h	ε
90	51.5		
91	11.2		
92	17.1		
93		1.5 My	β^-
94	17.4		
95		64.0 d	β^-
96	2.80		
97		16.9 h	β^-
98		31 s	β^-

When purified from hafnium, its low neutron absorption and its retention of mechanical properties at high temperature make it useful fuel cladding in nuclear reactors. See **zircaloy**.

Zr Symbol for **zirconium**.

zyg-, zygo- Prefixes from Gk. *zygon*, yoke.

zygote The cell that results from the fusion of two gametes.